Earth and Environmental Sciences Library

The *Earth and Environmental Sciences Library* (EESL) is a multidisciplinary book series that showcases innovative approaches and in-depth reviews, aimed at enhancing the role of the Earth and Environmental Sciences communities. The series also offers valuable guidance for stakeholders, decision-makers, policymakers, international organizations, and NGOs.

Topics of interest include oceanography, the marine environment, atmospheric sciences, hydrology, soil sciences, geophysics, geology, agriculture, environmental pollution, remote sensing, climate change, water resources, and natural resource management.

To further advance these critical areas, the *Earth and Environmental Sciences Library* invites contributions from the Earth and Environmental Sciences communities in the form of edited volumes, monographs, textbooks, conference proceedings, and more. The series fosters knowledge exchange and collaboration across disciplines, encouraging comprehensive discussions and solutions to pressing environmental challenges.

Faiza Khebour Allouche
Editor

Environmental Remote Sensing and GIS in MENA Region

GIS Urban and Rural Applications Geospatial Technologies and Risks

Springer

Editor
Faiza Khebour Allouche
High School of Agronomic Sciences
Sousse University
Sousse, Tunisia

ISSN 2730-6674 ISSN 2730-6682 (electronic)
Earth and Environmental Sciences Library
ISBN 978-3-031-86573-2 ISBN 978-3-031-86574-9 (eBook)
https://doi.org/10.1007/978-3-031-86574-9

This Springer imprint is published by the registered company Springer Nature Switzerland AG
The registered company address is: Gewerbestrasse 11, 6330 Cham, Switzerland

Preface

This volume came into conception to highlight the *Environmental Remote Sensing and GIS in MENA Region* and succeed the book *Environmental Remote Sensing and GIS in Tunisia*. In this volume, different experts from the MENA region present the result and findings of their research work. The volume consists of four main parts/ sections, including the "Introduction" and "Conclusion," written in two chapters. The main four parts of this volume contain 14 chapters written by more than 40 authors. The four main sections are:

- Overview of Remote Sensing and GIS Applications and Strategies (written in four chapters).
- Geospatial Technologies and Risks (written in four chapters).
- Urban Applications (written in three chapters).
- Rural Applications (written in three chapters).

This volume covers contributions from several countries from MENA regions including Jordan, Saudi Arabia, Egypt, Tunisia, Algeria, and Morocco. It explains how the application of remote sensing and GIS techniques can benefit RWH studies and render further analyses possible that provide deeper insights into RWH practices and changes over time. It emphasizes using GIS and the Vessel Monitoring System to map the fishing areas' evolution between 2003 and 2021, revealing a great expansion in small pelagic fish harvesting. Also, it applies GIS tools and spatial interpolation to estimate the distribution of freshwater fish species between the reservoirs of the northern sub-basins of Medjerda (Kasseb) and those of the southern sub-basins (Siliana). Additionally, it explores GEE by applying two radiometric indexes in Algeria.

Moreover, this volume proposes a methodology to assess and monitor the eutrophication using remote sensing and GIS tools applied to a coastal region in Tunisia. It explains the output for some works using statistical, and different remote sensing and geospatial techniques to define the flood-prone area within the high Oum Er Rbia basin, Middle Atlas region, Morocco. It uses an empirical and simple approach developed by NEPSEN-E6 and combine UHI score and permeability factors in a GIS tool to create the current and projected UIH maps. It provides a methodology with some stochastic models and the structured part of data formulated into a mathematical model to elaborate the prediction.

Furthermore, this volume proposes a methodology to assess and monitor the eutrophication using remote sensing and GIS applied to a coastal region in Tunisia. It demonstrates the output for some works using statistical, and different remote sensing and geospatial techniques to define the flood-prone area within the high Oum Er Rbia basin, Middle Atlas region, Morocco. It uses an empirical and simple approach developed by NEPSEN-E6 and combines UHI score and permeability factors in a GIS tool to create the current and projected UIH maps.

Furthermore, the volume proposes a methodology to evaluate the sptio-temporal variation of vegetation by using satellite images, radiometrix indexes, and supervised classification of the PCA component. It provides information on urban landscape using Google Earth images to design an urban park in Bouficha region, Tunisia. Also, it explains that through the visual interpretation of satellite imageries and GIS tool, coastal development is not in line with the concept of the green and blue belts to the coastal side in Algeria.

On the other hand, the volume highlights the importance of change detection methods for understanding urbanization patterns and assessing their impacts on the oasis ecosystem tested in Saudi Arabia. It presents how GIS can be used to study and map flood risks, based on identifying the relationship between the flood hazard and the hydro-morphometric characteristics (drainage density, hydrological density, compactness index) of order 1 impluviums, lithology, and slope values. Additionally, this volume explains how to assess land degradation in Egypt, different steps are done: assessment of water quality by using piper diagram; assessment in terms of salinity and sodicity hazard according to USDA (1954); applying geostatistical analysis and producing the IWQI map by overlapping of the thematic maps (EC, Na^+, Cl^-, HCO_3^- and SAR).

Overall, this unique volume provides a valuable tool for decision support and a valuable guide for policy planners, decision-makers, stakeholders, researchers, and graduate students.

Last but not least, special thanks are due to all authors who contributed to this volume, without their efforts and patience, it would not have been possible to produce this unique volume on Environmental RS and GIS in MENA region. Also, thanks should be extended to the Springer team including the series editor, project manager, publishing editor, and all who helped in the production of the book. Their help largely supported the volume editor and the authors of chapters during the production of this volume.

Sousse, Tunisia
October 2024

Faiza Khebour Allouche

Contents

List of Figures

GIS and Remote Sensing for a Better Understanding of Ancient and Modern Rainwater Harvesting Strategies and the Identification of Current Potentials and Improvement Possibilities

The National Strategy for Small Pelagic Fisheries' Development in the Northeast of Tunisia (Central Mediterranean): Status and Prospects

The Use of Geographic Information System (GIS) Data to Assess and Rebuild Stocks in Tunisia's Largest Freshwater Reservoirs

Remote Sensing Applications with Google Earth Engine: Case from Algeria

Coastal Eutrophication Detection and Assessement Using Sentinel-2 MSI Imagery

Performance of Modeling Based on Statistical Methods in Spatial Flood Susceptibility Mapping: Case of the Upper Oum Er-Rbia Watershed

Urban Heat Island in Algiers: Case of the Belcourt District

Design Urban Park. Role of Geoinformatics in Urban Planning

Impact of Human Activities on the Bay of Algiers (Algeria)

GIS Contribution for Mapping of Flood-Hazard Areas in Arid Tunisia: Application to the Maknassy Basin (Lban-Ouadran Wadi Watershed, Tunisia)

Assessment of Underground Water Quality for Irrigation Proposes Using GIS and Remote Sensing Techniques

List of Tables

Design Urban Park. Role of Geoinformatics in Urban Planning

Impact of Human Activities on the Bay of Algiers (Algeria)

Using RS and GIS to Map Land Cover Changes in the East of Saudi Arabia: The Al-Ahsa Metropolitan Area

GIS Contribution for Mapping of Flood-Hazard Areas in Arid Tunisia: Application to the Maknassy Basin (Lban-Ouadran Wadi Watershed, Tunisia)

Assessment of Underground Water Quality for Irrigation Proposes Using GIS and Remote Sensing Techniques

Introduction to "Environmental Remote Sensing and GIS in MENA Region"

Faiza Khebour Allouche

Abstract The main focus of this chapter is to present summaries of the chapters provided in the book "Environmental Remote Sensing and GIS in MENA region". The book consists of 15 chapters including the introduction and the conclusion chapters. The 13 chapters are exposed under four themes (i) Overview of Remote Sensing and GIS applications and strategies; (ii) Geospatial technologies and risks; (iii) Urban applications and (iv) Rural applications.

Keywords Geo-information · Remote sensing · GIS · Google Earth · Radiometric indexes · Strategies · Planning · Monitoring · Flood · Degradation · Management · Environmental · MENA

F. K. Allouche (✉)
Department of Horticultural Sciences and Landscape, High Institute of Agronomic Sciences of Chott Mariem, Sousse University, Sousse, Tunisia
e-mail: allouchekhebour@yahoo.fr

Lr GREEN TEAM (LR17AGR01), National Agronomic Institute of Tunis, Carthage University, Tunis, Tunisia

F. Khebour Allouche (ed.), *Environmental Remote Sensing and GIS in MENA Region*, Earth and Environmental Sciences Library,
https://doi.org/10.1007/978-3-031-86574-9_1

1 Background

The review done by [8] to the research conducted on earth observation (EO) in the Eastern Mediterranean, Middle East, and North Africa region shows a gap in fundamental research in the big earth data thematic area. Future needs were identified by this study like the limited availability of very high-resolution and near-real-time remote sensing data, the lack of harmonized methodologies and the need for further development of models, algorithms, early warning systems, and services. Recently [21], has published a book who provides theory, methodology, and future perspectives of remote sensing and GIS techniques applied to peri-urban modelling, analysis and sustainability through the use of spatio-temporal geospatial datasets. However, the increasing availability of satellite EO data and trends of innovations in geospatial technology demonstrated through different applications mentioned in this book will play an important role in the transfer of techniques to make the right decisions in development, planning, and reduction of risks that threaten the MENA regions. These applications will be useful for researchers, decision-makers and students to achieve sustainable development goals.

Scientists from Jordan, Egypt, Saudi Arabia, Algeria, Morocco and Tunisia put their great efforts for about two years to produce this unique book for all who are interested in MENA regions and its unique features.

The book's intention is, therefore to improve and tackle the following main theme.

- Overview of Remote Sensing and GIS applications and strategies
- Geospatial technologies and risks
- Urban applications
- Rural applications

The summary of the objectives of the chapters under each theme will be presented in the following section. However, this book comes as a continuation of the previous efforts to support the environmental remote sensing and GIS in MENA region. It is the fourth book to be published in "Springer Water" which is follower to the book "Environmental Remote Sensing and GIS in Tunisia" which was edited by Faiza Khebour Allouche and Abdelazim Negm (2021); "Environmental Remote Sensing and GIS in Iraq" which was edited by Ayad Al-Quraishi and Abdelazim Negm (2020) and the book "Environmental Remote Sensing in Egypt" which was edited by Elbeih et al. (2020). The follower book will focus on applications in the MENA regions.

2 Themes of the Book

Therefore, the book intends to address in more detail the following main themes.

- Overview of Remote Sensing and GIS applications and strategies
- Geospatial technologies and risks
- Urban applications
- Rural applications

3 Chapters’ Summary

The next subsections present the main technical elements of each chapter under its related theme.

3.1 Overview of Remote Sensing and GIS Applications and Strategies

This theme is covered in four chapters from 2 to 5. The first chapter assessed “GIS and Remote Sensing for a Better Understanding of Ancient and Modern Rainwater Harvesting Strategies and the Identification of Current Potentials and Improvement Possibilities.” In recent times, there has been an increased interest in the adoption of old or traditional RWH strategies. This renewed interest can be seen as part of the quest for solutions to the water shortages, which represent an episodic or permanent problem in many parts of the world such as in the Middle East, who the situation is further aggravated by population growth, which entails a higher water demand. In the field of RWH, a common task is the improvement of decision-making concerning the selection of suitable RWH sites, which can be accomplished with GIS-based spatial decision support systems [1] and [3]. Furthermore, GIS tools and extensions are often used to model and estimate surface runoff, and develop a map of the RWH potential of a given area or region [24] and [26].

Thus, this chapter presents a pioneering approach to assessing and understanding rainwater harvesting (RWH) strategies, drawing from a case study from central West Jordan (Karak Plateau). By blending traditional fieldwork methods and modern remote sensing technology, particularly the analysis of very high-resolution (VHR) satellite imagery like GeoEye-1 data, firstly, efficient data collection and the compilation of a comprehensive, area-wide inventory of RWH structures becomes possible. Integrating the latter with collected on-the-ground data, particularly regarding the dimensions of underground cisterns, facilitates the estimation of the total water storage potential available with the existing cisterns in the area. Additionally, outcomes from satellite imagery interpretation, including rooftop area and land use/land cover (LU/LC) information, allow for estimating the overall RWH potential and the rooftop rainwater harvesting (RRWH) potential for each settlement. The data enable a spatially explicit comparison between storage and harvesting potential, thus making the identification of discrepancies, untapped opportunities, and suitable actions for development possible. Moreover, the chapter explores spatial patterns in RWH structure distribution concerning potential relations to environmental factors and the historical development of the area. Overall, these analyses offer valuable insights into the development of RWH strategies and steps for their adoption and improvement in modern contexts. Notably, the findings are applicable beyond the studied region, while the methodological approach holds promise for addressing research questions beyond RWH assessment.

Chapter “The National Strategy for the Development of Small Pelagic Fisheries in the Northeast of Tunisia (Central Mediterranean): Status and Prospects”, discuss a review of the small pelagic fisheries' status in the Northeast of Tunisia (GFCM-GSA 12 & 13) before and after implementing of the National Strategy for the Development of Small Pelagic Fisheries (NSDSPF) is presented. Then, the Geographic Information System and the Vessel Monitoring System were used to map the fishing areas' evolution between 2003, when the national strategy for the development of bluefish capture (NSDSPF) was established, and 2021, revealing a great expansion in small pelagic fish harvesting. This strategy was based on the results of the research conducted by INSTM, which revealed the availability of an important exploitable potential of small pelagic species in Tunisian waters [4, 9]. A proactive approach to resource management is demonstrated by the emphasis on continued support, monitoring and modernization. In addition, proposals for the exploration of new fishing areas and the introduction of a biological rest period demonstrate a commitment to long-term sustainability. Overall, the chapter paves the way for the continued growth and development of the sector by presenting innovative solutions to the challenges it faces.

The next chapter is about "The use of Geographic Information System (GIS) Data to Assess and Rebuild Stocks in Tunisia's Largest Freshwater Reservoirs". Water resources in Tunisia, are unequally distributed in time and space due to the country's arid climate. However, there was a proposal to exploit such water resources for aquacultural, economic, and ecological purposes since certain types of fish can maintain an ecosystem's balance in these water bodies thanks to their feeding behaviour [15, 17]. However, freshwater fishery resources in Tunisia have just been tapped compared to sea fishing activities [7, 27]. In this context, this chapter presents a pioneering approach to resource management that combines cutting-edge technology with conservation efforts by using Geographic Information System (GIS) data to assess and reconstruct stocks in Tunisia's largest freshwater reservoirs. GIS tools and spatial interpolation known as Inverse Distance Weighting (IDW) is used to estimate the distribution of freshwater fish species between the reservoirs of the northern sub-basins of Medjerda (Kasseb) and those of the southern sub-basins (Siliana). This provides an original perspective for understanding freshwater ecosystems and developing sustainable fisheries strategies. In addition to improving the accuracy and efficiency of stock assessment, this new use of GIS will provide a framework for informed decision-making, which is essential for the conservation and restoration of freshwater resources in Tunisia's reservoirs.

The last chapter focuses on "Remote Sensing Applications with Google Earth Engine: Case from Algeria." Due to the continuous increase in the volume of geospatial data and the creation of storage tools and the cloud to process data. The objective of this chapter is to describe how a method of analysis of remote sensing data put into practice by deploying interactive applications that rely on the resources of Google Earth Engine without the need to be an expert in web programming or HTML. This platform is a powerful cloud-based platform that allows researchers to analyze massive amounts of geospatial data for various applications [19, 20]. The chapter aims, then, to introduce some basic function used to treat data of remote sensing for many applications and to calculate the Normalized difference vegetation index NDVI, and Normalized difference salinity index NDSI from Sentinel and Modis images for the North of Algeria.

3.2 Geospatial Technologies and Risks

This theme is covered by four chapters, 6 to 9. The first chapter deals with "Coastal Eutrophication Detection Using Sentinel-2 and Impact of Climatic Parameters". In coastal waters, eutrophication has become more frequent over the last ten years on a global scale [10] and can have serious effects on the marine ecosystem and human health, as well as negative economic consequences affecting fisheries and pleasure areas [14]. This is the case of the northern part of Monastir Bay, located on Tunisia's eastern coastline, who shows signs of eutrophication, with spectacular algal growth beginning during the winter season (green tide) and leading to chronic anoxia [22]. Thus, the main objective of this chapter is to assess the phenomenon of eutrophication in the Monastir region. The specific objectives include: (i) the study of water bodies; (ii) characterization of eutrophication using radiometric indexes and (iii) monitoring this phenomenon according to different factors. 100 dates on which eutrophication appeared on the Tunisian coast were collected, from 1988 to 2021, covering the Tunisian coastline from Mahdia (Rejiche) to Djerba/Zarzis. GIS data containing vector and raster shapefiles was implemented. Based on sentinel-2 MSI imagery, the NDWI and NDCI radiometric indices applied to assess eutrophication phenomenon and the monitoring of it is based on the exploration of a temporal sentinel series from EO browser. In another part of the chapter, the characterization of the eutrophication phenomenon according to physical and anthropogenic factors is presented.

The "Performance of Modelling Based on Statistical Methods in Spatial Floods Susceptibility Mapping: Case of the Upper Oum Er-Rbia Watershed" is discussed in the chapter. Flood risk management is nowadays part of most governmental and UN organizations strategies for disasters risk reduction. However, in the MENA region where countries represent different climatic, hydrological, land use, storms characteristics and observational capacities, these strategies are not yet integrating flood risk issues [16]. Then, to define the flood-prone area within the high Oum Er Rbia basin, Middle Atlas region, Morocco, the authors establish a methodology using statistical methods such as frequency ratio, weight of evidence, and weighting factor based on the mathematical relationship between observed flood and controlling factors. The FR, WF, and WoE method was integrated with remote sensing and geospatial techniques. Then, detailed hazard maps were reclassified into five categories: very low, low, moderate, high, and very high susceptibility to flooding.

The third chapter "Urban Heat island in Algiers: case of the Belcourt District", addresses this phenomenon in a region who is densely populated and dominated by the residential urbanization that represent about 75% including the Hamma botanical garden (32 ha). So, it is one of the MENA regions who temperatures are predicted to increase by 3 °C by 2050 [13]. However, to assess UIH in the study area, authors used an empirical and simple approach developed by NEPSEN-E6, that consists of attributing a UHI score to each element of the ground surface: plant, buildings, parks and roads. The UHI score tool proposes to classify different parameters in nine temperature bands from 0 for coolest island to 1 for heat island. It takes into account the urban surfacing materials, their position in the shade or the sun, the presence of vegetation, water, and various factors. Nevertheless, since it is considered that the permeability of certain urban elements plays a role as heat or cool islands like albedo and the color of materials, another parameter is added to the UHI score tool: the water permeability score. Thus, the integration and combination of these factors (UHI score and permeability) in a GIS tool allowed authors to create the current and projected UIH maps. This research can provide a decision-making tool for the municipality and ensure effective action to reduce the impact of UHIs.

The fourth chapter deals with "Stochastic Modeling and Prediction of Four Annual and Monthly Stream Flow Time Series in Northern Algeria". To cope with water resources lack and scarcity, under climate change and increasing needs for drinking, developing agronomy, industry and environment, the new Algerian policy is oriented towards the integrated management of water resources [11]. Besides desalination and water storage infrastructure, that are expensive and time consuming solutions, one of the cheapest, easiest and most sustainable solution is the optimal management that relies principally on predictive models. The major problem in Algeria is that hydrological variables like stream flows, which are stochastic in nature, have always been considered as completely random so not predictable.

The present work provides with some stochastic models that help dissociate the structured part of data from the random one inside the observed noisy stream flow time series. The extracted structured part is then formulated into a mathematical model, hence prediction is made possible. The first part of this chapter exposed the Box-Jenkins theoretical background and its application to the annual and monthly stream flows observed at four hydrometric stations in northern Algeria, all located on operating dams. The second part of this chapter was rather interested to demonstrate that these models allow to dissociate the random variability from the structured one within a time series. This can be done taking into account the stochastic nature of the hydrological variable of concern. The recommended method is a procedure that is essentially based on the internal dependence structure of the data, and its theoretical foundations are based on the assumptions of normality and stationarity. Indeed, the contribution of groundwater flow can be considered as the auto-regressive (AR) part, while random precipitation can be considered as the moving average (MA) one.

3.3 Urban Applications

Part III has the theme "Urban applications" and is covered in three chapters. The first chapter is titled "Spatio-Temporal Variation in Urban Vegetation in Maghrebian Cities Since the XXe Century Using Remote Sensing". In the Maghreb, vegetation cover, a crucial factor for the sustainability of urban ecosystems, is undergoing significant change in cities. After the French colonial period giving key consideration to public vegetation in the new towns built such as in Sousse, Algiers or Marrakesh, the importance attached to greenery in cities is not clearly defined in independent Tunisia, Algeria and Morocco, and today consideration of urban vegetation in Maghrebian spatial planning policy is still nascent [25]. The aim of the chapter is to document the spatio-temporal variations in urban vegetation in the three cities using a spatialized approach involving satellite remote sensing and to determine the factors of variation. The comprehension of the social, economic and political forces driving these trends and its effects on the urban space of the cities is then deduced. However, the methodology adopted in this research has three main steps: The first step is to determine the extent of the green spaces across the three cities and its variation over time from a set of Landsat satellite images dated 1980's and 2020's. During the second step, built-up areas (artificial surfaces) have been determined for a spatial comparison for the year 2020 using Landsat satellite images. For Marrakesh, as an example of Maghrebian cities, the method have been developed in order to have details about the evolution of the city and its vegetation. The method has been applied to Landsat satellite images from 1973 to 2020 and a last step involves the use of NDVI data at 8-day intervals from 2000 to 2020, to confirm the results and to compare the variation in vegetation with rainfall rates for Marrakesh.

The second chapter is devoted to "Design Urban Park Role of Geoinformatics in Urban Planning". Since, GIS is important to analyze and assess urban landscapes [12]. This chapter focuses on the use of a method combining landscape analysis and spatial organization of units to understand the evolution of urban spaces. The methodology developed in this research comprises three steps. First, the evolution of landscapes and environments in the study area was examined using Google Earth images. Then, a survey was performed to find the concept that would drive the development project. Third, the urban park's design reflected the unique qualities of the region. Solutions that use geoinformatics in urban planning can have an impact on the development of these areas' designs and planning. The chapter proposes the advantages of Google Earth application in urban design, which is useful to get a full-fledged geographic information system.

The last chapter treats the "Impact of human activities on the Bay of Algiers". The Mediterranean coastal zone is one of the most fragile and densely urbanized areas in the world [23]. From this observation, the authors of this chapter applied a diachronic study on the bay of Algiers using Geographic Information System (GIS) and imagery to assess the expansion of the mineral element often to the detriment of the plant element, and the necessity of opening the bay of Algiers and its infrastructures onto the sea. From 2001 to 2022, the study area lost 2.55 km^2 of vegetation, this represents 6.76% of the total surface area for the year 2022. Moreover, the cartographic results show that coastal development is not in line with the concept of the green and blue belts to the costal and sea. Developments are conceived and built without integrating notions of ecological connectivity and the city's current dynamic development reinforces the feeling of enclosure and encourages people to turn their backs on the sea.

3.4 Rural Applications

The last theme in the book contains three chapters, from 13 to 15 chapter. The first chapter is titled "Land Cover Transformation Mapping using GIS and Remote Sensing in Saudi Arabia: Case of Al-Ahsa Oasis". Because of severe natural and climatic conditions, a few areas of the Kingdom of Saudi Arabia (KSA) are populated: towns, the oasis, cultivated areas, or artificial communities built for the petroleum industry or at crossing points on the main roads [6]. Thus, this chapter aims to map the recent evolution of land use and land cover (LULC) in Al-Ahsa Metropolitan Area (Saudi Arabia) between 1990 and 2022. To understand this complex theme, an integrative, global, and hierarchical methodological approach, combining multi-source and multi-date data was adopted. This approach involves the creation of a prototype land-use map based on the strict rules of graphic semiology. The main cartographic results present the various changes in land use and land cover. However, the application aims to highlight the importance of change detection methods for understanding urbanization patterns and assessing their impacts on the oasis ecosystem. Post-classification comparison of land-use maps revealed a growth in built-up areas to the detriment of the old oasis center and bare soil. The Remote Sensing (RS) and Geographic Information System (GIS) approaches are highly effective and offer real added value in visualizing land occupation and land usage. All these documents are indispensable and highly important for rational planning.

The second chapter focuses on "GIS Contribution for Mapping of Flood-Hazard Areas in Arid Tunisia: appLication to the Maknassy Basin (Lban-Ouadran Wadi Watershed, Tunisia)". Flooding is a scourge that threatens the economy in the MENA region, especially as the number of flash floods and people affected or killed has doubled over the last ten years [5]. In this way, this research forms part of the economic development effort in the MENA region by mobilising soil and water resources and preserving and mobilising water resources on a regional and local scale. Furthermore, understanding the flood hazard helps to prevent this risk and to combat this scourge. Especially, this chapter focuses on the use and contribution of the GIS to the study and management of flood hazards in the Maknassy basin (the Lban-Ouadran wadi watershed) in Tunisia. The innovative approach used lies in the use of GIS to study and map flood risks, based on identifying the relationship between the flood hazard and the hydro-morphometric characteristics (drainage density, hydrological density, compactness index) of order 1 impluviums, lithology and slope values.

Following Chapter 14 comes Chapter 15 to discuss the "USE of Remote Sensing in Monitoring and Evaluation of Land Degradation Processing". Groundwater in arid/semiarid regions plays crucial roles in providing water supply, supporting irrigated agriculture, and sustaining ecosystems (El-Rawy et al. 2021). Nowadays, Egypt pays attention to new land reclamation projects to achieve self-sufficiency from strategic crops and to establish new settlements. West-West El-Minya area is considered one of the promising areas for future development plans in Egypt [18]. The innovative aspect can be highlighted in the current chapter through the observed relation between increased population growth and geographic practices. Such relation could be confirmed by focusing on the resulted impact of climate change in precipitation patterns, water quality and supply, to be considered major global issues playing an important role in agricultural planning and reclamation. Innovation extends to employing the Piper diagrams in assessing water quality, where evaluation in terms of the risk associated with salinity and sodicity were considered.

Acknowledgements The writers who wrote this chapter would like to acknowledge the authors of the chapters for their efforts during the different phases of the book including their inputs into this chapter.

References

1. Al-Adamat R, Diabat A, Shatnawi G (2010) Combining GIS with multicriteria decision making for siting water harvesting ponds in Northern Jordan. J Arid Environ 74:1471–1477
2. Al-Houri Z, Al-Omari A (2022) Assessment of rooftop rainwater harvesting in Ajloun, Jordon. Water Reuse 12(1):22–32
3. Al-Khuzaie MM, Janna H, Al-Ansari N (2020) Assessment model of water harvesting and storage location using GIS and remote sensing in Al-Qadisiyah, Iraq. Arab J Geosci 13:1–9
4. ANPE (National Agency for Environmental Protection) (2005) Sustainable fishing: Tunisian observatory for the environment and sustainable development. Ministry of Environment and Sustainable Development, 61p
5. Banerjee A, Bhavnani R, Burtonboy CH, Hamad O, Linares-Rivas Barandiaran A, Safaïe S, Tewari D, Zanon A (2014) Natural disasters in the Middle East and North Africa: a regional overview (English). Global Facility for Disaster Reduction and Recovery (GFDRR). World Bank Group, Washington, DC. http://documents.worldbank.org/curated/en/211811468106752534/Natural-disasters-in-the-Middle-East-and-North-Africa-a-regional-overview
6. Dhieb M (2022) A bivariate dasymetric population map of Saudi Arabia. Curr Urban Stud 10:673–696. https://doi.org/10.4236/cus.2022.104040
7. Djemali I (2005) Evaluation of fish biomass in Tunisian freshwater bodies: analytical and acoustic approach. Doctoral thesis in Agronomic Sciences at the National Agronomic Institute of Tunisia 250 p
8. Eliades M, Michaelides S, Evagorou E, Fotiou K, Fragkos K, Leventis G, Theocharidis C, Panagiotou CF, Mavrovouniotis M, Neophytides S (2023) Earth observation in the EMMENA region: scoping review of current applications and knowledge gaps. Remote Sens 15(17):4202. https://doi.org/10.3390/rs15174202
9. Fezzani S (2023) Ecobiological and dynamic study of the horse mackerel Trachurus trachurus from the northern and eastern coasts of Tunisia. Doctoral thesis of the National Agronomic Institute of Tunisia, 248 p
10. Glibert PMS, Heil SCA, Burkholder JM, Parrow MW, Codispoti LA, Kelly V (2005) The role of eutrophication in the global proliferation of harmful algal blooms. Oceanography 18(2):198–209. https://doi.org/10.5670/oceanog.2005.54
11. Kherbache N (2020) Water policy in Algeria: limits of supply model and perspectives of water demand management (WDM). Desalin Water Treat 180(2020):141–155. https://doi.org/10.5004/dwt.2020.25009
12. Khebour Allouche F, Hamaideh A, Hammami S, Gatta, E & Khasawneh A (2021) Envisioning sustainable landscape management in the environmentally sensitive areas of Ksour Essef, TUNISIA. Int J Res Granthaalayah 9(1):85–98. https://doi.org/10.29121/granthaalayah.v9.i1.2021.2715

13. Lelieveld J, Proestos Y, Hadjinicolaou P, Tanarhte M, Tyrlis E, Zittis G (2016) Strongly increasing heat extremes in the Middle East and North Africa (MENA) in the 21st century. Clim Change 137(1):245–260. https://doi.org/10.1007/s10584-016-1665-6
14. Lopez-Rodas Huertas IE, Rouco MV, Costas E (2008) Warming will affect phytoplankton differently: evidence through a mechanistic approach. Proc Roy Soc Biol Sci 278:3534–3543
15. Loss GF, Nau W, Winter M (1991) The development of freshwater fishing in northern Tunisia. Study carried out within the framework of the Tunisian-German technical cooperation GTZ Gmbh: Tunisia 403 p
16. Loudyi D, Kantoush SA (2020) Flood risk management in the Middle East and North Africa (MENA) region. Urban Water J 17(5):379–380. https://doi.org/10.1080/1573062X.2020.1777754
17. Mili S, Ennouri R, Laouar H, Khedhri I, Zarrouk H, Chargui T, Romdhane N (2021) Freshwater fish farming and fishery management in Tunisian reservoirs: limitations and opportunities. Agriculture Productivity in Tunisia Under Stressed Environment Springer Water, 34 p
18. Morsy SM (2023) Planning for groundwater management using visual MODFLOW model and multi-criteria decision analysis, West–West Minya, Egypt. Appl Water Sci 13:72. https://doi.org/10.1007/s13201-023-01881-x
19. Mutanga O, Kumar L (2019) Google earth engine applications. Remote Sens 11(5):591
20. Zhao Q, Yu L, Li X, Peng D, Zhang Y, Gong P (2021) Progress and trends in the application of Google Earth and Google Earth Engine. Remote Sens 13(18):3778
21. Sahana M (2024) Remote sensing and GIS in peri-urban research. Perspectives on global change, sustainability and resilience. Modern cartography series, vol 11. Elsevier, p 752
22. Sassi R, Souissi F, Soussi N, Boukaaba M, Belayouni H (1998) Geochemistry of organic matter for the analysis of coastal degradation of Monastir-Ksibet el Mediouni (eastern Tunisia). C.R. Acad. Sci. Paris. Earth and Planetary Sciences, pp 303–308
23. Satta A, Puddu M, Venturini S, Giupponi C (2017) Assessment of coastal risks to climate change related impacts at the regional scale: the case of the Mediterranean region. Int J Disaster Risk Reduct 24:284–296
24. Ouessar M, Bruggeman A, Abdelli F, Mohtar RH, Gabriëls D, Cornelis WM (2009) Modelling water-harvesting systems in the arid south of Tunisia using SWAT. Hydrol Earth Syst Sci 13:2003–2021
25. Taïbi AN, El Hannani M (2019) The colonial landscape in Africa. Plants in the public space of the colonial cities of Marrakech, Algiers, Antananarivo and Toliara. Africana Studia: Int J Afr Stud 32:27–42. ISSN 0874–2375
26. Yousif M, Bubenzer O (2015) Geoinformatics application for assessing the potential of rainwater harvesting in arid regions: case study: El Daba'a area, northwestern coast of Egypt. Arab J Geosci 8(11):9169–9191
27. Zaouali J (1981) Aquaculture problems: brackish waters and aquaculture potential. Arch Pasteur Inst Tunis 58:93–103

GIS and Remote Sensing for a Better Understanding of Ancient and Modern Rainwater Harvesting Strategies and the Identification of Current Potentials and Improvement Possibilities

Barbara Brilmayer Bakti, Arwa Hamaideh, and Olaf Bubenzer

Abstract Based on a regional case study from central West Jordan (Karak Plateau), the applicability, benefits and limitations of high-resolution remotely sensed imagery, semi-automatic image analysis with Feature Analyst, and GIS analyses are explored with respect to rainwater harvesting (RWH) practices. GeoEye-1 satellite imagery and derived land use/land cover information facilitated the mapping of existing RWH structures and the differentiated estimation of the regional RWH potential by the use of semi-automatic analyses. Hydrological analyses in ArcGIS, in combination with digital elevation model data (ASTER GDEM), allowed the detection and quantification of spatial patterns. The results, particularly those of the site characteristics and proximity analyses, provided valuable insights into the relationship between water resources, landscape, and settlements, and the changes over time in the management of water resources. Additionally, a comprehensive assessment of the regional RWH potential was achieved. In the around 160 km^2 big study area, over 1,400 individual structures associated with RWH were identified. This figure includes almost 400 cisterns of various age, many of which were probably built centuries to millennia ago. Their total storage capacity was estimated at around 21,270 m^3, which corresponds to approximately 0.38% of the calculated total annual on-site runoff from potentially suitable pervious areas. Additionally, a rooftop RWH potential of almost 154,000 m^3 was computed for years with average precipitation. In general, RWH is seen as a valuable part of an integrated water management strategy, in which it helps securing water supply and enhances resilience by increasing the diversity of available water resources.

B. Brilmayer Bakti · O. Bubenzer (✉)
Institute of Geography, Heidelberg University, Heidelberg, Germany
e-mail: olaf.bubenzer@uni-heidelberg.de

A. Hamaideh
Water, Energy, and Environment Center, University of Jordan, Amman, Jordan
e-mail: Hamaideh.arwa@ju.edu.jo

F. Khebour Allouche (ed.), *Environmental Remote Sensing and GIS in MENA Region*, Earth and Environmental Sciences Library,
https://doi.org/10.1007/978-3-031-86574-9_2

Keywords GIS · GeoEye-1 · Feature analyst · Rainwater harvesting · Cisterns · Jordan · Image analysis · Object detection

1 Introduction: What is Rainwater Harvesting and How Are GIS and Remote Sensing Used in This Context?

Rainwater harvesting (RWH) is an umbrella term used to describe a variety of methods to collect, store and apply rainwater and resulting runoff for different purposes (e.g. [28]). RWH strategies can include, for example, the use of shallow walls, bunds, ridges, dikes, channels, and check dams to redirect, hold back or spread surface runoff, as well as cisterns or open reservoirs for water storage and structures for the artificial recharge of groundwater aquifers. Many of the RWH practices known today have a long tradition and were developed millennia ago. Depending on the region and structures considered, the earliest traces of RWH are thought to date back to the Neolithic or even earlier periods (e.g. [13, 23, 29]). In many places, RWH activities probably began with sedentism, fueld by the need to secure sufficient water resources at settlement sites or nearby. This particularly applies to the semiarid to arid areas of the world, such as the MENA region (Middle East and North Africa). Similar RWH techniques were developed and employed in many different parts of the world to overcome temporary and/or local water shortages and manage the available water resources most efficiently (e.g. [1, 4, 11, 19, 28, 29]). Due to their long history and ancient roots, many RWH practices often are also described as traditional or indigenous water management.

In recent times, there has been an increased interest in the adoption of old or traditional RWH strategies, while occasionally new or modified RWH techniques have also been proposed (e.g. [3, 17, 18, 26, 44, 49]). This renewed interest can be seen as part of the quest for solutions to water shortages, which are an episodic or permanent problem in many parts of the world. Water deficits are expected to increase under climate change conditions leading to rises in temperature, more erratic precipitation patterns, and often decreases in annual precipitation amounts. In many of these regions, such as e.g. the Middle East, the situation is further aggravated by population growth which entails a higher water demand. Typically, RWH practices are seen as ways of managing and exploiting local to regional natural water resources efficiently and sustainably, in contrast to e.g. the exploitation of nonrenewable, fossil ground water resources. Moreover, RWH is usually based on decentralized, low-tech strategies which also makes it suitable for less developed and/or remote regions. Although typically associated with and practiced in the dry regions of the earth, RWH efforts are not restricted to those. Most RWH schemes include temporal intermediate storage of collected water, e.g. in cisterns, reservoirs or soil profiles, or otherwise delay, spread, or reduce the surface runoff resulting from rainfall in a given area and time. Thereby, RWH helps to reduce floods and other detrimental effects of intense rainfall and runoff events. Thus, RWH can also benefit areas with high annual precipitation amounts, particularly when those are associated with intense rainfalls.

Geographic information systems (GIS) and remote sensing techniques have been employed in many RWH studies, at least in the last two decades, while the purposes they serve can largely be divided into two broad categories. Remote sensing was typically used to acquire necessary auxiliary information, such as e.g. data on land use/land cover (LU/LC) or the terrain via digital elevation models (DEMs), or even climate or weather data. For many areas of the world, this information is still not readily available from other sources, e.g. local government institutions and ground sensor data, or the quality of the available data did not suit the requirements of the study objectives, e.g. in terms of spatial resolution, currentness of data or similar. Yet, data on the environment is indispensable for the investigation of RWH schemes. Therefore, this information often is obtained through the interpretation of visual satellite imagery, from DEMs derived from suitable stereoscopic satellite imagery, and through other remote sensing products and techniques. GIS has mostly been established as the standard environment for data processing and integration. Moreover, its capabilities are employed for hydrological analyses and runoff modeling, based on DEMs and auxiliary data where the latter are available, and further spatial analyses. In the field of RWH, a common task is the improvement of decision-making concerning the selection of suitable RWH sites, which can be accomplished with GIS-based spatial decision support systems (e.g. [5, 8, 27, 34, 37, 50]). Furthermore, GIS tools and extensions are often used to model and estimate surface runoff, and develop a map of the RWH potential of a given area or region (e.g. [21, 35, 39, 48, 25]). To conduct these analyses and achieve a realistic estimation of the RWH potential, a number of environmental input data is necessary, which often can only feasibly be obtained through remote sensing.

In the case study presented in the following sections, GIS and remote sensing were applied in ways similar to those described above. Yet, compared to other, previous studies, some modifications were made to the methodological approaches and procedures, and the study design in parts differed with regard to the perspectives and research objectives. Firstly, the interpretation of visual satellite imagery was used not only to acquire LU/LC information but also to map the existing RWH structures directly and thus compile a comprehensive inventory. This data formed the basis for the subsequent GIS analyses which were conducted to gain a better understanding of current and former RWH strategies, as well as the site selection process. In this sense, an inverted version of a GIS-based decision support system was set up. Here, thorough analyses of the characteristics of actual RWH sites should provide deeper insights into the factors that led to the observed decisions. Additionally, RWH potential maps were created and evaluated by comparing the estimated potentials to the calculated storage capacities of the mapped existing cisterns and their distribution. Finally, based on the collected data and analyses outcomes, recommendations could be derived for possible suitable further improvements and/or expansions of RWH efforts in the area.

2 Case Study: Investigating RWH in Jordan—Environmental and Cultural Circumstances

When studying RWH, Jordan provides an ideal setting, due to two main factors: its natural water shortage and its long settlement history. Together, these two factors have spawned a variety of different solutions for the effective management of the scarce local natural water resources. Traces and remains testifying to Jordan's rich history of water management, including RWH, are still found in many places throughout the kingdom. Over time, Jordan or some of its territory have formed parts of the Nabataean, the Roman, the Byzantine, and the Ottoman Empire. The continuous or recurrent interest in the territory throughout history can at least partly, if not mostly, be explained with its geostrategical position, being located between the shores of the Mediterranean Sea with its harbors in the west and the Arabian hinterland with its deserts in the east. Moreover, important trade and migration routes were running parallel to the Dead Sea and the Jordan Valley, connecting Africa and the Red Sea region in the south with Westasia and southeast Europe in the north. Hence, Jordan is situated at an important crossroads position between east and west, and north and south.

The above described also applies to the Karak plateau, a highland area directly east of the Dead Sea in the central western part of Jordan. In the northern half of the plateau, an area of approximately 160 km^2 extending from the central part of the tableland to the east was selected for the case study (see Fig. 1). Altitudes there range from around 1,000 to 700–800 m above sea level, and the relief is softly undulating. The study area is bordered by steep slopes leading down to Wadi al-Mujib in the north and a southward trending branch of Wadi al-Mujib, Wadi Nukhayla, in the east. In the central parts of the plateau, a semi-arid Mediterranean climate prevails with over 300 mm annual precipitation on average, a warm to hot, dry summer season, and moderately cold to cold, humid winter months. Towards the east, as continentality increases, the climate becomes drier and transitions into a steppe climate and finally into the cool desert climate of the Badia or Eastern Desert of Jordan. The transitional character reflected by the climate can also be observed in the landscape. The relatively flat terrain, together with the semi-arid Mediterranean climate and the mostly deep, fertile soils make the western and central parts of the Karak plateau highly suitable for rainfed agriculture. In contrast, towards the east, and particularly in the southeastern part of the study area, the climate becomes more arid, the terrain more rugged around the valley of Wadi Nukhayla and its numerous smaller tributaries, and the soil cover is mostly shallow. Therefore, this part of the study area is predominantly used as grazing land.

In general, the study area and the Karak plateau as a whole can be characterized as a rural, agricultural highland area at the desert margin. The area is not densely populated, with a total of around 23,600 people living in 12 individual settlements in 2015 [15]. More than half of the local population are residents of the two largest towns, ar-Rabba and al-Qasr. Most settlements (nine out of the twelve) are located in the central to western part of the highland, along or near the King's highway,

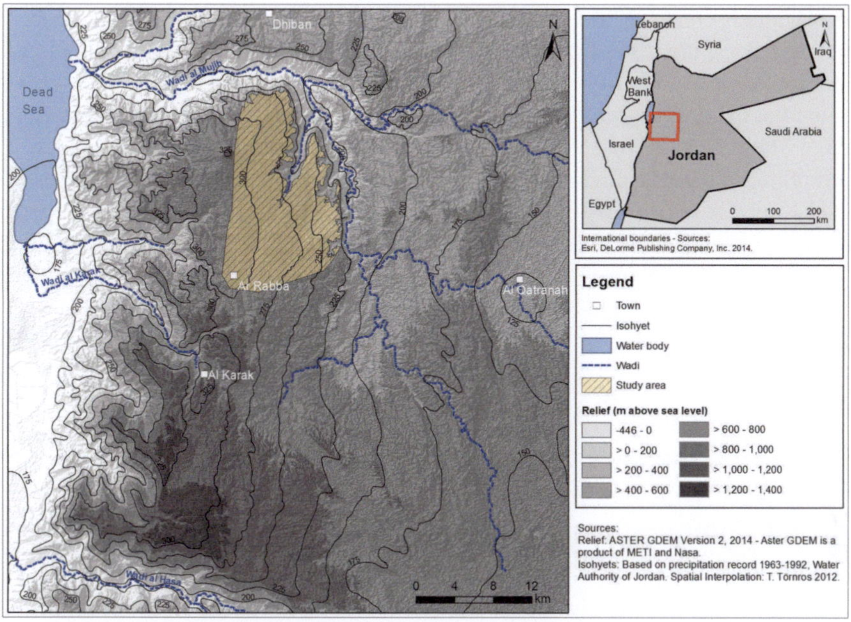

Fig. 1 Location of the study area in Jordan, altitudes, depicted by different shades of gray for each altitude class, and precipitation in the region, indicated by isohyets labeled with mean annual values according to the collected and interpolated data

the main traffic route leading over the plateau. The King's highway and many of the modern settlements have ancient roots. For example, today's King's highway roughly follows the course of the Roman Via Nova Traiana over the Karak plateau, while the modern town of ar-Rabba was built in the place of Roman Rabbathmoab or Areopolis and the settlement of al-Qasr developed around a Nabataean temple. In both aforementioned places, parts of the buildings from Nabataean and Roman times have been preserved and can still be seen today. With respect to its history, the area experienced its greatest prosperity during Nabataean, Roman, and Byzantine times (ca. 300 BC to 600 AD), which is reflected by numerous finds and remains from these periods at many different sites.

Since ancient times, settlers and travelers on the Karak plateau had to rely either on some form of RWH or on transporting water over considerable distances to cover their essential daily water demand. This can be deduced from the absence of lakes, ponds, streams, and springs on the tableland itself. Thus, natural sources of freshwater, such as springs and streams, are only available around the fringes of the plateau, in lower laying areas. Therefore, precipitation and resulting runoff in the rainy season were a valuable source of freshwater which needed to be collected and stored to ensure a sufficient water availability also during the dry summer months. This at least applies to the time before the advent of a public water line and water pumping which allowed the distribution of water resources from various places over great distances. Today, most buildings on the Karak plateau are connected to the public water line and people mainly rely on tap water for most of their needs. The water distributed comes from various sources, primarily groundwater (renewable and nonrenewable) and surface water. For the collection and storage of the latter, dams have been constructed along major streams, such as e.g. the Mujib dam north of the Karak area. Yet, due to intermittent supply through the public water system, water storage is still a necessity and is typically fulfilled with storage tanks on the rooftops of buildings. The restricted supply, a growing demand, and the overexploitation and partial reliance on declining, non-renewable water resources, such as fossil groundwater reserves, have rekindled the interest in RWH practices on various levels in Jordan (e.g. [6, 9]).

3 Materials and Methods: Interpretation of GeoEye-1 Satellite Imagery for Data Collection and Mapping, Spatial Analyses in a GIS Environment, and Computation of the RWH Potential

The study design has already been outlined briefly in the introduction. In the following sections, details on the methodological approaches and the materials employed to realize the study's research objectives are given. An overview of the methodological approach and the procedures, materials, and respective outputs involved can be found in Fig. 2.

3.1 Mapping and Investigating Existing RWH Structures

For the first study objective, the collection of data on existing RWH facilities in the area, a combination of remote sensing and field work was used. During the first field work campaign, primarily cisterns that were labeled as ancient or very old ("Roman wells") by representatives of the local communities were mapped with GPS devices. This data was then used as training examples for the detection of further cisterns in the following remote sensing-based mapping process. The latter was based on two GeoEye-1 satellite imagery scenes recorded on July 21 and 29, 2010, which together covered the study area. GeoEye-1 imagery was chosen due to its very high spatial resolution which allows the detection of smaller objects and structures. The images were selected as suitable since they had almost no cloud cover and were readily available from the GeoEye-1 image repository at the time of purchase. GeoEye-1 satellite images have four multi-spectral (MS) and one pan-chromatic (pan) band that cover the range of the optical light and the near-infrared spectrum. The images are recorded with a spatial resolution of 1.65 m in the MS bands and 0.41 m in the pan-chromatic band, and were resampled to 2 m (MS) and 0.5 m (pan) resolution due to legislative provisions at the time of purchase. After preprocessing (see section "Semi-Automatic Analysis of Satellite Imagery for the Extraction of LU/LC Information"), the imagery was interpreted visually-manually to detect further cisterns and other RWH structures, such as bunds, ridges, dikes, shallow walls made of earth and/or stones, and similar. Here, a visual-manual approach was necessary, in contrast to a (semi-) automatic image analysis procedure, as the target objects varied largely in shape, size, and visual appearance and often were very small, hardly discernible, in the case of cistern manholes. Then, in a second field work campaign, a ground check was conducted for a small subset of the RWH structures that were mapped based on the satellite imagery.

During the fieldwork campaigns, also, additional information was gathered on the RWH facilities, primarily the old cisterns. This included the written and photographic documentation of their condition, the respective setting, construction details, and similar. To gain further insights and collect additional data, a small subset of allegedly old cisterns was also examined from the inside. During inside inspections, measurements were taken at crucial points to capture the approximate size and shape of the respective cistern. Additionally, further construction details, e.g. concerning the plaster lining used for waterproofing of the cistern, were documented with photographs.

The storage capacities of the inspected cisterns were calculated based on the measurements from the inside and three-dimensional figures approximately matching the cisterns' shapes. The findings on common dimensions of cisterns could then be extrapolated using the data resulting from the remote sensing-based mapping process, i.e. the total number of cisterns identified for the area. Thus, the overall storage potential available with the existing cisterns could be estimated and the spatial distribution of storage potentials could be assessed.

3.2 Semi-automatic Analysis of Satellite Imagery for the Extraction of LU/LC Information

The GeoEye-1 satellite imagery was preprocessed to ensure the best possible outcomes in the subsequent image analysis process. First, the raw imagery was orthorectified in ERDAS IMAGINE using a DEM with 5 m spatial resolution, resampled to 2 and 0.5 m resolution to match the imagery. A substantial number of cisterns mapped with GPS in the first field work campaign served as ground control points to further enhance the location accuracy of the resulting orthorectified image. Subsequently, atmospheric corrections were made with ATCOR in ERDAS IMAGINE. Finally, a panchromatic sharpening procedure was applied and the two image scenes were merged into one image using the MosaicPro module in ERDAS IMAGINE.

The preprocessed GeoEye-1 image was then interpreted semi-automatically with Feature Analyst (FA) to create a LU/LC map and identify rooftop areas. FA is available as an extension for ArcGIS. In contrast to many other object-based methods,

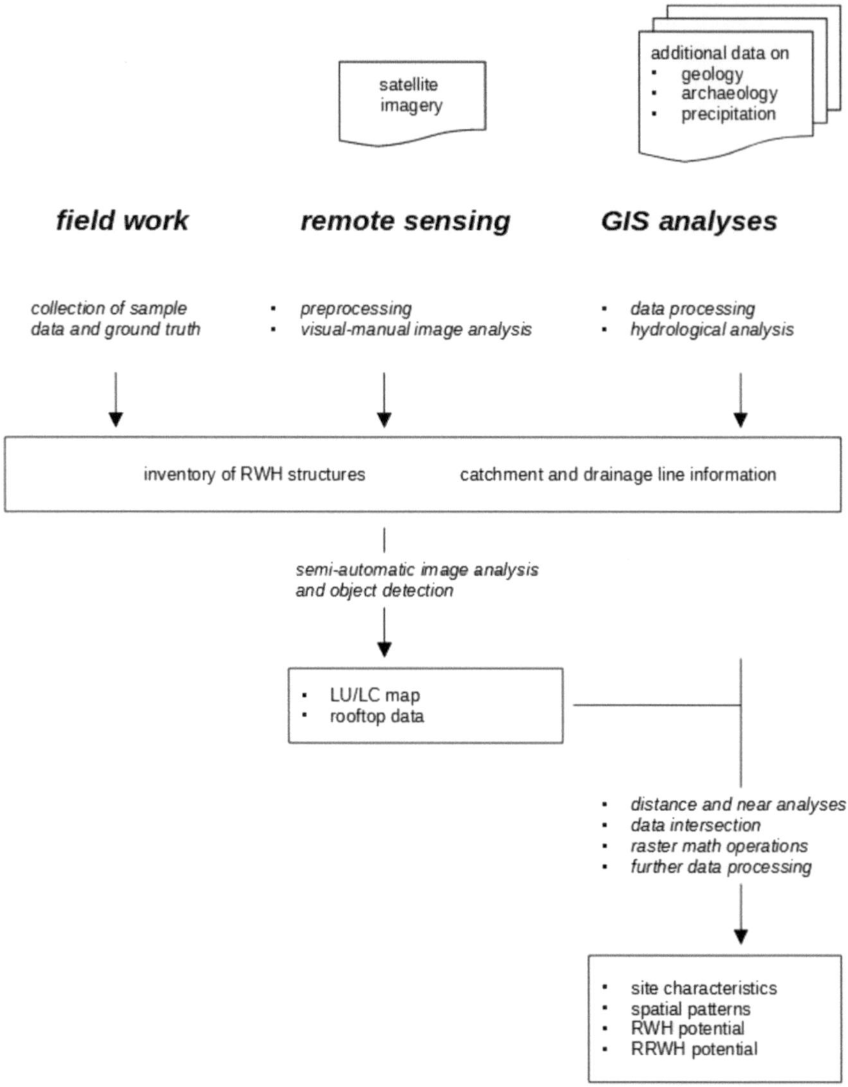

Fig. 2 Overview of the methodological concept showing the input materials (top row), the involved procedures (in italics), and the respective (intermediate) outcomes (information in boxes)

the FA concept does not include a segmentation step and instead individual pixels are classified directly. However, in the classification procedure, not only the spectral signature of the pixel under examination but also a sample of the surrounding pixels is considered. The sampling pattern, i.e. distance, arrangement, and number of pixels included, can be configured individually by the user. Thus, FA is able to capture and consider shapes, spatial patterns, and context, in addition to spectral values. The software mimics human visual perception and image interpretation and, thereby, is able to deliver highquality classification results. Another of FA's assets is that it relies on machine learning in which, for each classification task, an optimal ensemble of common algorithms is developed based on the user input.

At least when a supervised classification is conducted, the procedure in FA is not fully automated but semi-automatic, as the user needs to provide templates, i.e. manually digitized examples of the target objects or areas. Therefore, the FA concept has also been referred to as a template-matching approach (e.g. [20]). In this study, only the supervised classification mode of FA was employed since it is the only procedure that allows the definition of particular LU/LC classes and the detection of specific target objects. Each supervised classification starts with the user delineating areas on the image that serve as examples (training data) of each LU/LC class or group of target objects. Subsequently, the learning parameters can be adjusted as desired or settings that have been predefined for specific classification tasks can be selected. Next, the software analyzes the provided data (image, training data, and, if provided, auxiliary data) and performs the actual classification. The user can then examine the output, and if not satisfied, proceed to further classification cycles in which the results can be improved. Within FA, this is called "hierarchical learning". It is an iterative, optional, multi-step procedure that allows the user to refine the initial outcomes based on further examples. Thereby, either the amount of false positives can be reduced ("clutter removal") or missed areas can be added and thus the detection results be broadened. For a cycle of clutter removal, the user delineates parts or selects features in the outcomes and flags them as correctly or incorrectly classified. To add missed features, the user digitizes a few examples of areas that were not included in the output of the previous classification. In each case, afterward, a learning or classification cycle is performed. Based on the quality, the user can then choose to accept the results or, conduct another pass of clutter removal or adding missed features until outcomes are considered suitable. For all classification tasks in this study, hierarchical learning was applied after initial outcomes were examined. Typically, two or three cycles of hierarchical learning, i.e. clutter removal and adding missed features, were necessary to balance over- and underclassification and obtain satisfactory results.

Within the supervised classification mode, FA offers a multi- and a single-class approach. The multi-class procedure allows the simultaneous detection of two or more groups of target areas or objects. With the back-to-back classification option, a whole image or designated area can be classified, i.e. each pixel is assigned to one of the target categories, thus leaving no unclassified areas. This option was employed in the first step of the preparation of the LU/LC map in which a multi-class analysis was used to distinguish between sealed surfaces, that are impervious to water, and unsealed surfaces that are pervious to water. The impervious class included surfaces made up of concrete and other (building) materials, and mainly contained built-up areas and associated infrastructure (e.g. roads). All other types of LU/LC and related areas, e.g. vegetated parts, agricultural fields, bare soil, and range land, were assigned to the pervious class. Every square meter of the study area was classified as belonging to either one of the two categories. Initial classification results were improved with hierarchical learning. When using a multi-class approach, hierarchical learning has to be applied to the output for each class individually. Then, in a final learning pass, classes are joined again and detected simultaneously. Thereby, overlaps and "gaps", i.e. unclassified areas, can be avoided.

Subsequently, vegetation areas were detected separately with a single-class approach. The areas included in this class primarily corresponded to those with active, photosynthetic vegetation characterized by high reflectance values in the near-infrared (NIR) band. In the classification process, the batch processing option of FA was employed. It allows the application of a previously developed classification or automatic feature extraction (AFE) model to another image. Here, the image covering the whole study area was split into three parts to expedite the classification process. A classification model that was able to deliver satisfactory results was developed for the first image tile. The batch processing method was then applied to the second image tile. To enhance its suitability for this case and improve the quality of the classification results, the initial AFE model was then refined with further cycles of hierarchical learning based on examples from the respective image tile. This process was repeated for the third image part. There, the improved AFE model from the second image tile was applied and adjusted with one more hierarchical learning pass. A similar procedure employing the batch processing option was also used in the detection of the pervious and impervious areas described before.

For the preparation of the map, the classification results for the three categories of LU/LC had to be integrated. Before that, the outcomes for the pervious and the impervious class were slightly edited manually, primarily to remove a number of misclassified areas from the impervious class. Especially in the southeastern part of the study area, limestone and similar bare ground areas were identified as impervious surfaces by the software due to their strong resemblance with concrete surfaces with respect to their spectral values. To distinguish between these bare rock or ground areas and sealed surfaces covered by man-made materials, such as concrete, these parts were reassigned to the pervious category. Similarly, cloud-covered areas were also classified as impervious in the semi-automatic image analysis. These areas were manually assigned to a new category, which was excluded from further analyses. Subsequently, the results for all LU/LC classes were integrated by subtracting the vegetation areas from the other two classes of LU/LC, primarily the pervious areas. Finally, an aggregation procedure was applied to eliminate smaller patches of pervious or impervious areas resulting from the integration procedure.

The rooftop areas were then detected with a separate single-class object extraction procedure. For this purpose, subsets, each showing one of the 12 settlements in the study area, were created from the satellite image. For each of the three largest settlements, an individual classification model was developed. With the batch processing procedure, each of the three AFE models was then applied to the other two image subsets as well as three further image tiles that had not previously been used in the model development process. Thus, the respective suitability for the classification of other image subsets could be tested, evaluated and compared for each AFE model. The model that appeared to be the most appropriate overall, i.e., delivered the best classification results, was then employed to detect the rooftop areas in the remaining six settlements and image tiles.

All classification results were evaluated visually and with a specific accuracy assessment procedure. Reference data was produced through visual interpretation of the satellite image and manual on-screen digitization. For the three LU/LC classes of vegetation, pervious areas, and impervious areas, the accuracy assessment procedure was based on a stratified random sampling strategy and a pixel-wise comparison between classification results and reference data. A detailed description of the procedure can be found in [14]. All classification results were assessed prior to any possible manual editing, as applied to the impervious and pervious class. The outcomes of the rooftop detection were evaluated based on the data for the six largest settlements in the area. Here, no sampling strategy was employed and instead, the six datasets were compared entirely to the reference data (manually digitized rooftops). Through data intersection, areas of overlap and those existing only in the classification results (i.e. false positives) or in the reference data (i.e. false negatives) could be identified. In addition to this area-based approach, the detection results were also assessed on an object level. The number and percentage of objects exhibiting a partial overlap between classification output and reference data were expressed as "Feature Score" results. The outcomes of all individual assessment procedures were noted in error matrices. Based on the latter, then, the respective statistical accuracy indices were calculated, including the overall, the user's and the producer's accuracy, as well as Cohen's kappa.

3.3 GIS Analyses—Hydrological Analyses, Data Intersection, and Distance Analyses for the Detection of Spatial Patterns

A hydrological analysis was conducted with Arc Hydro for ArcGIS to identify and trace the main wadis and surface water basins of the wider region. The analysis was based on a freely available digital elevation model (DEM) with approximately 30 m spatial resolution, the ASTER GDEM V2. For the study area itself, a second, more detailed hydrological analysis was performed based on a custom-made DEM with a finer spatial resolution of 5 m. This DEM was derived from ALOS PRISM triplet satellite images, which had a spatial resolution of 2.5 m. Analysis outcomes, i.e. the computed drainage lines and their catchments, were evaluated visually with the help of the GeoEye-1 satellite imagery. Although the data was found to be in good accordance in general, a number of discrepancies were observed, primarily in relatively flat areas with little relief. Therefore, the computed stream network was improved with manual digitization of the drainage lines, i.e. incisions in the terrain that are discernible in the satellite image. Then, a third hydrological analysis was conducted in which the new, improved stream network was "burnt" into the terrain represented by the ALOS DEM, prior to the actual analysis steps. Thus, the number, position, and course of drainage lines could be predetermined and corresponding catchments could be computed. The procedure is described in detail in [14].

Different data layers were intersected to systematically gather information on the characteristics of RWH sites. The previously detected RWH structures served as the input layer with which the datasets containing the desired environmental information were sampled. For this purpose, all RWH objects that were not already represented by an x, y coordinate pair, e.g. linear structures such as bunds, ridges, dikes, and similar, were converted into point features whose location corresponded to the center point of the respective linear or areal object. Each point was considered to represent one RWH site of a specific type, according to the nature of the RWH structure. For every site, then, the corresponding value was extracted from the other data layers. Altitude and slope information were gathered from the ALOS DEM and the slope map derived from it. Information on the surface drainage basins and the subcatchments in which the RWH sites were located, were retrieved from the outcomes of the hydrological analyses. Additionally, data on the geology at each point was obtained from the digital version of the geological maps (scale 1:50 000) for the area. Annual precipitation data for every site were extracted from the precipitation maps described in the following section "Assessing the RWH Potential and Its Distribution".

Following the hypothesis that things in close vicinity to each other tend to be related more often, also distance or near analyses were conducted to discover potential spatial patterns and systematic relationships. The distance from each site to the nearest RWH structure of the same or a different type were determined as well as the number of other RWH sites in a specific radius. Moreover, the nearest settlement and vegetation area, and the nearest archaeological site, as well as the respective distances to those were determined for each RWH site. Information on archaeological sites were obtained from the MEGA (Middle Eastern Geodatabase for Antiquities) Jordan online database [45] which provides data on archaeological finds and remains including their respective (discovery) location. All places with relevant archaeological discoveries were represented as points in the analyses.

3.4 Assessing the RWH Potential and Its Distribution

To estimate the RWH potential of the area, the potential annual on-site runoff of different surfaces was calculated. In view of the purpose and the available data, an empiric formula (see Eq. 1) was chosen which expresses runoff as a function of precipitation and (land) surface characteristics. The latter were reflected in the runoff factor (RC) assigned to the specific type of surface (LU/LC category) or area.

$$Q = P * RC \quad (1)$$

in which: Q = annual runoff in mm

P = annual precipitation in mm

RC = (annual) runoff coefficient, dimensionless

The different classes in the prepared LU/LC map (see section "Semi-Automatic Analysis of Satellite Imagery for the Extraction of LU/LC Information") were assigned different, suitable runoff coefficients. Appropriate runoff factors were derived from literature, i.e. the reported findings or suggested values from previous studies, taking into account the specific circumstances in the study area, e.g. common soil types and properties, specifics of the local climate such as expectable typical precipitation intensities, and similar. Unsealed land surfaces, i.e. areas belonging to the two LU/LC classes vegetation and other pervious areas (mainly bare land), were subclassified according to slope. For this purpose, the following three categories of slope were defined and used to classify the slope map: flat terrain or land with small gradients (0–5% slope), terrain with moderate gradients (>5–15% slope), and steep sloping terrain (>15% slope). The resulting modified slope map was intersected with the LU/LC map to produce a RC map with the following classes and RC values assigned to them:

- impervious areas (sealed surfaces, including built-up areas, roofs, streets, asphalt and concrete areas, etc.)

independent of slope/all slope values	RC 0.80

- pervious areas (unsealed surfaces with no significant active vegetation cover, mostly bare ground, soil and rock areas, without specific treatment to enhance or reduce runoff)

0–5% slope	RC 0.12
>5–15% slope	RC 0.15
>15% slope	RC 0.18

- vegetation (areas with mostly active, photosynthetic vegetation)

0–5% slope	RC 0.06
>5–15% slope	RC 0.08
>15% slope	RC 0.10

Implementing Eq. 1, the RC map was then multiplied with the precipitation map(s). Here, the raster math function in ArcGIS was used to calculate the potential on-site runoff per raster cell which corresponded to one square meter. The precipitation maps used in this process were derived from the daily precipitation records of 28 weather stations in and around the study area. With regard to data gaps and the best possible coverage concerning the daily precipitation data available from the recording stations, the period 1963 to 1992 was selected as the most suitable 30 year observation period. For each day in the aforementioned period, a precipitation map was produced through spatial interpolation of the respective available data from the weather stations. A tool developed and described by [47] was employed for the interpolation (for details on the method, refer to [31] and [46]). One of the two main advantages of this approach was the automatic selection of the best suited interpolation algorithm, based on the properties of the available input data, e.g. number and distribution of stations with a record, for the given task, i.e. day. Additionally, latitudinal and longitudinal gradients, and orographic effects were taken into account. This required the availability of a DEM to represent the relief of the area. Here, the ASTER GDEM V2, resampled to spatial resolution of 250 m, was used in the interpolation process. Later, monthly, annual, and long-term (30 years) mean precipitation maps were derived from the daily precipitation maps for the period 1963–1992.

The hitherto described procedure was employed for the calculation of the overall RWH potential of the study area and the annual on-site runoff that could be collected from potentially suitable bare ground surfaces, e.g. unsealed, vegetation-free plots. With the same formula (see Eq. 1) but in a separate procedure, the rooftop rainwater harvesting (RRWH) potential was estimated for each settlement in the study area. All rooftop areas identified in the detection process (see earlier section "Semi-Automatic Analysis of Satellite Imagery for the Extraction of LU/LC Information") were assigned the same runoff factor of 0.80 which means that approximately 80% of the annual precipitation that falls on rooftops is converted into runoff. The respective annual runoff amount, i.e. the RRWH potential, was computed by intersecting and multiplying the rooftop areas with the precipitation maps.

The overall RWH potential or potential annual on-site runoff from ground catchments and the yearly RRWH potential were estimated for long-term mean precipitation conditions as well as dry and wet year conditions. The latter were exemplified by the years 1963 (dry year) and 1965 (wet year). Thus, the likely range of potentially collectible runoff or RWH potential in a given year can be illustrated.

4 Results

In the following sections, the outcomes of the mapping process, i.e. the database of RWH structures, are reported. Then, the results of the GIS near analyses and the identified characteristics of RWH sites are described. Finally, the overall RWH potential and the RRWH potential in the study area are presented. These sections also include details on the LU/LC map and the rooftop data, since those formed a main part of the input data in the estimation procedure and had a substantial influence on the results.

4.1 Existing RWH Structures in the Study Area

The database of the mapped RWH facilities in the study area contained two major categories with 1,477 objects in total. The locations of the individual RWH structures and their distribution are shown in Fig. 3. Linear RWH structures like bunds, dikes, ridges, and similar formed the largest group with 1,088 entries. The remaining 389 objects belonged to the category of cisterns which was subdivided into two classes: the allegedly ancient and most likely very old cisterns ("Roman Wells" as they are called by the locals) which amounted to 96, and the 293 other cisterns that were associated with a more recent construction and therefore labeled as modern. The ground check performed on a small sample of the modern cisterns confirmed their correct identification in the remote sensing-based mapping process.

In addition to the aforementioned RWH structures, one open reservoir was documented during field work in ar-Rabba. The object was also clearly visible in the satellite image. Based on its dimensions, the reservoir's capacity was estimated at ca. 5600 m^3. The object was also included in the MEGA Jordan database and it has been suggested that the reservoir probably dates back to Roman times. According to the MEGA Jordan database, there should be two more reservoirs in the study area. Yet, their existence could not be confirmed in the mapping process as fitting structures were not discovered, neither in the satellite image nor during fieldwork. In the remote sensing-based mapping process, 152 suspicious structures were identified that might indicate underground cisterns, possibly older ones, at the respective locations. However, ground checks would be required for each of these objects to clarify their nature.

In a number of the allegedly old cisterns, and particularly six of the seven cisterns inspected from the inside, construction details were found that parallel findings from other ancient cisterns elsewhere, at Gadara (Umm Qais) in northern Jordan (see e.g. [22]). The documented distinguishing features include half circle collar stones around the cistern entrances and different layers of plaster lining with fishbone and other patterns on the walls inside the cisterns. In view of the findings from other areas, this evidence together circumstantiates the assumption that several of the cisterns in the study area date back to ancient, e.g. Nabataean, Roman or Hellenistic times.

The storage capacities of the cisterns inspected from the inside which also showed signs of a probably ancient age, were calculated to amount to 54–189 m^3 per cistern, depending on the respective shape and dimensions. Another cistern that was examined but unlike the other ones appeared rather modern, without any features suggesting a very old age, and had an estimated capacity of 106 m^3. Based on the aforementioned findings, the average storage capacity of the 96 old cisterns was assumed to be ca. 130 m^3. Hence, the total water storage potential of this group of cisterns was estimated at 12,480 m^3. For the modern cisterns, no reliable average size could be determined from the data collected in this study. Therefore, here, a typical capacity of 30 m^3 was assumed, based on one of the author's experiences from cistern building projects in Jordan in recent years and the figures reported for modern cisterns in other studies (e.g. [10, 12]). Given the total of 293 mapped modern cisterns, their overall storage capacity thus was estimated at 8,790 m^3. For all, old and

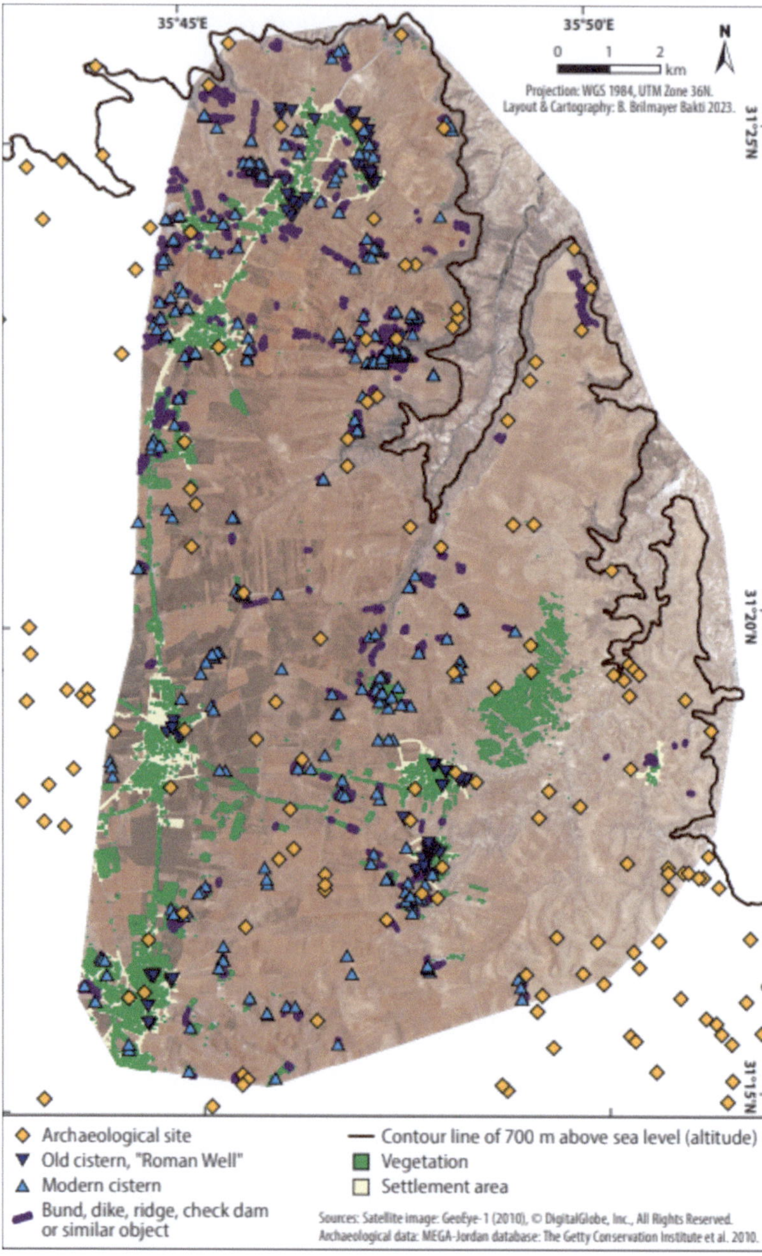

Fig. 3 Distribution of mapped RWH structures in relation to archaeological sites, and vegetation and settlement areas. Archaeological sites include all known places of archaeologically relevant finds and/or remains within the study area or the surrounding area on the Karak plateau (i.e. in areas with altitudes of at least 700 m above sea level). The surrounding area is defined as a zone of no more than 6 km distance to the study area. Modified from [14]

modern, cisterns detected in the study area, an available total water storage capacity of 21,270 m^3 was calculated.

4.2 Characteristics of RWH Sites, Results of Near Analyses, and Apparent Spatial Patterns

Sites belonging to the category of old cisterns exhibited the narrowest range of altitudes and the highest mean elevation (see Table 1). The elevation range was broader for the class of modern cisterns and the largest for the category of linear RWH structures. Along with that, the mean altitude values were lower in both these groups. The amplification of the elevation range resulted from sites in areas of lower as well as higher altitudes. It parallels differences in the distribution pattern of the sites in the different categories. In contrast to the old cisterns, which are found only at a comparatively limited number of places, modern cisterns and linear RWH structures are located in a broader variety of areas, including the (south-) eastern part of the study area and around the fringes of the plateau where altitudes are lower. However, the group of old cisterns showed the broadest range of slope values and the highest mean gradient. Modern cisterns and linear RWH structures such as bunds, check dams, ridges and similar, in contrast, typically seemed to be sited in rather flat or more gently sloping terrain (see Table 1).

According to their type, the discussed RWH structures are connected in different ways and to varying degrees to the geological situation of a given location. Whereas cisterns as underground water storage excavations strongly depend on suitable strata in appropriate depths, above ground structures, like bunds, dikes, ridges, and similar, are largely independent from a site's geology. In these cases, the geological setting may rather determine the availability of suitable local building materials. The analyzed linear RWH structures were distributed over all geological units present in the study area. In this group, the percentage of sites associated with each geological unit roughly corresponded to the respective unit's extent or share in the study area, with slightly higher percentages of sites in calcrete and basalt areas and a slightly lower share in soil (over bedrock) areas (see Table 1). According to the geological map, soil (over bedrock) was the most extensive unit, covering over 70% of the study area. Beneath the soil cover, strata of the Al-Hisa Phosphorite unit, calcrete or basalt can be expected. In each category of RWH sites, around or more than two thirds of the sites were located in soil (over bedrock) areas. The rest of the old cisterns were associated with the Al-Hisa phosphorite unit and calcrete areas. Modern cisterns were also found in areas of these units, however, a considerable share (14%) of this group was located in basalt areas, in contrast to the old cisterns.

The distribution of annual precipitation values largely paralleled the findings for the sites' general distribution over the study area and their altitudes, as described above. Annual precipitation in the range of 273–321 mm (mean 294 mm) was observed at the sites of old cisterns. For the group of modern cisterns and that of bunds, dikes, ridges, and similar, this range was broader and the mean value was

Table 1 Characteristics of RWH sites

	Category of RWH structures		
	Old cisterns (96)	Modern cisterns (293)	Linear structures (1,088)[a]
□ *Height above sea level*			
Min.–max	822–958 m	800–975 m	765–1,008 m
Mean	884 m	876 m	871 m
□ *Slope*			
Min.–max	2–30%	1–18%	0–25%
Mean	10%	7%	8%
○ *Geological unit (share in study area)*			
Soil (76%)	73 (76%)	190 (65%)	763 (70%)
Calcrete (3%)	4 (4%)	33 (11%)	110 (10%)
Basalt (all) (6%)	0	41 (14%)	132 (12%)
Al-Hisa Phosphorite (7%)	19 (20%)	29 (10%)	75 (7%)
Qatrana Phosphorite (4%)	0	0	8 (1%)
□ *Mean annual precipitation (in mm)*			
Min.–max	273–321 mm	260–318 mm	234–320 mm
Mean	294 mm	292 mm	291 mm
○ *Major surface drainage basin (share in study area)*			
Mujib basin (total) (97%)	96 (100%)	288 (98%)	1074 (99%)
Lower Wadi Mujib (9%)	9 (9%)	61 (21%)	382 (35%)
Wadi Nukhayla (total) (88%)	87(91%)	227 (77%)	692 (64%)
Lower Wadi Nukhayla (4%)	24 (25%)	14 (5%)	58 (6%)
Wadi Balu (48%)	28 (29%)	171 (58%)	468 (43%)
Middle Wadi Nukhayla (2%)	0	0	24 (2%)
Wadi Ghuwayta (30%)	35 (37%)	39 (13%)	121 (11%)
Upper Wadi Nukhayla (3%)	0	0	7 (1%)
Wadi ad-Dabba (1%)	0	3 (1%)	14 (1%)
Dead Sea basin (total) (3%)	0	5 (2%)	14 (1%)
Wadi Ibn Hammad (3%)	0	5 (2%)	14 (1%)
□ *Catchment information*[c]			
Total no. of catchments affected	38	135	288

(continued)

Table 1 (continued)

	Category of RWH structures		
	Old cisterns (96)	Modern cisterns (293)	Linear structures (1,088)[a]
Total area of catchments affected	8.138 km^2	36.840 km^2	37.260 km^2
Sizes of catchments affected			
Min.–max	0.012–2.828 km^2	0.009–7.942 km^2	0.002–2.828 km^2
Mean	0.214 km^2	0.198 km^2	0.129 km^2
Features per catchment			
Min.–max	1–12	1–14	1–23
Mean	2.5	2.0	3.3

The collected data of individual objects were accumulated according to the three defined categories of RWH structures. For the extent of the study area, see Fig. 1

[a] Information in this column refers to center points of individual objects. Linear structures are bunds, dikes, ridges, check dams, and similar features

[c] Only catchments of drainage lines within the study area and features located in these catchments were considered. This led to a reduced number of objects in the categories of modern cisterns (264) and linear RWH structures (969). Only catchments without signs of truncation by the boundary of the study area were included

□ Quantitative attribute: value(s) comprised by features in the respective category of RWH structures (column)

○ Qualitative attribute: absolute number and corresponding percentage (in parentheses) of features in the respective category (column)

slightly lower (see Table 1). This finding results from a considerable number of objects located in the eastern and southeastern parts of the study area where annual precipitation is lower.

Differences in the spatial distribution of the different types of RWH structures also become apparent with regard to the major surface water sub-basins in the area. The majority of the old cisterns are situated in the watershed of Wadi Ghuwayta, Wadi Balu, and the lower Wadi Nukhayla. In contrast, most of the modern cisterns and the linear RWH structures were found in the sub-basin of Wadi Balu, that of the lower Wadi Mujib, and that of Wadi Ghuwayta.

With regard to individual catchments within the major surface water sub-basins, the number of catchments affected varies considerably between the three classes of RWH structures. While the old cisterns are distributed over 38 catchments, modern cisterns were found in 135 catchments. This number is the highest for the group of bunds, check dams, ridges, and similar, as objects of this type were identified in 288 different catchments. Along with an increase in the number of host catchments, a decrease in mean catchment size was observed. However, the differences between the two groups of cisterns, were less stark compared to the category of linear RWH structures (see Table 1). Similarly, the typical number of objects per individual catchment was rather comparable for the old and the modern cisterns whereas it was distinctly higher in the category of bunds, dikes, check dams, and similar.

The outcomes of the near analyses showed that for each RWH structure typically another one of the same type is the closest object, except for the modern cisterns (see Table 2). In the latter category, in more than two third of the cases, a linear RWH structure formed the nearest object while for the rest of the modern cisterns, another object of the same group was the closest one. However, in the group of bunds, dikes, ridges, and similar, only 13% of the objects had a modern cistern closest to them whereas in almost all other cases, another linear RWH structure was identified as the nearest feature. Distances to the next closest RWH object varied significantly from less than 10 to over 700–900 m. However, on average, the nearest object was found at a distance of 20–60 m. In general, the distances between neighboring cisterns tended

Table 2 Outcomes of the analysis of distances between individual RWH structures according to their category

Near features—RWH category		Input features—RWH category		
		Old cisterns	Modern cisterns	Linear structures
Nearest feature according to category and corresponding distances[a]				
Old cisterns		88 (92%)	3 (1%)	1 (<1%)
Distance	Min.–max	9–325 m	33–479 m	4 m
	Mean	54 m	194 m	–
Modern cisterns		1 (1%)	92 (31%)	141 (13%)
Distance	Min.–max	33 m	4–290 m	5–145 m
	Mean	–	61 m	22 m
Linear structures		7 (7%)	198 (68%)	946 (87%)
Distance	Min.–max	4–304 m	5–993 m	0–757 m
	Mean	96 m	38 m	27 m
Near features according to type in a max. distance of 250 m[b]				
Old cisterns		90 (94%)	11 (4%)	59 (5%)
Modern cisterns		18 (19%)	252 (86%)	722 (66%)
Linear structures		28 (29%)	257 (88%)	1,082 (99%)
Number of near features according to type in the 250 m zone[c]				
Old cisterns	Min.–max	1–15	1–7	1–7
	Mean	5.5	2.7	2.1
Modern cisterns	Min.–max	1–3	1–20	1–24
	Mean	1.5	5.4	4.7
Linear structures	Min.–max	1–19	1–47	1–49
	Mean	4.2	10.5	11.6

[a] Number (percentage) of input features whose nearest RWH structure belongs to this category, and corresponding distances
[b] Number (percentage) of input features which have a RWH structure of the respective type within a max. distance of 250 m
[c] Number of near features of the respective category which are located at a max. distance of 250 m to individual input features

to be slightly higher than those between cisterns and nearby linear RWH structures, or between nearby individual objects of the latter type (see Table 2).

The analysis of the 250 m radius around each RWH site revealed that nearly all of the old cisterns had another object of the same type in this zone while almost 20% also had a modern cistern and around 30% a linear RWH structure within the aforementioned radius. Of the modern cisterns, more than 80% respectively had a modern cistern and a bund, dike, ridge, or similar feature within a maximum distance of 250 m. In the category of linear RWH structures, in almost all cases, another object of the same class was found in the analyzed radius. Around two-thirds of the sites in this group had a modern cistern observed nearby. Yet, only very few of the linear RWH structures and the modern cisterns had an old cisterns in the described radius. The number of nearby objects, i.e. those within a distance of no more than 250 m, generally varied largely. The cisterns typically had 1–20 and on average around five nearby sites. The figure was similar for the linear RWH structures with regard to nearby modern cisterns. However, of the modern cisterns that had a linear RWH structure in their vicinity, each had on average almost 11 nearby sites of this type. The bunds, dikes, check dams, and similar structures had on average almost 12 nearby sites of the same type and thus, apparently formed larger clusters.

On average, the distance to the nearest archaeological site was the lowest for the old cisterns while it was considerably higher for the modern cisterns and the linear RWH structures (see Table 3 and Fig. 3). For roughly one-quarter of the old cisterns, the nearest archaeological site could be found within a distance of no more than 250 m. In the other two groups of RWH structures, only slightly more than 10% of the objects were located in similarly close vicinity to archaeological sites. Similar results were observed for distances of up to 500 m. Here, for almost two-thirds of the old cisterns, the nearest archaeological site was found in this zone. In contrast, only about one-third of each the modern cisterns and the linear RWH structures had an archaeological site within the aforementioned maximum distance.

Concerning the proximity to settlement areas, the majority of old cisterns were located within or in close vicinity to settlements (see Table 3 and Fig. 3). The mean distance to these areas was only 59 m in this group of RWH structures while 95% of the old cisterns were sited in distances of no more than 250 m to the nearest settlement. In contrast, the average distance to the nearest settlement area was around 20 times higher for the modern cisterns and the bunds, ridges, check dams, and similar. Both classes of RWH structures were also very similar in regard to the share of sites located within a maximum distance of 250 m to settlements: only about one third or less of the objects were found in this zone. Similar outcomes were registered for the distances to vegetation areas, which in this case typically were trees (e.g. olive groves, orchards) or irrigated gardens. Here, the range of identified distances was already significantly greater for the modern cisterns and the linear RWH structures than that observed for the old cisterns. In both of the aforementioned groups, the identified mean distance was around five times higher than that of the old cisterns. Moreover, nearly all of the old cisterns were located in distances of 250 m or less to the nearest vegetation patch, whereas the same applied to only about two-thirds of the modern cisterns and the linear RWH structures (see Table 3). Hence, the results consistently show that the old cisterns are usually located in close proximity to settlement and vegetation areas,

Table 3 Distances between RWH structures of different types and archaeological sites, settlements, and vegetation areas

Near features	Input features—RWH category		
	Old cisterns	Modern cisterns	Linear structures
Archaeological sites[a]			
□ *Distance to nearest on plateau*			
Min.–max	103–1,640 m	25–1,559 m	5–1,785 m
Mean	535 m	692 m	659 m
○ Nearest in up to 250 m [b]	23 (24%)	35 (12%)	159 (15%)
○ Nearest in up to 500 m [b]	60 (63%)	99 (34%)	404 (37%)
Settlement areas (3.19 km^2)[c]			
□ *Distance to nearest*			
Min.–max	0–2,661 m	0–3,318 m	1–4,917 m
Mean	59 m	1,119 m	1,073 m
○ Nearest in up to 250 m [b]	91 (95%)	79 (27%)	357 (33%)
Vegetation areas (1.44 km^2)[d]			
□ *Distance to nearest*			
Min.–max	0–390 m	2–1,482 m	0–1,524 m
Mean	44 m	259 m	293 m
○ *Nearest in up to 250 m*[b]	93 (97%)	198 (68%)	706 (65%)

[a] Archaeological sites, represented by point features, fulfilling three criteria: dated to periods before 1918 AD or not dated yet (applies only to water structures), located within or in a zone of up to 6 km distance around the study area, and situated on the Karak Plateau, i.e. at altitudes of at least 700 m above sea level
[b] Number (percentage) of input features whose nearest site/area is found within the specified maximum distance
[c] Impervious surfaces and areas assigned to individual settlements
[d] Mostly photosynthetically active vegetation (summer half year)
□ Quantitative attribute: value(s) comprised by features in the respective category of RWH structures
○ Qualitative attribute: absolute number and percentage (in parentheses) of input features to which the statement applies

while the modern cisterns as well as the bunds, dikes, ridges, and similar features, are mostly found in similarly greater distances (see Fig. 3).

4.3 Estimated RWH Potential

Since the estimated RWH potential or annual on-site runoff highly depends on the LU/LC, the corresponding results will be presented first. The extent of cloud-covered areas that were excluded from the prepared LU/LC map amounted to only 0.039 km^2 or roughly 0.02% of the study area. The final LU/LC map covered 163 km^2 of which 4.2 km^2 or approximately 3% was assigned to the impervious class while the remaining 158.8 km^2 or ca. 97% were classified as pervious to water (see Fig. 4). The vegetation area amounted to 1.4 km^2 or roughly 1% of the pervious area. From the impervious areas, relatively compact regions of continuous or adjacent impervious patches were delineated and defined as settlement areas (see Fig. 4). The latter comprised about 3.2 km^2 and thus constituted ca. 76% of the land classified as impervious or 2% of the study area. The delineated regions were assigned to 12 individual settlements. The extent of their respective settlement area varied between 0.84 km^2 for the largest town, al-Qasr, and 0.07 km^2 to 0.06 km^2 for the two smallest settlements, Rashadiya and Misr.

The accuracy assessment conducted for the pervious and the impervious classes revealed an overall accuracy of 90% and a value of 0.8 for Cohen's Kappa. Here, the outcomes of the semi-automatic image classification procedure were assessed prior to manual editing. With the latter, the suitability of the pervious and particularly the impervious data may even have improved. For the vegetation class, a separate accuracy assessment was conducted, based on which a user's accuracy of 95% and a producer's accuracy of 92% was calculated for this class. An overall accuracy of 94% and a Kappa of around 0.9 was identified. More specifically, the unadjusted value for Cohen's Kappa was 0.88 while a Kappa of 0.91 was found when the value was adjusted with regard to the maximum possible Kappa for the analyzed sample data.

The computed amounts of annual precipitation and resulting on-site runoff that could potentially be harvested are given in Table 4. The results for three different precipitation scenarios are presented, including long-term average (mean values based on the period 1963–1992), dry year (1963) and wet year (1965) conditions. Overall and for the study area as a whole, local runoff has been estimated to be around 16% of the precipitation amount in a given year. In dry years, precipitation and on-site runoff are decreased to about 50% of their long-term mean, whereas in wet years, amounts are almost doubled and equal ca. 180% of the long-term mean, according to our findings. Most of the runoff comes from pervious areas as these make up 97% of the studied land (see above). Despite their much higher runoff efficiency, impervious areas contribute only around 14% to the total figure of annual on-site runoff (see Table 4), due to their limited extent. The spatial distribution of on-site runoff generally closely reflects the pattern of the terrain characteristics, primarily the LU/LC, and the precipitation distribution. Therefore, the lowest runoff values were always found in vegetated, rather flat areas while the highest runoff amounts occurred in impervious areas. Moreover, generally lower runoff values were observed in the (south-) eastern part of the study area where annual precipitation was also lower compared to the western half and central part of the Karak plateau. Within the limits of this general pattern, the spatial distribution of on-site runoff varied locally from year to year, particularly with regard to maxima and minima. For example, our results showed that under normal (long-term average) precipitation conditions and in dry years, the lowest annual on-site runoff occurs near Rashadiya in the southeastern part of the study area. In wet years, in contrast, the lowest runoff amounts can be expected near Ariha in the northern part of the plateau. These findings result from

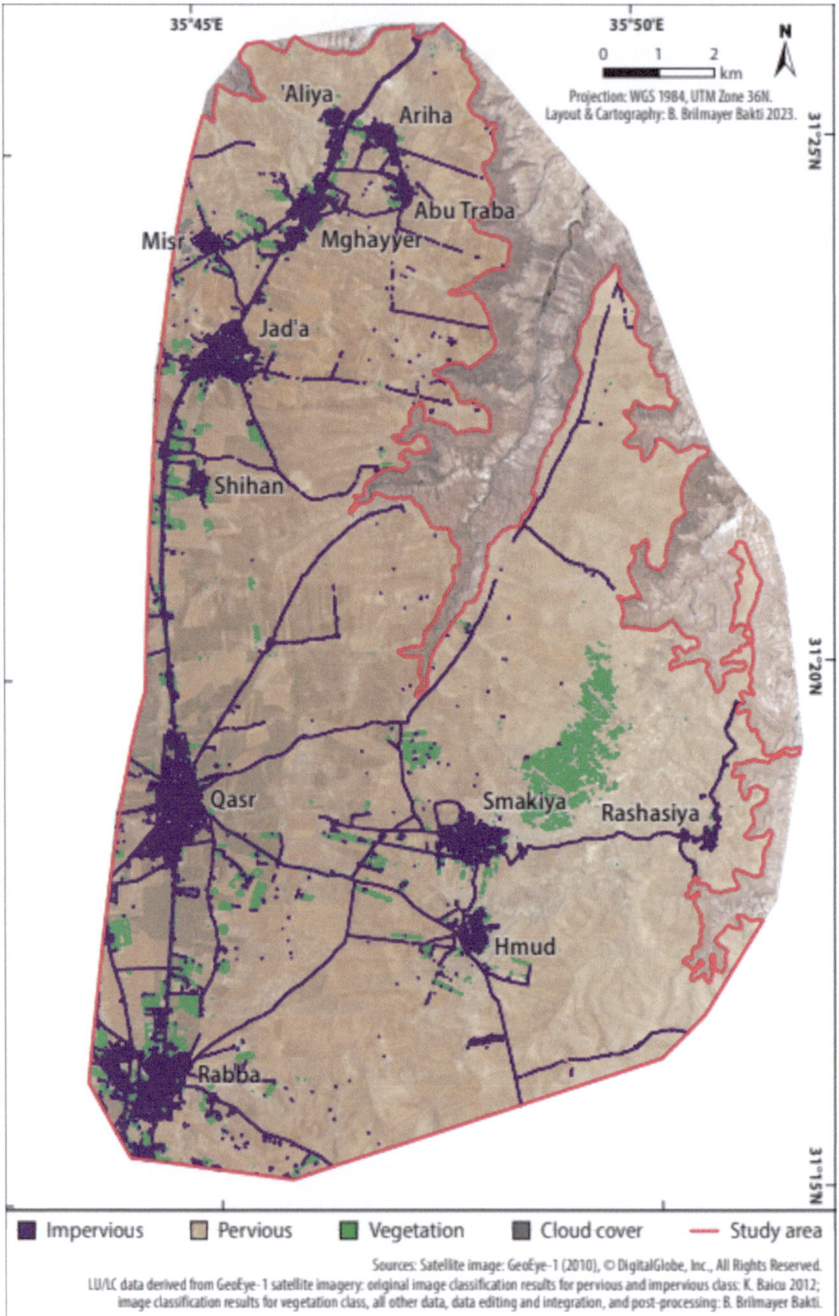

Fig. 4 LU/LC map for the area. The settlement areas of the 12 towns and villages in the study area are labeled with their respective names. Cloud-covered areas were excluded from the LU/LC analyses. Modified from [14]

interannual variances in the spatial distribution pattern of precipitation which may be even more pronounced in years when precipitation is abnormally high or low.

Based on the computed runoff estimates, the catchment area was determined that would be necessary to collect enough water to fill an old or a modern cistern once a year. The extent of the required catchment varies throughout the study area depending on local circumstances, i.e. LU/LC characteristics and annual precipitation amounts. Therefore, in Table 5, the likely range of required catchment sizes under three different precipitation conditions is presented. In general, in wet years, only about half of the catchment area necessary in years with average precipitation would be needed. In dry years, in contrast, the catchment area would roughly have to be doubled to collect sufficient runoff. However, not all surfaces are considered equally suitable for RWH purposes. Commonly, impervious ground surfaces, e.g. parking lots, yards, streets, and similar, are regarded as unsuitable as catchment areas due to water quality concerns. Similarly, vegetated areas typically produce very little runoff and, therefore usually also represent inadequate catchment areas. Additionally, steep sloping terrain is commonly excluded from RWH schemes as runoff from these areas is more dynamic and thus more difficult to control. It may also carry a higher sediment or particle load leading to an increase in water impurity and filtration requirements. Based on these considerations, basic criteria were defined for potentially suitable catchment areas. The latter had to be ground surfaces classified as pervious, without detected vegetation, and low to moderate gradients of no more

Table 4 Estimated annual precipitation and resulting on-site runoff in the study area

	Long-term mean conditions (1963–1992)		Dry year conditions (1963)		Wet year conditions (1965)	
		Total		Total		Total
Precipitation						
Min	224 mm	46.18 MCM	97 mm	21.20 MCM	395 mm	84.89 MCM
Max	321 mm		159 mm		677 mm	
Mean	273 mm		128 mm		523 mm	
Runoff						
Pervious areas						
Min	14 mm	6.22 MCM	6 mm	2.86 MCM	26 mm	11.44 MCM
Max	58 mm		29 mm		121 mm	
Impervious areas						
Min	186 mm	1.02 MCM	82 mm	0.47 MCM	334 mm	1.85 MCM
Max	257 mm		127 mm		542 mm	
Overall						
Mean	62 mm	7.24 MCM	29 mm	3.33 MCM	123 mm	13.30 MCM

In the column "total", the annual precipitation and on-site runoff are stated in million cubic meter (MCM, 10^6 m^3) for the three different precipitation scenarios. Additionally, the associated minima, maxima, and mean values are given in mm per m^2

Table 5 Estimated annual on-site runoff and respective catchment size required to collect enough runoff to fill an old/modern cistern once per year, when considering only pervious areas without vegetation, with slopes of up to 15%, and a minimum extent of 769 m^2

	Results under the respective precipitation conditions		
	Long-term mean (1963–1992)	Dry year (1963)	Wet year (1965)
Estimated annual on-site runoff/RWH potential in the study area			
Total amount	5.547 MCM	2.549 MCM	10.188 MCM
Min.–max. per m^2	27–48 mm	12–24 mm	47–102 mm
Mean per m^2	37 mm	17 mm	71 mm
Required catchment area			
For an old cistern (130 m^3)	2,708–4,815 m^2	5,417–10,833 m^2	1,275–2,766 m^2
For a modern cistern (30 m^3)	625–1,111 m^2	1,250–2,500 m^2	294–638 m^2

than 15%. Additionally, a minimum extent of 769 m^2 was defined. This threshold was determined according to the average catchment area required to fill a modern cistern under normal precipitation conditions. The resulting refined RWH potential in the study area is presented in Table 5. More than 90% of the pervious land or 88% of the study area (144 km^2) were identified as potentially suitable for RWH based on the described minimum criteria. Moreover, distance analyses showed that almost all modern cisterns (290 of 293) and bunds, dikes, ridges, and similar (1,070 of 1,088), as well as a substantial number of old cisterns (84 of 96) were located either within or adjacent (10 m tolerance) to pervious areas identified as potentially suitable catchments.

4.4 Estimated RRWH Potential

In the description of the RWH potential so far, runoff that could be harvested from rooftops, i.e. the RRWH potential, had been excluded. However, rooftops are commonly regarded as excellent catchment areas which are usually associated with high runoff efficiencies and a good water quality. Therefore, the findings regarding the RRWH potential in the study area are presented in detail below. Since the potential estimates strongly depend on the detected rooftop area, first, the results of the image analysis process, i.e. the characteristics of the semi-automatically detected rooftop data, are described.

For the 12 settlements in the study area, a total rooftop area of 627,654 m^2 or roughly 0.63 km^2 was found according to the outcomes of the semi-automatic image analysis. The rooftop area detected in the individual settlements varied between 182,329 m^2 in Qasr, the largest town in terms of the identified rooftop area, and only 7,451 m^2 in Rashadiya, the smallest village (see Fig. 5). Based on the results of the comprehensive accuracy assessments of the classification outcomes for five of the six larger settlements, a likely error margin of ca. 16% under- to ca. 15% overestimation was identified in regard to the detected rooftop area (see Table 6). According to the evaluation, this accuracy can be assumed for the output when the developed AFE model is applied to detect rooftops in a given image subset of the study area (batch processing procedure). Higher accuracies were observed for the classification outcomes from the original image subset showing the town of Qasr, based on which the AFE model had been developed. The accuracy findings for the latter case and the determined characteristics of the AFE model regarding the accuracy of outcomes obtained through batch processing are summarized in Table 6.

In general, the outcomes of the batch processing procedure were characterized by similar or lower accuracy values than the original classification results from the primary image subset. In all cases, the overall accuracy was very high at around 98%. This partly results from the correct identification of the vast background areas as non-rooftop regions (see Fig. 6). In contrast, the user's and the producer's accuracies are significantly lower with around 63%, respectively, in the outcomes of the original classification results. In the assessment of the batch processing outcomes, even lower values of around 40–66% and about 56% on average were observed. These figures reveal a significant amount of false positives (overclassification) and false negatives (underclassification) in the detection results compared to the reference data. Since the user's and producer's accuracies were characterized by comparable values in all cases, both types of classification error apparently were present at similar levels. With regard to the total rooftop area, under- and overclassification tendencies often balanced, resulting in an on-average good match between classification results and reference data. However, the evaluation of the individual batch processing

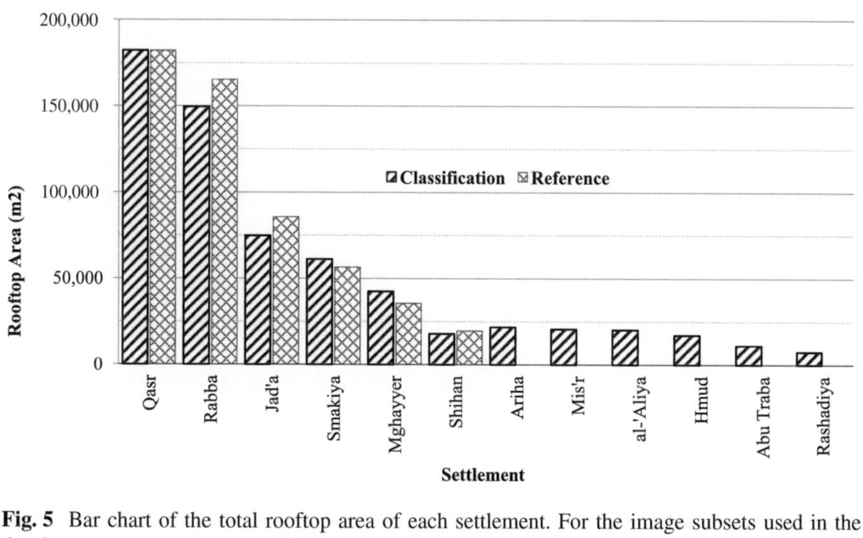

Fig. 5 Bar chart of the total rooftop area of each settlement. For the image subsets used in the development or the evaluation of the AFE model, the total rooftop area according to the reference data (columns with cross-hatching) are also given. Modified from [14]

Table 6 Accuracies derived from the assessment of the rooftop data obtained with the developed and applied classification model

	Original image classification	Batch processing	
		Min.–max	Mean
Area difference (FA/ reference data)	+0.08%	−12.69%–+19.90%	−0.54%
95% confidence interval[a]		−16.09%–+15.01%	
Feature score user's accuracy[b]	95.34%	92.09–98.58%	95.09%
Feature score producer's accuracy[c]	79.21%	42.34–78.46%	66.99%
User's accuracy (rooftops)	62.81%	41.80–65.69%	56.66%
Producer's accuracy (rooftops)	62.86%	38.26–66.19%	56.28%
Overall accuracy	98.03%	96.60–98.56%	97.52%
Kappa[d]	0.62	0.41–0.66	0.59

Accuracy findings of classification outcomes for the original image subset for which the AFE model was developed and those of the results obtained through batch processing (five datasets) are presented separately

[a] Range of likely over- and underestimation in classification results in comparison to reference data

[b] Percentage of features in the classification results which (partly) overlap with those in the reference data

[c] Percentage of features in the reference data which (partly) overlap with those of the classification results

[d] Here observed Kappa as proportion of maximum possible Kappa value for individual datasets in order to compensate biases from cases in which maximum possible Kappa is not 1

outputs showed a relatively broad range concerning differences in the identified total area between classification outcomes and reference data. Finally, the comparison of classification results and reference data on an object (feature) level generally led to higher accuracies than the area-based evaluation (compare feature scores versus normal user's and producer's accuracies in Table 6). Particularly the feature score user's accuracy was very high with values of around 95%. This means that regions labeled as rooftops in the classification results could nearly always, at least partly, be confirmed as rooftops based on the reference data. In contrast, the feature score producer's accuracy was significantly lower, with around 40–80% and about 67% on average, indicating that many of the rooftops in the reference data were missed by the semi-automatic classification procedure.

For the study area and the 12 settlements within it, the annual runoff from rooftops that could potentially be harvested, i.e. the RRWH potential, was estimated at 153,937 m^3 under long-term average precipitation conditions. In dry years, such as 1963, the potential can drop by more than 50% to 70,859 m^3 while in wet years, such as 1965, it may increase by 80% to 277,329 m^3, according to our findings. The estimates given

Fig. 6 Semi-automatically detected rooftops in the village Misr. Satellite image: GeoEye-1 ©DigitalGlobe, Inc. All Rights Reserved. Modified from [14]

here and in the following were based on the rooftop data acquired through semi-automatic image classification. Therefore, a tolerance of 15–16% under- or overestimation should be considered in all cases in accordance with the error margin identified for the rooftop detection results (see above). The RRWH potential of the individual settlements is presented in Fig. 7. Under normal, i.e. long-term average, precipitation conditions it ranged from 46,187 m^3 in Qasr to only 1,437 m^3 in Rashadiya. The annual collectible runoff primarily depended on the size of the settlement in terms of the identified total rooftop area. For the six larger settlements in the study area together, a RRWH potential nearly six times that of the six smaller villages was computed. The mean annual runoff per square meter of rooftop area varied substantially between individual settlements and directly reflected spatial differences in annual precipitation amounts. In years with average precipitation, an annual runoff of 0.253 m^3 of roof area can be expected in Qasr and ar-Rabba while in Abu Traba and Rashadiya only 0.226 and 0.193 m^3 are likely to occur. In some cases, e.g. in the village Mis'r, differences in annual precipitation led to a higher RRWH potential despite a slightly smaller total rooftop area in comparison to another settlement, here, the village of Ariha.

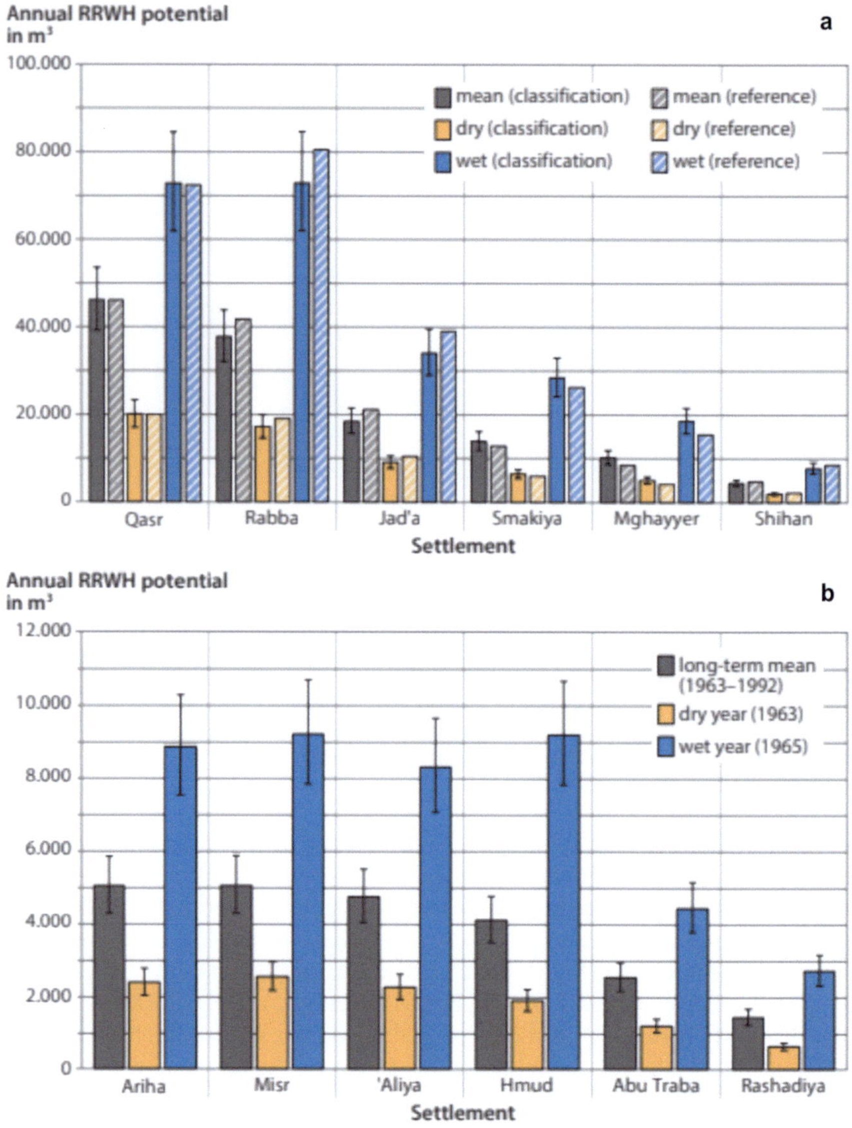

Fig. 7 RRWH potential of each settlement in the study area. For each settlement, the estimated runoff from all rooftops, under three different precipitation conditions is illustrated: long-term mean (grey bars), dry years exemplified by 1963 (yellow bars), and wet years represented by 1965 (blue bars). Part (**a**) includes the results for the six larger settlements and part (**b**) displays the outcomes for the six smaller villages. The hatched bars in part (**a**) show the potential calculated based on the reference data. The plain bars in both diagrams visualize the potential estimated based on the image classification results. Error bars indicate the range of likely over- or underestimation according to the error margin associated with the extracted rooftop data. Modified from [14]

The respective estimated RRWH potential of each settlement in dry and wet years is also included in Fig. 7. Additionally, Fig. 8 shows the percentage differences in potential compared to the long-term average. For 1963, an example of a very dry year, a rather uniform reduction in runoff by 50–57% was found for all towns and villages in the study area. For 1965, an exceptionally wet year, in contrast, differences compared to the long-term average were larger overall and varied more between individual settlements. In Smakiya and Hmud, two villages in the eastern half of the study area, the estimated RRWH potential in 1965 was almost double that of a normal year, whereas in Qasr, situated in the central part of the Karak plateau, only an increase of 58% in the potential was observed. The remaining settlements in the study area showed increases of around 75–90% in their RRWH potential in the same year (see Fig. 8). This illustrates the range and interannual variations that can be expected in regard to the RRWH potential. Moreover, the influence of interannual variations in the spatial distribution of precipitation also becomes apparent. According to our results, in the wet year 1965, the highest RRWH potential was found in ar-Rabba, in contrast to years with average precipitation, in which Qasr ranks first due to the town's larger total rooftop area. Similarly, in the same period, the RRWH potential of Misr and Hmud surpass those of the larger villages Ariha and al-'Aliya. In the dry year 1963, in contrast, the ranking of the individual settlements in regard to their total runoff from rooftops paralleled that described above for years with average precipitation amounts.

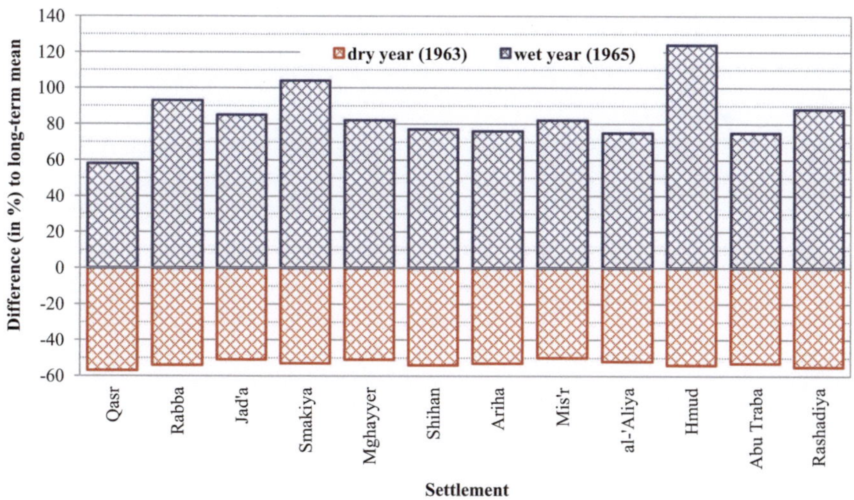

Fig. 8 Percentage differences in annual RRWH potential in years with very low or high precipitation, compared to long-term mean. For each settlement, the corresponding bar shows the deficit (red part) and surplus (blue part) compared to the respective RRWH potential in years with average precipitation. Dry and wet year conditions are exemplified by the data for the years 1963 and 1965. Modified from [14]

Finally, the RRWH potential in each settlement and the study area as a whole was compared to the storage potential available with the existing cisterns. In this evaluation, only a subset of the mapped cisterns and the detected rooftops was considered, which would potentially be suitable for integration into RRWH schemes. Here, for the cisterns, a distance of no more than 250 m to the nearest detected rooftop (edge) was determined as the selection criterion. For the rooftops to be eligible, they had to be located, at least partly, within a radius of 250 m around an existing cistern. These prerequisites were defined to ensure that rooftops as water collection areas and cisterns as storage facilities could be connected with reasonable efforts in the form of piping and construction works. In total, 155 cisterns were identified as potentially suitable for RRWH. This figure included 91 or 95% of the old cisterns as well as 64 or 22% of the modern cisterns. For the defined subset of objects, the mean distance to the nearest rooftop was 47 m for the old cisterns and 126 m for the modern cisterns (see Table 7). In all settlements but the smallest one, Rashadiya, at least one eligible cistern was found. In three settlements, in Jad'a, Shihan, and Mis'r, no potentially suitable old cisterns were identified, whereas in two other villages, in Smakiya and Abu Traba, no eligible modern cisterns were detected.

Overall, for the whole study area, the findings presented in Table 7 showed a considerable surplus in runoff from eligible rooftops compared to the total storage potential of the cisterns in their vicinity. For the selected rooftops, an annual runoff of 76,397 m^3 was computed under average precipitation conditions. The storage potential of all 155 eligible old and modern cisterns amounted to 13,750 m^3 or 18% of the aforementioned runoff, thus indicating a quite large untapped potential. However, with regard to individual settlements, the proportion between RRWH and storage potential varied largely. For Qasr, for example, an annual runoff of 15,066 m^3 from

Table 7 Comparison of RRWH and available storage potential

Study area/total of 11 settlements	Old cisterns	Modern cisterns
Roof area required, per cistern	544 m^2	126 m^2
Potentially eligible cisterns	91	64
Distance to nearest rooftop (edge)		
Min.–max	0–199 m	0–243 m
Mean	47 m	126 m
Total area of eligible rooftops	236,605 m^2	77,655 m^2
Annual runoff from eligible rooftops	57,370 m^3	19,027 m^3
Estimated total capacity of eligible cisterns	11,830 m^3	1,920 m^3
Eligible cisterns with sufficient roof area	74	44

All figures refer to normal, i.e. long-term average precipitation conditions. The required roof area is the average extent necessary to collect enough runoff to fill an old/modern cistern once per year. For a description of the criteria defining (potentially) eligible cisterns and rooftops, see text above. Cisterns with sufficient roof area are those which have rooftops with a total extent at least as big as the specified threshold (see roof area required) in their vicinity (250 m radius). All results were computed based on semi-automatically detected rooftop data all potentially suitable rooftops was calculated, whereas the storage capacity of the six old and two modern eligible cisterns was estimated at 840 m^3. The latter corresponds to only around 6% of the estimated runoff and thus suggests a considerable storage deficit with respect to the exploitation of the RRWH potential. In Abu Traba, 17 potentially suitable old cisterns were found leading to an estimated total available storage capacity of 2,210 m^3. The runoff from eligible rooftops was estimated at 2,536 m^3 yearly. Hence, with the existing cisterns, 87% of the computed runoff from nearby rooftops could be stored.

Under typical precipitation conditions and with the annual precipitation amounts occurring in the study area, on average, a rooftop area of 500–600 m^2 is necessary to collect enough runoff to fill an old cistern once per year to its capacity (see Table 7). For the same purpose, for a modern cistern, 120–130 m^2 of rooftop area are sufficient due to its much lower estimated capacity (see 4.1). An evaluation based on these requirements revealed that 118 or 76% of the 155 cisterns that were identified as potentially suitable for RRWH, had rooftops with an adequate total extent in their vicinity (250 m radius). This finding suggests a spatial mismatch between RRWH and storage potential on a local scale within individual settlements. In several cases, it may result from cisterns being clustered around a few buildings, while in the vicinity of other buildings, no cisterns are found. With regard to differences between old and modern cisterns, evaluation outcomes showed that most (over 80%) of the old cisterns had a sufficiently large rooftop area in their proximity despite their typically bigger capacity. In the group of eligible modern cisterns, the share of objects with a sufficiently large rooftop area in the analyzed radius was smaller (less than 70%). However, again, circumstances varied largely between individual settlements. In the village Hmud, with 23 old and 4 modern ones, the largest number of cisterns potentially suitable for RRWH was found. Yet, only 50–60% or 13 of the old and 2 of the modern cisterns had a rooftop area of adequate size in their surroundings. In Jad'a, no old but 12 modern cisterns were identified as eligible for RRWH, all of which had rooftops with suitable extents in their proximity. In Mghayyer, in contrast, only one of the 12 potentially suitable modern cisterns had a sufficiently large rooftop area in its vicinity.

5 Interpretation and Discussion of Findings

The compiled comprehensive inventory of RWH structures can be assumed to be fairly accurate in regard to the location and nature of the individual included objects, as the ground check on a subset of objects mapped based on the satellite image confirmed their correct identification. This also suggests that the applied remote sensing-based mapping procedure is highly specific since no false positives were discovered in the evaluation. The sensitivity of the method could not be determined as false negatives were not detected and they could not be excluded based on the assessment. The remote sensing-based mapping approach is prone to overlook some objects, e.g. those obscured from direct view by overhanging trees, clouds or similar, whereas the limitations of the field survey result from its dependence on the knowledge of local guides and the accessibility of locations, and its time-consuming nature. Here, the combination of both approaches allowed for the reaping of the benefits and counterbalancing the drawbacks of each method. The remote sensing-based mapping approach, in particular, facilitated the gathering of comprehensive data which otherwise would not have been possible. In combination with information that could only be collected through field surveys, such as cistern capacities, this data enabled the estimation of the total water storage potential available with the existing cisterns. Findings from field surveys, such as construction details of several cisterns, and from analyses concerning the proximity to places with archaeological finds or remains, also corroborated the previously tentative assumption of an old, probably ancient, age for the cisterns in the corresponding category. Hence, the collected evidence supported and justified the definition of two separate categories of cisterns, old and modern ones, and the comparative analyses based on these.

Outcomes from site characteristics and distance analyses highlighted differences between the defined groups of RWH structures and their sites. Differences in the characteristics of sites between linear RWH structures and cisterns may be attributed to the respective nature and functionality of the two types of structures. In contrast, discrepancies in the findings for old and modern cisterns most likely reflect changes in site preferences over time and emphasis on different factors in the site selection process. The old cisterns are almost all located within modern settlements or in the immediate surroundings of those. Since most of the modern settlements in the study area were built upon ancient foundations or near places used during earlier times as settlement sites or for other purposes, the observed spatial relation to settlement areas and proximity to buildings can be assumed as a rather consistent factor over time. This may reflect the role RWH and water storage played during former times when these cisterns were built: in the absence of other, year-round available water resources, water from precipitation in the winter half year and its collection and storage for the dry summer months was of utmost importance for survival on the Karak plateau. Since the construction of underground cisterns was labor intensive and costly, they were probably founded where they could be most beneficial and the water was most needed: within settlements, near residential buildings or at other central places of heightened importance (e.g. palaces or temples). Today, most of these places and the buildings in settlement areas are connected to the water line which provides a central distribution of water resources over long distances. Therefore, nearly all of the modern cisterns are found outside of settlement areas and often at larger distances to those. There, the water stored safely underground is used to water animals such as sheep and goats which are kept as livestock by local farmers and (semi-) nomadic people. The advent of modern drilling equipment and excavation machinery apparently led to the construction of smaller cisterns at more diverse places, even in basalt areas which previously were probably considered unsuitable due to the usually hard rock. Hence, the findings suggest, when interpreted holistically, that anthropogenic factors such as socio-economic and technological circumstances are probably more decisive in the site selection process than purely natural environmental factors. In other words, which places are regarded as most suitable for RWH may be subject to changes, depending on the respective setting during that time and the weighting of different aspects. The apparent differences between old and modern cisterns with regard to the characteristics of their sites and their spatial distribution pattern most likely reflect the shift in the role of RWH over time, from the primary, essential source of water to a complementary one that is used only in the absence of other solutions, e.g. in remote areas, and/or for specific purposes such as farming.

The LU/LC data acquired through the semi-automatic image interpretation process was of high quality as indicated by accuracy values of 90% and above and a Kappa coefficient of at least 0.80. The latter value is commonly interpreted (see e.g. [30]) as evidence for a substantial or strong agreement (between classification results and reference data) or good reliability of the data, here the classification results. More specifically, for the vegetation class, an even higher reliability can be assumed, based on the index values identified for this class. These included an overall accuracy of 94% and a Kappa of around 0.90 which indicates an almost perfect agreement or an only marginal, about 10%, chance of error. The evaluation of the semi-automatically acquired LU/LC data also revealed typical sources of error and thus allowed for the narrowing down of potentially problematic areas that might require special attention, visual-manual re-evaluation or similar corrections. Here, an overestimation of the impervious area was observed which resulted from the inclusion of larger bare (lime) stone or other bare ground areas primarily in the southeastern part of the study area. Due to similarities especially in reflectance values, and in shape, extent, and environment, these areas strongly resemble impervious surfaces like concrete areas, (dust-covered) asphalt roads, and similar. On the other hand, the otherwise good reliability and quality of the semi-automatic classification results, here can at least partly be attributed to the definition of three broad classes with distinct characteristics and sufficient differences based on which the corresponding areas could reliably be distinguished in the image interpretation process. At the same time, these classes were characterized by larger differences in runoff (coefficients) and thus, were most relevant for the scope of this study. In regard to the aforementioned difficulties concerning the correct classification of a specific subset of bare ground areas, the integration of additional, auxiliary data, such as e.g. a texture layer, index layer, or other remote sensing images with a higher spectral resolution, could be beneficial and lead to enhanced results. However, the necessary data might not be available in all cases and associated costs can outweigh the expected benefits. Here, for the objectives of this study and the acquisition the necessary data, the semi-automatic image analysis procedure and its applicability were largely confirmed. Generally, chances for a successful semi-automatic detection with suitable and reliable outcomes are highest, when target areas or features are relatively homogeneous in appearance, sufficiently large in size (group of pixels versus individual pixels), and exhibit noticeable differences with their surroundings (i.e. contrast with the immediate environment). Since the RWH structures studied here, in particular the above-ground features of cisterns, did not fulfill these criteria, the semi-automatic procedures were limited to the acquisition of the LU/LC and rooftop data.

In comparison to the LU/LC data, the semi-automatically extracted rooftop data showed considerably lower accuracy values, particularly in regard to the user's and the producer's accuracy of the individual datasets. To a large part, this results from the comparably narrow scope of the category and the overlap in surface materials, primarily cement, between target objects and other areas like parking lots, roads, and similar. Hence, rooftop areas have to be identified and distinguished from these other regions based on not only spectral information but also additional criteria such as shape, size, and surroundings. In view of the complexity of the classification tasks, unsurprisingly, (semi-) automatic rooftop or building footprint detection results are often associated with a higher error margin compared to other classes of target areas, as findings from previous studies have illustrated (e.g. [38, 40]). Yet, for the rooftop data in our study, a consistently very high overall accuracy was observed which contrasts with the low user and producer accuracies. This finding can be explained with the employed comprehensive assessment procedure which included vast background (non-rooftop) areas that were mostly correctly identified as such in the image classification process. In combination with the feature score results, consistently exhibiting very high user and much lower producer accuracies on an object level, this suggests that the developed AFE model and rooftop detection procedure were quite specific but less sensitive. Depending on the settlement or the subregion, a considerable number of rooftops were not recognized in the classification process. Additionally, the differences between feature score values and area-based user's and producer's accuracies point to difficulties in the precise delineation of individual rooftop areas. Often, the rooftop edge was not clearly discernible, making tracing its exact outline difficult even for the trained user in a visual-manual image interpretation approach. This particularly applies to areas in which rooftops showed very poor contrast with surrounding areas. Thus, based on the synoptical interpretation of the findings resulting from the evaluation of the rooftop data, the two main sources of observed errors could be identified. Here, again, the integration of additional, auxiliary data into the (semi-) automatic building (footprint) or rooftop detection process is suggested. As previous studies (e.g. [16, 42]) have shown, especially the use of very high resolution (VHR) elevation data, in combination with VHR satellite or aerial images, can lead to improved classification outcomes for the building class. Despite the difficulty in capturing the exact shape and boundary of individual rooftops correctly, the total area of rooftops of our classification results approximately matched that of the reference data in most cases. However, the reliability and suitability of the semi-automatically extracted data varied markedly between individual settlements. Upon comparison of the evaluation results, apparently, the best outcomes can be achieved with a classification (AFE) model that has been specifically developed for or adjusted to the respective subregion and image subset. Overall, the rooftop as well as the LU/LC data produced through semi-automatic image analysis in this study, in view of their limitations, are not recommended for planning purposes on the highest level of detail, such as the design and dimensioning of individual (rooftop) RWH systems. Yet, the data are suitable for use with a medium level of detail such as the area-wide estimation of the general RWH and the RRWH potentials and their spatial distribution, which were the objectives of the present study.

The estimates of the annual on-site runoff revealed a considerable, yet mostly undeveloped, RWH potential. This potential varied throughout the study area, depending on local circumstances such as LU/LC, slope, and local climate conditions. At the same time, with the mapped existing cisterns in the area, a significant storage potential already exists. In view of the costs associated with the construction of cisterns or the installation of storage tanks (e.g. [2]), already existing objects are particularly valuable. This especially applies to the older cisterns, many of which apparently date back to ancient times, since they are usually larger than their modern counterparts. Yet, as noted during field surveys, many of the old cisterns are in some state of decay and, depending on individual circumstances, would need at least a removal of debris, sediments, and rubbish accumulated inside the cistern, minor repairs, or larger refurbishment efforts, in order to be fully functional again. Additionally, many old cisterns today are not connected to adequate runoff collection areas and are therefore often merely used for storing tap water from intermittent public water supply. According to our findings, most of these cisterns are also located near modern buildings and, therefore, appear particularly eligible for an integration into modern RRWH systems in which the rooftops provide a usually excellent runoff collection area. This also illustrates the particular value of the old cisterns and the possible benefits from their rehabilitation where necessary. Based on the study outcomes, such as the comparison of the RRWH and the storage potential on the level of individual settlements, priority areas and suitable measures can be defined, like cistern rehabilitation programs or the construction of new cisterns. Similarly, this also applies to the RWH potential of the area in general and other RWH schemes besides RRWH, for which suitable areas and strategies matching local conditions can be determined based on the collected data.

Given the estimated overall storage capacity of 21,270 m^3, all cisterns identified in the study area would allow storing around 0.38% of the computed total annual on-site runoff from potentially suitable areas, i.e. pervious, mostly vegetation-free land with gradients of no more than 15%. However, as the proximity analyses revealed, most of the old cisterns are located inside settlement areas and would not be suitable for RWH schemes relying on ground catchments. Considering only the modern cisterns, which in view of their location in most cases are more suitable for these RWH strategies, only 0.16% of the estimated yearly local runoff from potentially eligible pervious areas could be stored. The aforementioned figures refer to runoff estimates for years with long-term average precipitation, while the share of readily storable runoff would roughly double or be halved in exceptionally dry or wet years. However, the total capacity of the existing cisterns could be expanded with (partial) multiple fillings, meaning that at least part of the stored water is used during the winter half-year when precipitation and water harvesting occur. According to [12], over the year, this could lead to an increase of 30% in the total amount of available stored cistern water from RWH. On the other hand, other factors, mainly concerning insufficient cistern maintenance, may reduce the quantity of annually available cistern water. Sediments and other materials accumulating inside cisterns over time can decrease the remaining capacity for water storage to a minor or a major degree. Similarly, cracks, tree roots reaching the cistern body, or deficient plaster lining of cistern walls can lead to leakages, thereby reducing the amount of available water.

In regard to the RWH potential, the area actually available for RWH schemes and, thus, the applicable potential can be expected to differ considerably from the estimated hypothetical potential, primarily due to competing purposes and (land) uses in the eligible areas. For the pervious regions that were mapped as vegetation-free in the summer, cultivation and other agricultural purposes during the winter half year may be in opposition to many types of RWH practices, e.g. the collection and storage of surface runoff in cisterns for a variety of purposes. However, these areas could benefit from the application of RWH practices, e.g. with bunds, dikes, ridges, and similar, which decrease surface runoff, thereby increasing infiltration and soil moisture, and thus can enhance plant growth. The different interests and (land) uses competing with RWH were not included in the present study, although they can influence the available area and the spectrum of applicable RWH techniques considerably. However, the availability of areas for RWH strongly depends on the priorities assigned to the different purposes or uses and the availability of alternatives. Besides the factors decreasing the actually applicable RWH potential, there are also aspects that could increase it. For example, the preparation of land surfaces can increase runoff efficiency (e.g. [24, 41]), thus leading to a reduced required collection area or a larger quantity of harvested water.

The surface runoff harvested with the existing linear RWH structures, such as bunds, ridges, check dams, and similar, could not be quantified here since the harvested water is stored directly in the soil profile or contributes to groundwater resources. Hence, quantification of these effects requires studies with a different focus than that of the present study. The collected comprehensive data on existing RWH structures in the area could serve as the basis for further studies regarding these effects and their dimensions. The high number of mapped bunds, ridges, check dams, and similar objects may already point to their relevance and the associated benefits. These have been documented in previous studies (e.g. [43]) and encompass effects like increased soil moisture, enhanced plant growth, and reduced erosion.

The water demand of the local population in the study area is estimated at 1,889 m^3 per day or 689,529 m^3 per year. This results from a population of 23,614 people in 2015 as estimated by the Jordanian Department of Population Statistics [15] and an assumed water demand of 80 l per day per person, as suggested e.g. by [12] and the Jordanian Ministry of Water and Irrigation [32, 33]. To cover this demand, around 13% of the calculated, long-term average, overall RWH potential or annual on-site runoff from pervious areas would have to be harvested. If all of the computed long-term mean annual runoff from rooftops was collected and made available, ca. 22% of the aforementioned demand or the annual water supply required by 5,272 residents could be covered. At the same time, the storage potential of all cisterns together corresponds to 3% of the annual water demand of the local population or that of 728 people. While the typical storage capacity of a modern cistern is sufficient to meet the water demand of one person per year, the estimated mean capacity of an old cistern allows for storing enough water for at least four persons each year. With partial multiple fillings, these figures can be increased to 4% of the annual water demand of the local population or the necessary yearly water supply for 947 residents. These figures and the assumed demand of 80 l per day per person include the water needed for domestic purposes in a broader sense, such as drinking, cooking, washing and cleaning, as well as watering of livestock and plants (gardening). However, municipal and mainly domestic purposes account only for less than 40% of the water demand in Jordan, while the remaining water supply is allocated to primarily (about 60%) agricultural purposes (e.g. irrigation) and industrial uses (ca. 3%) [7, 32, 33]. In the Karak governorate, which receives around 8% of the country's water supply, municipal and domestic purposes make up an even smaller share of about 22% of the water demand [33]. Not all purposes require the same water quality, and therefore, harvested rainwater, or actually surface runoff, may be more or less suitable, depending on the purpose or use case, the type of RWH practice and the design of the respective system. RRWH is typically associated with a high quality of the collected water that, therefore, can be utilized for a variety of purposes, including human consumption, after appropriate treatment (e.g. [2]). However, as the figures above have shown, the RRWH potential is limited at least by the available rooftop area. In contrast, unsealed ground catchments may be available for RWH in vast areas, but the harvested water may rather be suitable for agricultural purposes only, especially for supplemental irrigation, e.g. for trees during crucial growing phases. This illustrates that not all water supply from other sources can readily be replaced by RWH.

6 Recommendations for Decision-Makers, Policy Planners, Stakeholders, and Future Studies

1. Enhance RWH studies and planning processes with remote sensing and GIS technologies:
   (a) The case study exemplified how the application of remote sensing and GIS techniques can benefit RWH studies. Satellite and aerial images with very high resolution allow deriving detailed information and thus can be employed in mapping processes, even when target objects are rather small, such as RWH structures. Remote sensing-based mapping procedures not only speed up the data collection process, they also facilitate the assessment in difficult terrain or remote areas, which otherwise would be hard to accomplish.
   (b) The resulting comprehensive data, such as the RWH inventory, then render further analyses possible that provide deeper insights into RWH practices and changes therein over time. They also form the necessary basis for thorough subsequent planning processes.
2. Integrate field surveys with remote sensing efforts:
   (a) Combine remote sensing with field surveys for optimal data accuracy and ground truth validation. Insights from field surveys, such as typical cistern capacities, can also be extrapolated by combining them with data gained

through remote sensing-based approaches, such as the total number of existing cisterns. This upscaling provides new insights, in this case, the well-founded estimate of the total storage capacity available in the area.
    (b) Encourage community engagement by incorporating crowdsourcing and citizen science to gather substantial on-the-ground data. Local participants equipped with smartphones and mapping applications can be valuable contributors to data collection efforts.

3. Leverage advanced remote sensing for environmental data:
    (a) Remotely sensed imagery, recorded by a variety of different sensors aboard numerous satellite missions, by drones, and within the scope of aerial surveys, becomes increasingly available with further technological advances. In particular, this also applies to image products with a high or very high spectral or spatial resolution, such as the GeoEye-1 imagery employed in this study. From these, critical environmental data, such as land use / land cover (LU/LC) information, can be derived, which are essential for planning and optimizing RWH systems.
    (b) Invest in big data and image analysis automation, incorporating artificial intelligence (AI) and machine learning, to handle large datasets and enhance the efficiency of data processing and interpretation.

4. Adopt GIS for holistic spatial planning and analysis:
    (a) GIS technology should be the cornerstone of spatial analyses for RWH initiatives, as it provides essential tools for data processing, integration, and visualization.
    (b) GIS should be utilized in all stages of RWH research and planning projects to facilitate site selection, design, and analysis of RWH schemes, thereby enhancing spatial planning and stakeholder engagement.

5. Support infrastructure investments for sustainable water management:
    (a) Policy planners and decision-makers should promote and fund the modernization and expansion of existing RWH infrastructure, as well as ensure ongoing maintenance to maximize water harvesting potential.
    (b) Recognize the limitations of RWH as a stand-alone solution, since it cannot expand local or regional water resources to an indefinite degree. Therefore, RWH has to be combined with other water supply and management strategies, such as greywater reuse, water recycling, and, in particular, demand management, to strengthen local water security and resilience.

6. Promote RWH as a part of integrated water management strategies:
    (a) Policy planners should consider RWH as a tool within a broader portfolio of integrated water management strategies. This integration can help mitigate the overexploitation of groundwater resources by offsetting demand and contributing to a more resilient water supply system.

   (b) Encourage widespread adoption of RWH practices across regions to make a substantial impact on water resource sustainability, particularly in areas where groundwater depletion is a pressing issue.

7. Expand research into RWH potentials and water quality assessments:

   (a) Future studies should explore further how remote sensing and GIS can be applied to assess not only the spatial potential for RWH but also the water quality, depending on the qualities of the runoff collection areas and surfaces, such as rooftops (e.g. [36]). This could provide actionable insights for safe RWH implementation.

Acknowledgements The research was funded by Heidelberg University, the federal state of Baden-Württemberg, Germany, and the University of Jordan in Amman, Jordan. The research was carried out at the Institute of Geography of Heidelberg University, the Institute of Geography of the University of Cologne, and the Water, Energy, and Environment Center of the University of Jordan. The authors would like to thank the staff of the National Center for Agricultural Research and Extension (NCARE) in Jordan and the Water Authority of Jordan, in particular, Refaat Bani-Khalaf and Randa Shaker Tuffaha, for supporting the research, as well as Dr. Tobias Törnros for his assistance with the interpolation of the precipitation data, Ahmed Ebaid Saadallah for the on-screen digitization of data, and the former students Isabell Rühl, Alexandra Vögele (Stein), and Keren Baicu who wrote their theses in the context of this research project. For their time, efforts, practical help during field surveys, and hospitality, the authors would like to thank Mahmoud Alamr and his family, Youssef Tarawneh and his family, Mohamed Ali Majali, Rony Zumot, and Amjad Saleh Hijazin. Thanks are also due to Prof. Dr. Georg Bareth and his working group at the University of Cologne and Prof. Dr. Lucas Menzel and his working group at Heidelberg University.

References

1. AbdelKhaleq RA, Alhaj Ahmed I (2007) Rainwater harvesting in ancient civilizations in Jordan. Water Sci Technol: Water Supply 7(1):85–93
2. Abu-Zreig M, Ababneh F, Abdullah F (2019) Assessment of rooftop rainwater harvesting in northern Jordan. Phys Chem Earth 114:102794
3. Aklan M, de Fraiture C, Hayde LG (2025) Why we should revitalize indigenous water harvesting systems: lessons learned. Int Soil Water Conserv Res 13(1):152–163. https://doi.org/10.1016/j.iswcr.2024.05.004
4. Akpinar Ferrand E, Cecunjanin F (2014) Potential of rainwater harvesting in a thirsty world: a survey of ancient and traditional rainwater harvesting applications. Geography Compass 8(6):395–413
5. Al-Adamat R, Diabat A, Shatnawi G (2010) Combining GIS with multicriteria decision making for siting water harvesting ponds in Northern Jordan. J Arid Environ 74:1471–1477
6. Al-Houri Z, Al-Omari A (2022) Assessment of rooftop rainwater harvesting in Ajloun, Jordon. Water Reuse 12(1):22–32
7. Al-Kharabsheh A (2020) Challenges to sustainable water management in Jordan. Jordan J Earth Environ Sci 11(1):38–48
8. Al-Khuzaie MM, Janna H, Al-Ansari N (2020) Assessment model of water harvesting and storage location using GIS and remote sensing in Al-Qadisiyah, Iraq. Arab J Geosci 13:1–9

9. Al-Qawasmi O (2021) Feasibility of rainwater harvesting from residential rooftops in Jordan. Appl Water Sci 11(2):30
10. Ali A, Oweis T, Salkini AB, El-Naggar S (2009) Rainwater cisterns: traditional technologies for dry areas. ICARDA, Aleppo
11. Antoniou G, Kathijotes N, Spyridakis DS, Angelakis AN (2014) Historical development of technologies for water resources management and rainwater harvesting in the Hellenic civilizations. Int J Water Resources Develop 30(4):680–693
12. Assayed A, Hatokay Z, Al-Zoubi R, Azzam S, Qbailat M, Al-Ulayyan A, Abu Saleem M, Bushnaq S, Maroni R (2013) On-site rainwater harvesting to achieve household water security among rural and per-urban communities in Jordan. Resour Conserv Recycl 73:72–77
13. Bienert H-D (2000) Water shortage in Jordan since prehistoric times—lessons from history. ICID J 49(4):17–37
14. Brilmayer Bakti B (2020) Rainwater harvesting in rural Jordan: a GIS and remote sensing-based analysis of ancient and modern adaptations to water scarcity in a changing environment. Heidelberg University, Heidelberg
15. Department of Population Statistics (DoPS), Population and Social Statistics Directorate, The Hashemite Kingdom of Jordan (2015) The population of the kingdom by administrative divisions, according to the general census of population and housing results 2015. Department of Statistics, The Hashemite Kingdom of Jordan, Amman
16. Dinis J, Navarro A, Soares F, Santos T, Freire S, Fonseca A, Afonso N, Tenedório J (2010) Hierarchical object-based classification of dense urban areas by integrating high spatial resolution satellite images and LiDAR elevation data. In: Addink EA, Van Coillie FMB (eds) GEOBIA 2010: geographic object-based image analysis, Ghent, June 29–July 2, 2010. International Society for Photogrammetry and Remote Sensing (ISPRS). (The International Archives of the Photogrammetry, Remote Sensing and Spatial Information Sciences, XXXVIII-4/C7), pp 280–286
17. Ertop H, Kocięcka J, Atilgan A, Liberacki D, Niemiec M, Rolbiecki R (2023) The importance of rainwater harvesting and its usage possibilities: Antalya example (Turkey). Water 15(12):2194. https://doi.org/10.3390/w15122194
18. Feloni E, Nastos PT (2024) Evaluating rainwater harvesting systems for water scarcity mitigation in small Greek islands under climate change. Sustainability 16(6):2592. https://doi.org/10.3390/su16062592
19. Fiaz A (2019) Rainwater harvesting in Jordan: potential water saving, optimal tank sizing and economic analysis.Urban Water J 17(5):1–11
20. Güneralp İ, Filippi AM, Hales B (2014) Influence of river channel morphology and bank characteristics on water surface boundary delineation using high-resolution passive remote sensing and template matching. Earth Surf Processes Landforms 39(7):977–986
21. Hadadin N, Shawash S, Tarawneh Z, Banihani Q, Hamdi MR (2012) Spatial hydrological analysis for water harvesting potential using ArcGIS model: the case of the north-eastern desert, Jordon. Water Policy 14:524–538
22. Keilholz P (2014) The ancient cisterns of Hellenistic Gadara/Umm Qais (Jordan). In: Schäfer T, Schön F, Gerdes A, Heinrichs J (eds) Antike und Moderne Wasserspeicherung: Internationaler Workshop vom 11. - 14.05.2011 in Pantelleria (Italien). Rahden: Marie Leidorf. (Tübinger Archäologische Forschungen, 12), pp 27–36
23. Laureano P (2004) Traditional techniques of water management: a new model for a sustainable town and landscape: from the first water harvesting surfaces to Paleolithic hydraulic labyrinths. Perspectivas Urbanas 4:1–7
24. Li X-Y, Xie Z-K, Yan X-K (2004) Runoff characteristics of artificial catchment materials for rainwater harvesting in the semiarid regions of China. Agric Water Manag 65:211–224
25. Loper SA, Zimmerman SA, Stoughton KL, Pamintuan BC, Kilgannon EM (2024) Geographic information system mapping tool for rainwater harvesting in the United States. J Water Resour Plan Manag 150(5). https://doi.org/10.1061/JWRMD5.WRENG-6236
26. Madgundi MM, Kumbhar AP, Lele GM, Komble SP, Marane YH, Mate AR (2023) Design and investigation on rain saucer: The technique of roofless rainwater harvesting. Mater Today: Proc 72:1084–1088. https://doi.org/10.1016/j.matpr.2022.09.169

27. Mahmoud SH, Mohammad FS, Alazba AA (2015) Delineation of potential sites for rainwater harvesting structures using a geographic information system-based decision support system. Hydrol Res 46(4):591–606
28. Maliva R, Missimer T (2012) Arid lands water evaluation and management. Environmental science and engineering: environmental science. Springer, Heidelberg
29. Mays LW (ed) (2010) Ancient water technologies. Springer, Dordrecht, Heidelberg
30. McHugh ML (2012) Interrater reliability: the kappa statistic. Biochemia medica 22(3):276–282
31. Menzel L, Koch J, Onigkeit J, Schaldach R (2009) Modelling the effects of land-use and land-cover change on water availability in the Jordan River region. Adv Geosci 21:73–80
32. Ministry of Water and Irrigation (MWI), The Hashemite Kingdom of Jordan (2016) National water strategy 2016–2025. Ministry of Water and Irrigation (MWI), The Hashemite Kingdom of Jordan, Amman
33. Ministry of Water and Irrigation (MWI), The Hashemite Kingdom of Jordan (2016) Water reallocation policy. Ministry of Water and Irrigation (MWI), The Hashemite Kingdom of Jordan, Amman
34. Mwenge Kahinda J, Taigbenu AE, Sejamoholo BBP, Lillie ESB, Boroto RJ (2009) A GIS-based decision support system for rainwater harvesting (RHADESS). Phys Chem Earth 34:767–775
35. Napoli M, Cecchi S, Orlandini S, Zanchi CA (2014) Determining potential rainwater harvesting sites using a continuous runoff potential accounting procedure and GIS techniques in central Italy. Agric Water Manag 141:55–65
36. Norman M, Shafri HZ, Mansor SB, Yusuf B (2019) Review of remote sensing and geospatial technologies in estimating rooftop rainwater harvesting (RRWH) quality. Int Soil Water Conserv Res 7(3):266–274
37. Odeh T, Sawaqed R, Murshid EA, Mohammad AH (2023) GIS-based analytical analysis for selecting potential runoff harvesting sites: the case study of Amman-Zarqa Basin. Sustain Water Resour Manag 9(3). https://doi.org/10.1007/s40899-023-00879-2
38. Opitz D, Blundell S (2008) Feature extraction with the VLS feature analyst system. In: Proceedings of the ASPRS annual conference "Bridging the horizons – new frontiers in geospatial collaboration", Portland (Oregon, USA), April 28–Mai 2, 2008. American Society for Photogrammetry and Remote Sensing (ASPRS), Bethesda (Maryland, USA)
39. Ouessar M, Bruggeman A, Abdelli F, Mohtar RH, Gabriëls D, Cornelis WM (2009) Modelling water-harvesting systems in the arid south of Tunisia using SWAT. Hydrol Earth Syst Sci 13:2003–2021
40. Psaltis C, Ioannidis C (2008) Simple method for cost-effective informal building monitoring. Surveying Land Inform Sci 68(2):65–79
41. Richardson L, Hairsine PB, Ellis TW (2004) Water farms: a review of the physical aspects of water harvesting and runoff enhancement in rural landscapes. Cooperative Research Centre for Catchment Hydrology, Monash University, Canberra
42. Santos T, Freire S, Fonseca A, Tenedório JA (2011) Producing a building change map for urban management. EARSel eProceedings 10(1):56–65
43. Strohmeier S, Fukai S, Haddad M, AlNsour M, Mudabber M, Akimoto K, Yamamoto S, Evett S, Oweis T (2021) Rehabilitation of degraded rangelands in Jordan: the effects of mechanized micro water harvesting on hill-slope scale soil water and vegetation dynamics. J Arid Environ 185:104338
44. Szabó Z, Pedretti D, Masetti M, Ridavits T, Csiszár E, Falus G, Palcsu L, Mádl-Szőnyi J (2023) Rooftop rainwater harvesting by a shallow well – impacts and potential from a field experiment in the Danube-Tisza Interfluve, Hungary. Groundw Sustain Dev 20:100884. https://doi.org/10.1016/j.gsd.2022.100884
45. The Getty Conservation Institute (GCI), World Monuments Fund (WMF), Department of Antiquities (DoA), The Hashemite Kingdom of Jordan (2010) MEGA-Jordan: the national heritage documentation and management system. [Online database]. http://megajordan.org. Access 5 Nov 2015
46. Törnros T, Menzel L (2014) Leaf area index as a function of precipitation within a hydrological model. Hydrol Res 45(4–5):660–672

47. Wimmer F, Schlaffer S, Menzel L (2009) Distributed modelling of climate change impacts on snow sublimation in Northern Mongolia. Adv Geosci 21:117–124
48. Yousif M, Bubenzer O (2015) Geoinformatics application for assessing the potential of rainwater harvesting in arid regions: case study: El Daba'a area, northwestern coast of Egypt. Arab J Geosci 8(11):9169–9191
49. Zhou W, Matsumoto K, Sawaki M (2024) Potential of traditional domestic rainwater harvesting systems: current trends and future directions. J Asian Arch Build Eng 23(1):344–354. https://doi.org/10.1080/13467581.2023.2214193
50. Ziadat F, Bruggeman A, Oweis T, Haddad N, Mazahreh S, Sartawi W, Syuof M (2012) A participatory GIS approach for assessing land suitability for rainwater harvesting in an arid rangeland environment. Arid Land Res Manage 26(4):297–311

The National Strategy for Small Pelagic Fisheries' Development in the Northeast of Tunisia (Central Mediterranean): Status and Prospects

S. Mili, R. Ennouri, M. Fatnassi, S. Agrbi, and S. Tounsi

Abstract Fisheries is a strategic sector for Tunisia and has great social and economic potential, with average production expected to reach 137,470 tonnes by 2020. With 32% of national landings, small-scale pelagic fisheries occupy a privileged position. These species are caught in the waters of Northeastern Tunisia (GFCM GSA 12 and 13) and account for 18% (8,673 tonnes) of the national production, 80% of which is landed at the port of Kelibia. This study reviewed the small pelagic fisheries' status in the Northeast of Tunisia (GFCM-GSA 12 & 13) before and after implementing the National Strategy for the Development of Small Pelagic Fisheries (NSDSPF). This process helped to identify, and ultimately recommend corrective measures to ensure long-term exploitation, both the benefits and the obstacles associated with this activity. The results showed that the production of small pelagic fish in Tunisia has experienced several fluctuations over the last 28 years (1991–2019) after the NSDSPF was set up. The successful establishment of small pelagic fisheries in the Northeast of Tunisia (Kelibia region) depends on several factors, such as the number of trips, spatial delimitation of fishing areas, characteristics of the purse seiners, and number of processing and conservation factories. In addition, we drew attention to the overexploitation of the traditional small-scale pelagic fishing areas in the Northeast of Tunisia and the orientation towards new offshore marine fishing areas. To optimise the exploitation of small pelagic fish in Northeastern Tunisian waters, we strongly recommend the implementation of a coherent development plan.

S. Mili (✉) · R. Ennouri · M. Fatnassi · S. Agrbi · S. Tounsi
University of Carthage, Higher Institute of Marine Sciences of Bizerte, Bizerte, Tunisia
e-mail: sami.mili@issmb.ucar.tn

M. Fatnassi
e-mail: manel.fatnassi@issmb.ucar.tn

S. Mili
Laboratory of Fisheries Sciences, National Institute of Marine Sciences and Technologies, Salammbô, Tunis, Tunisia

R. Ennouri
Laboratory of Ecology, Biology and Physiology of Aquatic Organisms, Faculty of Sciences of Tunis, University of Tunis El Manar, Tunis, Tunisia

F. Khebour Allouche (ed.), *Environmental Remote Sensing and GIS in MENA Region*, Earth and Environmental Sciences Library,
https://doi.org/10.1007/978-3-031-86574-9_3

Keywords Small Pelagic · Fisheries · NSDSPF · Northeast · Tunisia

1 Introduction

Fishing plays a crucial role as a global source of animal protein, especially for the most disadvantaged. In Tunisia, this fundamental element of food security has to be preserved and needs to continue providing foreign currency to the country [1]. As a result, the government paid special attention to the fishing industry, focusing on basic infrastructure development, rationalizing fisheries exploitation, and upgrading the sector.

The Tunisian fisheries production reached 137,470 tons in 2020, ensuring 50,621 direct jobs and contributing to self-sufficiency and food security (9.5 kg/capita/year). The small pelagic fish sector mobilizes 5.568 workers and contributes 47.592 tons, i.e., 32% of the national production [2].

The National Institute of Marine Sciences and Technology (INSTM) and the General Directorate of Fisheries and Aquaculture (DGPA) have a major interest in small pelagic species (sardine, anchovy, sardinella, bogue, etc.) that offer great potential for the growth and sustainability of the Tunisian fishing activity. The acoustic and experimental fishing survey made by INSTM showed that small pelagic fish were under-fished in the 2000s and can be twice as highly exploited [3]. Hence, in 2001, the government set up the National Strategy for the Development of Small Pelagic Fisheries (NSDSPF), which aims to optimize the exploitation of these resources. To achieve this goal, a management plan was proposed, and several measures were taken to boost investments, enhance the processing industry, develop marketing, and upgrade this sector.

Fisheries' resources, although renewable, are exhaustible, and therefore, they must be effectively managed to be preserved. To maintain their nutritional, economic, and social contribution to a growing population, fish stocks must be monitored and managed sustainably. Therefore, the sustainability of fisheries through GIS application has been one of the key research priorities adopted by national research institutions. To contribute to these efforts, we aimed to gather and analyze data on the small pelagic fisheries in the Northeast of Tunisia (GFCM-GSA 12 & 13), using statistical and spatial analyses as implemented by the NSDSPF. This work also seeks to propose recommendations for sustainable management of these fisheries.

The present chapter consists of three parts: the first one is a review of the literature, including a delimitation and description of the study area, the state of exploitation of the resources studied, and the fishing techniques used, as well as the legislative framework governing the sector. The second part describes the methodology, scheduling of the surveys, design of the questionnaires, and analyses preceding data interpretation. The chapter is then wrapped up with a general conclusion that summarizes the most important results while emphasizing scientific recommendations and the potential for promoting responsible small pelagic fishing activities.

Table 1 Taxonomy of the most common small pelagic species in Tunisian waters [11]

Families	Species
Clupeidae	*Sardina pilchardus*
	Sardinella aurita
Engraulidae	*Engraulis encrasicolus*
Sparidae	*Boops boops*
Scombridae	*Scomber scombrus*
	Scomber japonicus
Carangidae	*Trachurus trachurus*
	Trachurus mediterraneus
	Trachurus picturatus

2 Overview

2.1 Introducing Small Pelagic Fisheries in Tunisia

In Tunisia, the targeted small pelagic fish include nine species of five families (Table 1). These species move freely along the bottom and are unaffected by the substrate. Their spatial and temporal distributions are determined by the physico-chemical and hydrographic parameters of the environment. They live in schools and move in search of optimal conditions [4–9]. The eco-biological studies on small pelagic fish show that these resources are characterized by a variable biogeographic distribution, an important migration, a rapid growth rate in the initial year of their existence, early sexual maturity, a relatively spread-out reproduction period and high fecundity [10, 42].

2.2 Small Pelagic Fishing Techniques and Gear

- **Light fishing**: This technique utilizes a surface or submerged light source to attract schools of fish, which respond positively to illumination. The gear is a rotating purse seine with a lead rope rigged with rings, through which a cable passes to aid in net hauling [12, 13].
- **Lampara net**: In this gear, the length of the head rope is greater than that of the foot rope. Therefore, as we haul up the net, the tightness of the footrope draws the net wall ahead of the head rope, trapping fish and preventing them from falling out. One or two boats can operate the net. A lampara net consists of a central bunt (small mesh size) and two lateral wings, that are which are dragged concurrently by attached ropes to facilitate fish encirclement. This gear is often used at night, and it deploys light to lure and aggregate small pelagic fish before the encirclement. This type of nets is deployed in areas along the coasts

to capture baitfish for hook-and-line fishing, particularly for tuna pole-and-line fishing. These nets are also not uncommon in lakes and rivers in catching small pelagic fish that are active close to the top [3, 14, 15].

- **Purse seines**: A purse seine is a netting wall that is used to enclose a shoal of pelagic fish at the surface. To boost the net's sinking speed and keep fish from escaping, weights, or lead lines are attached to the footrope, and thick netting materials such as polyamide (or polyester) are utilized. The distinguishing feature of a purse seine is a pursing cable woven through purse rings set along the net's lower border, that is drawn tight to shut the bottom of the net, resembling a drawstring purse, giving the name "purse seine". The central parts of the net are the most profound, and they begin to narrow near the wings and bunt, where the fish eventually gather. The bunt can also be positioned at the centre of the net, resulting in towing from the two wings [3, 15].
- **Small-scale seine fishing**: This technique is similar to light fishing for small pelagic fish, except that the operation is carried out during the day and no means of aggregating fish are used. The fishing unit comprises a master boat and a tender "Shkiff".
- **Hlig**: Beach seines are nets with elongated wings that can have a cod-end. Seining is particularly deployed in shallow coastal waters to capture schools of fish. We can set this type of gear from a boat or by hand and then pulled ashore. Spanning from the water's surface to the bottom, the net acts as a natural barrier, trapping the fish in the enclosed area. Bottomless seines are often equipped with a small-mesh bunt. Pulling beach seines can be accomplished from the beach manually, with the assistance of vehicles, winches, or a variety of machinery located on the beach. In the past, animals, particularly horses, were commonly deployed to haul beach seines [12, 15].
- **Mid-water trawls**: A mid-water trawl refers to a conical-shaped net that is pulled in the middle layer of the water column, typically by a single boat. This gear targets pelagic or semi-demersal fish, and it deploys components, which have no contact with the seabed during the fishing operation. Mid-water trawling targets species that live in schools; such as Clupeids and Scombrids, yielding substantial harvests. Hauling speed typically falls within the range of 3 to 5 knots, while 6 knots are necessary for fast fish. Mid-water trawls are generally larger compared to bottom trawls, particularly at their vertical opening. To minimize drag and maintain effective catch, the front section of the net is crafted using large meshes. The mouth of the net is kept open by using weights that are tied to the bottom wing extremities called clump weights [3, 15–17].

2.3 Assessment of Small Pelagic Stocks in Tunisia

The small pelagic fish of the Tunisian coasts, such as *Sardina pilchardus*, *Engraulis encrasicolus*, *Sardinella aurita*, *Boops boops*, *Trachurus* sp. and *Scomber* sp. represent 32% of the production in Tunisia. In 1996, the National Institute of

Marine Sciences and Technologies (INSTM) developed a long-term multidisciplinary research program dedicated to these species. The INSTM has been organizing OASIS surveys (Anchovy and Sardine Operations of the Salammbo Institute) to assess the stock of small pelagic species using acoustic techniques centred on echo-integration and accompanied by experimental fishing operations. This program has a twofold benefit: it allows us to assess the instantaneous biomass of fish stocks and to follow the evolution of their relative abundance in time and space. The most recent survey, OASIS 16, was conducted in 2018 and the estimated exploitable potential was 50 thousand tons. Additionally, Ben Abdallah et al. have shown that the sizes of the most available species (*Trachurus* sp. and *Sardina pilchardus*) are smaller than the sizes of first sexual maturity (Lm50) [18].

According to the OASIS 16 survey, the abundance of the small pelagic fish in Tunisian waters varied depending on sectors, seasons, and species (Fig. 1). The distribution is the result of one and/or several factors, such as the fish ecological needs. The diversity of the physicochemical and hydrographic conditions of Tunisian waters induced this variability. In comparison to previous studies, the current exploitable potential of small pelagic species in Tunisian waters is at its lowest level and it is dominated by horse mackerel [8, 18–21, 42].

2.4 National Strategy for the Development of Small Pelagic Fisheries

During the 2000s, a stock assessment of small pelagic resources in Tunisia showed significant exploitable potential. In the same vein, professionals confirmed that these resources are underutilised. Hence, the government encouraged the professionals to take an interest in these stocks by setting up a national strategy and encouraging fishermen to focus on this sector [22].

The objective of the NSDSPF strategy is to merge with various fisheries development plans. The general objectives of this strategy are.

- the development of an optimal investment environment and food security;
- the increase of the sector's productivity and the enhancement of its profitability, and
- the creation of jobs and the reinforcement of the sector's position.

The specific objectives of this strategy are

- the modernization of the fleet for better exploitation of the pelagic fisheries in the offshore areas;
- the improvement of the added value and competitiveness of Tunisian small pelagic fish; and
- the creation of new marketing channels and the promotion of fish export.

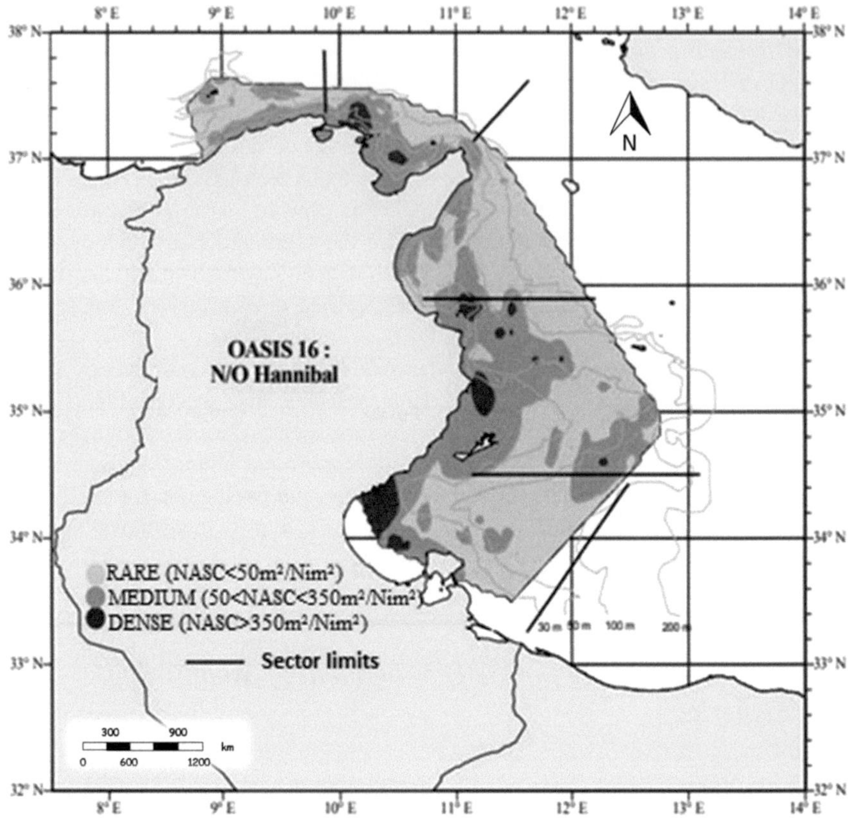

Fig. 1 Geographic distribution of small pelagic fish densities in Tunisian waters (2018) [18]

To boost production, NSDSPF planned to build 110 modern small pelagic fishing units. Specific measures have been taken to purchase these units, including an increase in the ceiling of investments in the "B" category from 300,000 TND to 1 million TND. This was accompanied by a subsidy on investment from 20 to 30%, with a first ceiling of 150.000 TND. In addition, it was mandated to install the Vessel Monitoring System (VMS) for fishing units exceeding 15 m to be able to track them, determine the fishing areas where they are operating, and identify the nature of the small pelagic fisheries in the Northeast of Tunisia (Central Mediterranean GFCM-GSA 12 & 13) [23]. Besides, the NSDSPF programmed the establishment of 10 units for container cleaning and 10 units for harvest transportation 'carriers or transport vessels'. Additionally, it was planned to create 15 new manufacturers of ice and 10 new facilities for fish processing. Furthermore, the existing canneries have been encouraged to join the upgrade program.

This was followed by specific marketing actions that included encouraging municipalities to upgrade commercial spaces; increasing the number of agents working at wholesale markets; and promoting the establishment of transportation facilities, such as refrigerated and isothermal trucks.

2.5 Tunisian Strategy Achievements

The following major accomplishments were made during the NSDSPF implementation in 2021: 89 fishing vessels have been built, 10 are under construction, and eight new construction authorizations have been granted. A total of 15 ice factories have been created, which is a completion rate of 100%. Concerning the units of conservation, six factories out of 10 have been established. However, no projects involving container-cleaning units have been carried out to date. In terms of marketing, this strategy has resulted in the establishment of approximately 500 retail outlets and 90 refrigerated trucks [24, 42].

3 Methodology

It is technically possible to obtain the required information for the small pelagic fish industry analysis from the administration. Administrative records have the benefit of mitigating much of the cost of field data collection and respondent burden. Results from surveys are also of interest because they contain most of the data [25]. To analyze the state of the small pelagic fisheries in the Northeast of Tunisia (GFCM-GSA 12 & 13), we have resorted to the collection of information from statistical resources on production (monthly/fishing method or technique/species) This information was gathered from organizations which are operating in this sector: INSTM, DGPA, Agency of Ports and Fishing Installations (APIP) and Inter-Professional Grouping of Fishing Products (GIPP). In addition, we conducted interviews with fishermen and crew members using specific survey forms.

The current work relied on analyzing 19 years of production statistics from 2000 to 2019. This relatively long interval elucidates the evolution of the profile of the small pelagic fisheries in the GFCM-GSA 12 & 13 areas.

Indeed, the analysis of the data on production, fleet, fishermen's population, and specific composition of fish gives us a deeper insight into the state of the small pelagic fisheries in this area. Besides, we have taken an interest in the spatial analysis of the fishing areas before and after the implementation of NSDSPF.

3.1 Study Area and Time Frame

Tunisia is endowed with a privileged location at the hinge of the Mediterranean's eastern and western basins. It is endowed with a maritime frontage, with one-third oriented towards the north (GFCM-GSA 12) and the other two towards the east (GFCM-GSA 13 & 14). With such a position, the ecosystem brings together a variability of biotopes and a great biocenotic diversity [17, 26].

According to the Minister of Agriculture's Order, dated September 28th, 1995, and regulating fishing activities [27], the Tunisian coasts have been divided into three fishing zones, delimited as follows (Fig. 2).

- The Northeast region (GFCM-GSA 12) extends from the Tunisian-Algerian border to the line passing through Burj Kelibia lighthouse;
- The eastern region (GFCM-GSA 13) extends from Burj Kelibia to the line passing through Rass Kapoudia (cape);
- The southern region (GFCM-GSA 14) is limited to the north by the line passing through Rass Kapoudia and to the south by the Tunisian-Libyan border.

The Northeast region is marked by rough landscapes, a low-amplitude coastal platform, and varied sediments [3, 17]. The western coast, which lies between Tabarka and Bizerte, is initially constituted by a recent coastline, then by a rocky coast which has numerous cliffs with rockslides, separated by small sandy bays. According to

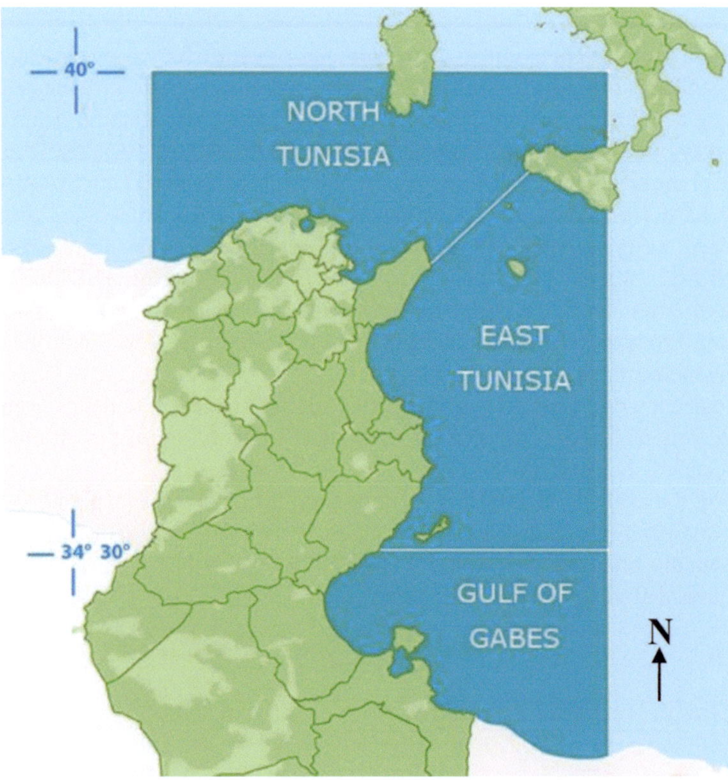

Fig. 2 Fishing areas in Tunisia (GFCM-GSA 12, 13 & 14) [28]

Kouki, Paskoff, Oueslati and Halouani et al., the coast of Cap Farina-Cap Bon is, in part, of an alluvial nature characterized by a sandy beach, mudflats, lagoons and deltaic structures and, in part, of a rocky nature interspersed with loose strings [29–32]. The slope of the coast becomes increasingly flatter from west to east. The coastal zone corresponding to this area is known for its humid to sub-humid and windy climate. It is also reputed for its rough landscape, which makes it sometimes difficult to access, due to its rocky and indented shores, which offer particularly varied marine landscapes, its extensive dune areas, as well as its steep bathymetry, which favors the agitation of marine waters [33]. The bathymetry of the coastal plateau is vast with a lower limit, which does not exceed 100 m depth [3, 17].

The dominant winds are those of the WNW sector in the spring and summer seasons and the west sector in the autumn and winter seasons [34]. The average water temperature in the studied area (the Northeast of Tunisia, GFCM-GSA 12 & 13) oscillates between 14.9° and 29.1 °C [28, 35]. In addition, according to Ben Ismail et al., three superimposed layers of waters with different densities and varying salinities co-exist: Surface waters salinity <37‰; deep waters salinity >38‰; middle layer salinity [37.5–38‰]; and salinity mix layer [37–37.5‰] [33].

The Currentology study showed that two water layers circulate in opposite directions in the Mediterranean: an intermediate layer (Levantine Water) moving from east to west and a surface layer (Atlantic Water) that circulates from west to east [36].

This study focuses on the Northeast of Tunisia (GFCM-GSA 12 & 13), specifically the region of Kelibia, as the small pelagic fishing activity is of paramount importance in this region. Regarding infrastructure, Kelibia Port offers more than 1,150 m^2 of quays and wharves for the landing of small pelagic fishing products. The port has a wholesale market that meets sanitary quality standards and ensures the traceability of products. It also has proper hoisting equipment and important refrigerating facilities for storage and conservation. Besides, the Kelibia region is home to 12 units of fish processing and transformation.

3.2 Data Collection

To enhance data collection and validity, we tailored the investigation tools to the fishermen's literacy level and degree of cooperation. We used a questionnaire, an interview and field notes (at Kelibia Port), which were supplemented by direct observations concerning the number of crews, the fishing areas, the production, etc. It is important to note that the preparation of the investigation was conducted in six stages:

- **Stage 1**: **Research Design**

At this preliminary stage, we defined the objective of the study and the questions it is expected to answer, namely, the state of the small pelagic fisheries in North-east Tunisia associated with the fleet, the socio-economic condition, and the fishing activity in Kelibia Port, along the state of the stock of these resources.

- **Stage 2: Identification of Background Information**

This step allowed us to assess the existing information, through the literature and the data provided by the administration, to avoid duplication and identify the major information gaps. It also helped specify the questions to fill these gaps and supply additional data. To design the investigation, it is necessary to gather preliminary information on the characteristics of the crews as well as the specificities of the fishing areas.

- **Stage 3: Additional Information**

After gathering background information, several aspects still needed to be clarified before data analysis could begin. Once the data were collected, a checklist was created to highlight the information gaps. This is to ensure that additional data and information on the small pelagic fishing activity in the Northeast region of Tunisia can be collected while still on site.

- **Stage 4: Survey Management**

Survey management is based on the availability of the target community. The fishermen leave the port around 5 p.m. and return after the fishing trip at 4 a.m. Once the fish landed, they left the port and returned only to prepare for the next trip in the afternoon. To get hold of them, we had to pick bad weather days or the full moon period.

- **Stage 5: Involvement of the Target Communities**

The survey could not be accomplished without the consent and cooperation of the involved fishermen. Thus, it was important to determine how they would be approached, kept informed and made aware that the collected data were theirs. Therefore, we contacted fishermen at the outset of work to inform them of the scope and objectives of the survey and the reasons for their selection as well as the form of the final results. We asked for their consent to participate in the study to gain their full support and commitment. We also familiarized ourselves with local customs and cultural practices and made sure to comply with them, i.e., not to disturb the preparations for the fishing trips or fish landing. We also considered potential issues related to the fishermen's age and level of comprehension of the questions.

3.3 Data Processing and Analysis

In the processing phase, we gathered information through a survey and *in-situ* observation. These were compiled and processed according to their nature and our needs. This was necessary to provide interesting results and to facilitate the investigation of the evolution of the production and fishing areas.

Data analysis was based on small pelagic fish landings in Kelibia Port, between 1991 and 2019, as well as data on the production of different fishing types in this region.

A correction was applied to these data according to each fishing type. This correction was based on the application of the DGPA coefficient, i.e., 1.42 for artisanal fishing, 1.17 for trawling and 1.15 for purse seining [37]. In addition, we collected and analyzed data related to the fishing effort expressed in the number of fishing units, the number of trips, and the fishing areas.

The methodology adopted is a combination of the following key and complementary steps:

- collection of data related to the landed quantities of small pelagic fish from 1991 to 2019;
- identification of the fishing areas and the collection of data associated with the fleet and crews;
- storage of the collected data according to a specific protocol;
- statistical analysis to diagnose the status of small pelagic fisheries status in the northeast of Tunisia (GFCM-GSA 12 & 13).

Kelibia Port comprises 48 small pelagic fishing units (seiners). During the present work, 30 units have been investigated, for which we have identified and recorded all the technical characteristics such as length, tonnage, engine power, etc. Then the similarity and homogeneity of these units were identified using multivariate analysis (PCA).

3.4 *Mapping of the Fishing Areas*

Following in-situ observations and interviews with the fishing operators, it was possible to map the fishing areas. The aim was to identify the small pelagic fishing zones before and after the NSDSPF was implemented. The obtained data map the areas frequented by the small pelagic fishing units.

The Geographic Information System (GIS), which was used for this study, is a type of geographic database associated with software tools that facilitate data entry and management in numerical form, analysis, modelling, and cartographic representation [38]. To make the cartography, we used the Vessel Monitoring System (VMS), which allows us to organize and standardize the information collected through surveys, observations and interviews. We analyzed the gathered data before and after the implementation of the NSDSPF.

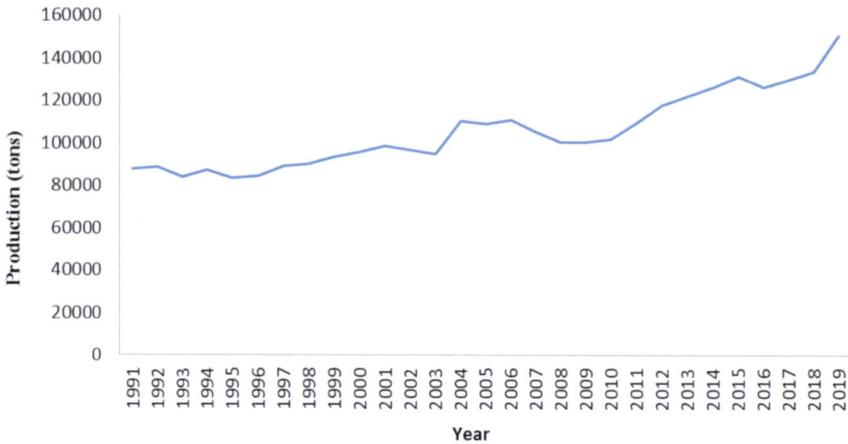

Fig. 3 Annual total fisheries production in Tunisia from 1991 to 2019 [39]

4 Results

The small pelagic fisheries help to feed many mouths, create jobs, and bring in foreign currency, all of which are the main objectives of the national economic policy.

4.1 *Diachronic Study of Fisheries Production*

The monitoring of Tunisian fisheries production indicates many fluctuations over the last 28 years (1991–2019), as shown in Fig. 3. The production differed significantly between years ($F = 3.54$; $P < 0.05$).

Between 1991 and 2019, the quantities that landed in Tunisian ports climbed continuously. The increase in fishing production was around 59%. An annual increase rate of 2% had been recorded until 2019, the year in which the maximum production reached 50.890 tons. This evolution resulted from the implementation of the NSDSPF [23].

4.2 *Evolution of the Small Pelagic Fish Production*

Until 2001, the production of small pelagic fisheries provided 35% of total national landings. From 2002 to 2015, the implementation of the NSDSPF induced an increase in the production of small pelagic fish, with a contribution to the national production exceeding 43%. Since 2016, small pelagic production has been decreasing and has

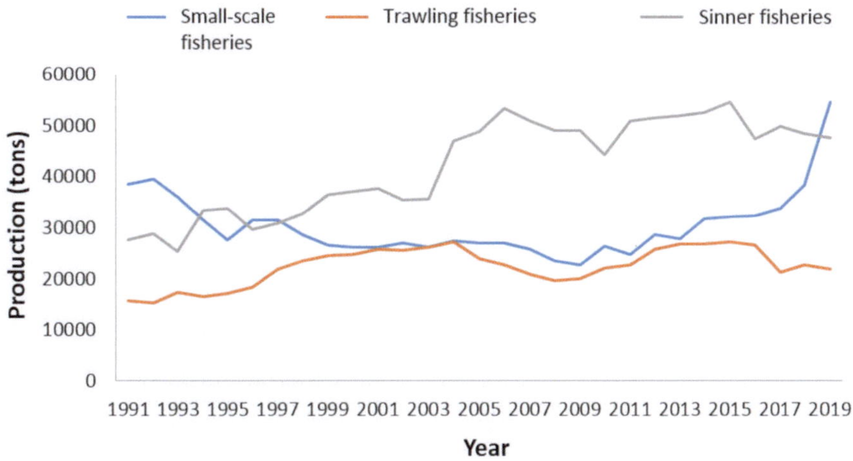

Fig. 4 Annual fisheries production in Tunisia between 1991 and 2019 [39]

not exceeded 36% (Fig. 4). The evolution of the national fisheries production remains proportional to the small pelagic production.

Between 1991 and 2002, the landing of small pelagic fish in Tunisian ports increased. During this time, the production growth was at a rate of 77%, with an annual increase of about 6%. When the NSDSPF was implemented, production climbed from 35.521 tons in 2002, to reach 53.406 tons in 2006, one of the highest levels of small pelagic fish production ever in Tunisia. This rise was the outcome of fleet modernization, the exploitation of small pelagic resources in offshore marine zones, and the creation of new marketing and conservation channels and factories.

Small pelagic fisheries play an important role in meeting the demand for marine products (Fig. 4). Between 1991 and 2019, two peaks of production were recorded: the first one in 2006 with 53.406 tons (48% of the national production) and the second in 2015 with 54.487 tons (41%). The production of these fisheries differed significantly between years ($F = 13.74$; $P < 0.05$).

4.3 *Identification of Small Pelagic Fisheries Areas*

Tunisia's five governorates that mostly generate small pelagic fish are Monastir, Nabeul, Mahdia, Medenine, and Gabes (Fig. 5). The production landed in the ports of Nabeul governorate accounts for 18% of the national production, and varies significantly among governorates ($F = 4.21$; $P < 0.05$). The remainder of the production (80%) is mainly landed in Kelibia Port.

Kelibia Port receives landings from several types of fishing activities, with a dominance of the small pelagic (63%), followed by benthic trawling (24%) and

Fig. 5 Variation of small pelagic fish production in Nabeul governorate

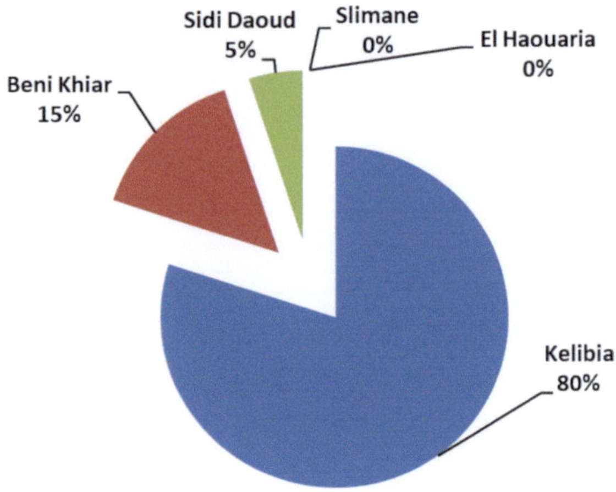

coastal small-scale fishing (13%). This qualifies Kelibia's harbor as a port with a vocation for small pelagic fisheries.

Fishing for small pelagic fish is one of the most ancient professions in the Northeast of Tunisia, especially in Kelibia region. The most commonly caught species are mackerel (*Scomber scombrus* and *Scomber japonicus*), sardine (*Sardina pilchardus*), sardinella (*Sardinella aurita*) and horse mackerel (*Trachurus* sp.) [3, 40–42].

The production of small pelagic fish landed in Kelibia Port underwent a significant evolution during the 2000s. 2006 was the year when production was at its highest (landings exceeded 15.231 tons). This rise in landing and profitability was the result of the implementation of the NSDSPF. After 2006, the production of small pelagic fish declined and did not exceed 6.689 tons in 2017 (Fig. 6). The production differed significantly between years ($F = 3.65$; $P < 0.05$).

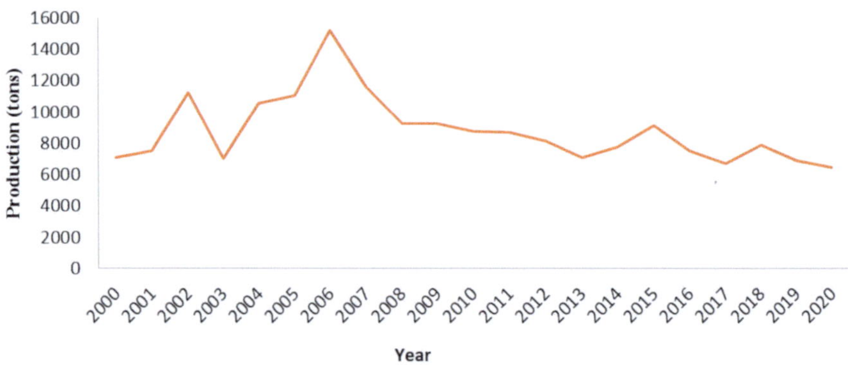

Fig. 6 Small pelagic landings in the Northeast of Tunisia (GFCM- GSA 12 & 13) (2000–2019)

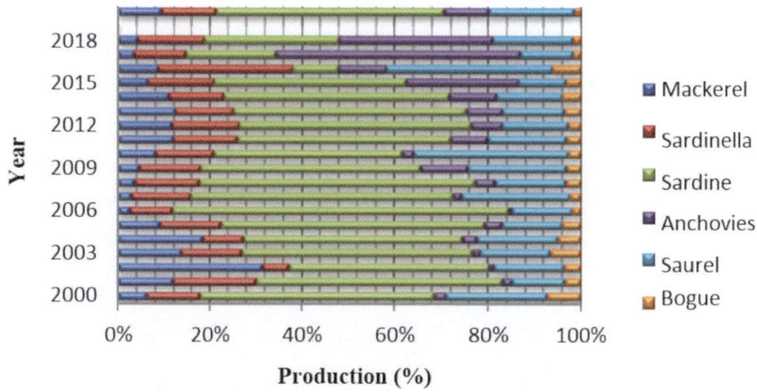

Fig. 7 Specific composition of small pelagic landings in the Northeast of Tunisia (GFCM-GSA 12 & 13) (2000–2019)

4.3.1 Specific Composition

The analysis of the production of pelagic species caught in the Northeast of Tunisia and those landed in Kelibia port shows that it is mainly composed of sardine (*Sardina pilchardus*), sardinella (*Sardinella aurita*), anchovy (*Engraulis encrasicolus*), bogue (*Boops boops*), mackerel (*Scomber* sp.) and horse mackerel (*Trachurus* sp.). The specific composition of the catches landed at Kelibia Port shows a significant variation over the years [3]. In 2000, sardines dominated pelagic fish production (51% of total landings), followed by horse mackerel (27%), sardinella (12%), bogue (7%) and anchovy (6%). That year, the specific production differed significantly ($F = 2.76$; $P < 0.05$).

The production in 2019 mostly contained sardines (29% of total landings), followed by anchovies (20%), horse mackerel (19%), mackerel (17%), sardinella (13%), and bogue (2%). Also in 2019, the specific production differed significantly ($F = 4.36$; $P < 0.05$).

During the period 2000–2019, the production of three species dropped continuously (sardine, horse mackerel and bogue), while three species witnessed an increase in their landing (anchovy, mackerel and sardinella) (Fig. 7).

4.3.2 Monitoring of the Small Pelagic Fish Landing

- **Sardine *Sardina pilchardus***

The harvest of sardine, *Sardina pilchardus,* in the Northeast of Tunisia (GFCM-GSA 12 & 13) fluctuated during the last 20 years (Fig. 8). Between 2000 and 2006, the quantities of sardines landed grew from 3.604 tons (2000) to 11.074 tons (2006). Between 2007 and 2019, landed production fell by 76% at a 6% yearly pace. This species' production varied significantly across the years ($F = 4.74$; $P < 0.05$).

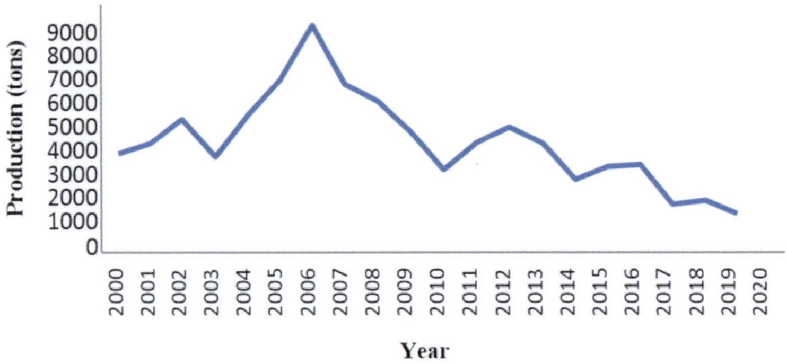

Fig. 8 Production of *Sardina pilchardus* in the Northeast of Tunisia (GFCM-GSA 12 & 13) (2000–2019)

- **Sardinella *Sardinella aurita***

The landings of *Sardinella aurita* in Kelibia Port changed substantially between 2000 and 2019 (Fig. 9). The production of this species reached 1,485 tons in 2005 and dropped to 690.3 tons in 2019. The production of *S. aurita* varied significantly between years ($F = 6.56$; $P < 0.05$).

- **Anchovy *Engraulis encrasicolus***

From 2000 to 2019, the quantities of anchovies landed in Kelibia port had been progressively climbing at an annual rate of about 14.5% reaching 2.477 tons (Fig. 10). The production of *Engraulis encrasicolus* differed significantly between years ($F = 3.46$; $P < 0.05$).

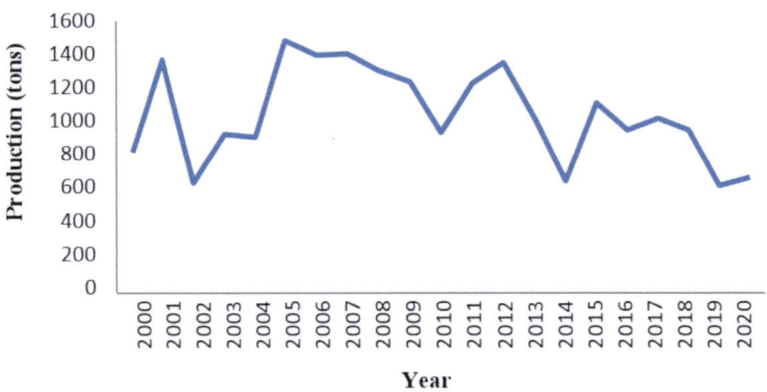

Fig. 9 Production of *Sardinella aurita* in the Northeast of Tunisia (GFCM-GSA 12 & 13) (2000–2019)

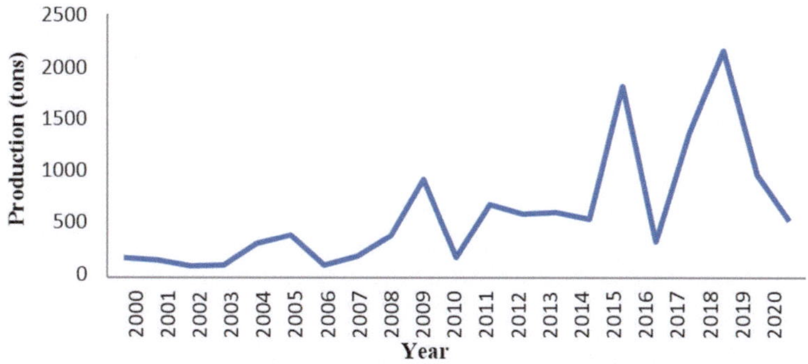

Fig. 10 Production of *Engraulis encrasicolus* in the Northeast of Tunisia (GFCM-GSA 12 & 13) (2000–2019)

- **Bogue *Boops boops***

Figure 11 depicts a continual decline in bogue production between 2000 and 2019. Landings peaked at 528 tons in 2004 before falling to 81 tons in 2019. *Boops boops* production varied considerably across the years ($F = 2.85$; $P < 0.05$).

- **Mackerel *Scomber* sp.**

The evolution of mackerel production is depicted in Fig. 12. Landing of this species has steadily declined over the last 20 years (2000–2019), falling to 267 tons in 2018, except for 2002, when production peaked at 3.548 tons. The production of *Scomber* sp. fluctuated significantly across the years ($F = 8.65$; $P < 0.05$).

- **Horse mackerel *Trachurus* sp.**

Landing of *Trachurus* sp. fluctuated between 2000 and 2010, with a peak in 2007 (2511 tons). Thereafter, we recorded a continuous decrease in production, which dropped to 774 tons in 2015 (Fig. 13). The production of *Trachurus* sp. differed significantly between years ($F = 2.64$; $P < 0.05$).

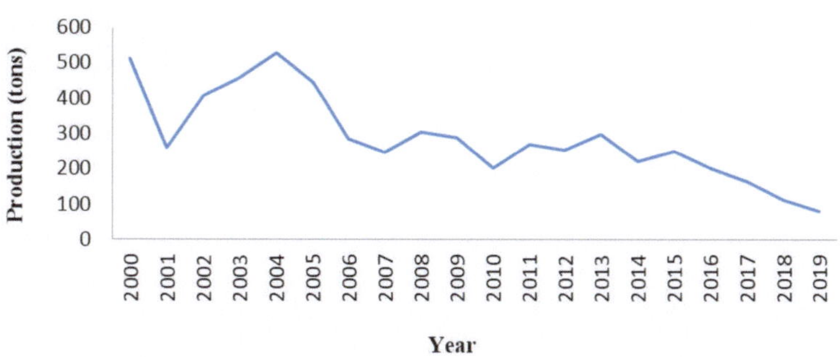

Fig. 11 Production of *Boops boops* in the Northeast of Tunisia (GFCM-GSA 12 & 13) (2000–2019)

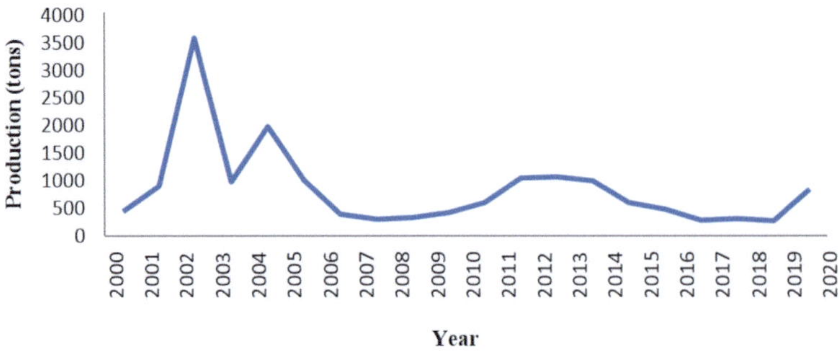

Fig. 12 Production of *Scomber* sp. in the Northeast of Tunisia (GFCM-GSA 12 & 13) (2000–2019)

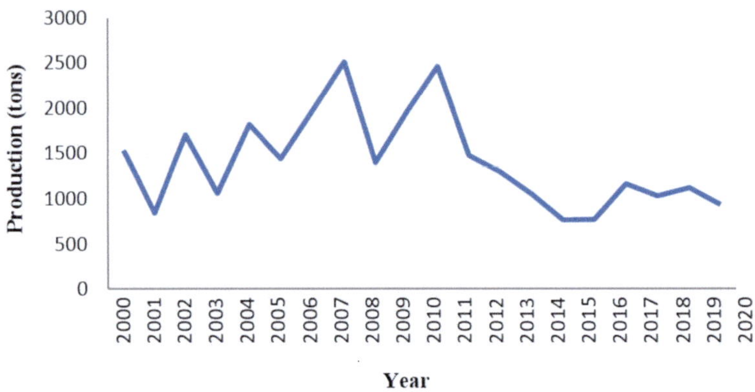

Fig. 13 Production of *Trachurus* sp. in the Northeast of Tunisia (GFCM-GSA 12 & 13) (2000–2019)

4.4 Factors Affecting Small Pelagic Fisheries

The evolution of small pelagic fisheries in the Northeast of Tunisia (GFCM-GSA 12 & 13) depends on several factors, including the number of active seiners, the number of trips per year, the workforce, the characteristics of the seiners, the fishing area, gasoil consumption, fish prices, the number of processing factories, etc.

4.4.1 Fishing Effort

The monitoring of fishing activities in the small pelagic fisheries of the Northeast of Tunisia (GFCM-GSA 12 & 13) shows a constant decrease in the number of working days since 2005 [24]. The small pelagic fishing fleet occupies second place, representing 24% of the vessels operating in this area. Additionally, Tounsi showed a downward trend in fishing efforts (79 operating vessels in 2000 compared to 50 in 2019), despite the encouragement to build new pelagic fishing units [24].

In this work, we used Principal Component Analysis (PCA) to analyze the data related to the small pelagic fishing fleet operating in the Northeast of Tunisia (GFCM-GSA 12 & 13).

This analysis investigated the characteristics of 30 fishing units including length, width, height, Gross Registered Tonnage (GRT), presence of sonar, and workforce number.

PCA was applied to analyze the characteristics of the different fishing vessels. The two factors of the PCA contribute 71.35% of the total variance. Figure 14 shows that tonnage, total length, engine power and height have a positive correlation. The width of the vessel has a positive correlation with the number of fishermen. Vessels number NA 454, NA 470, NA 389, NA 762, NA 460, NA 484 and NA 468, which were manufactured after the implementation of the NSDSPF, have the highest tonnage, engine power, height and overall length. These units are well-developed with modern equipment ensuring the highest production (Fig. 15). Despite having the greatest width and the greatest number of fishermen, the vessels NA 464, NA 362, NA 272, and NA 379 refused to join the NSDSPF and remained undeveloped.

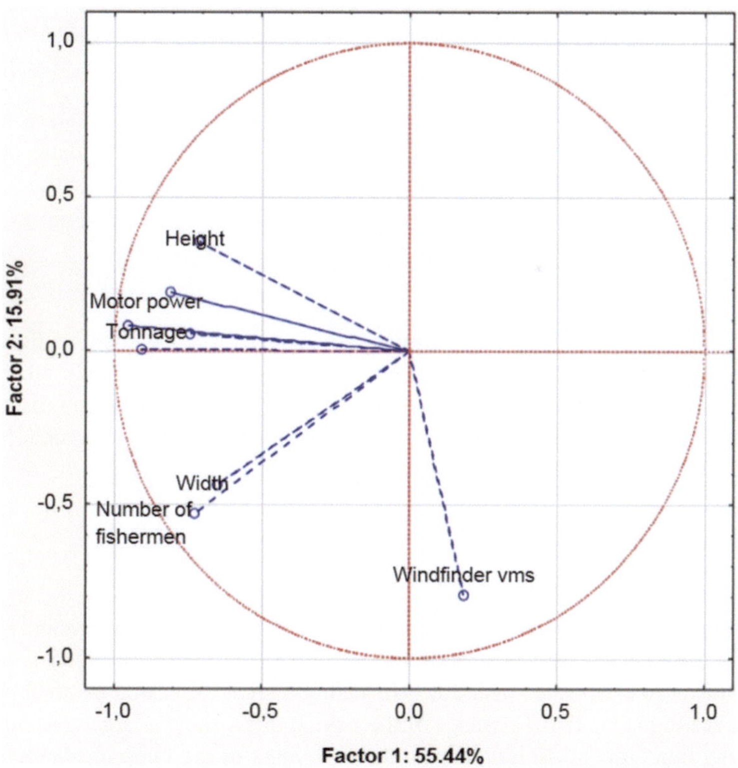

Fig. 14 PCA analysis of the characteristics of small pelagic fishing units in Kelibia region

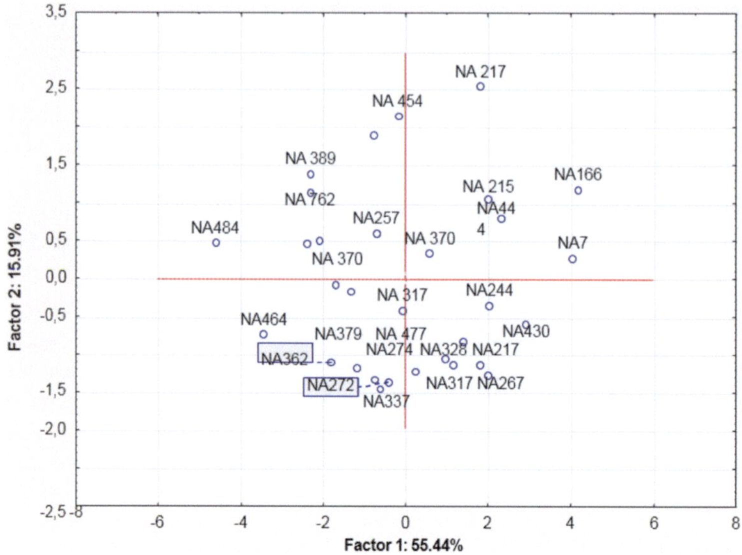

Fig. 15 PCA Analysis of the correlation between small pelagic fishing units in Kelibia region

The number of powerful units targeting small pelagic species in offshore fishing areas is not important compared to the rest of the fleet. The latter remains old and undeveloped, targeting traditional fishing areas, which are overexploited.

4.4.2 Workforce

Both the number of small pelagic fishing vessels and the total count of fishermen had been falling. The workforce dropped from 1.314 in 2000 to 810 in 2019. This decrease can be explained by the modernization of the fishing units and mainly the purchase of fish detection equipment, such as the echosounder and sonar, which drastically affected the need for fishermen.

4.4.3 Fishing Areas

The fishing areas' evolution was tracked using the Vessel Monitoring System (VMS) between 2003, when the national strategy for the development of bluefish capture (NSDSPF) was established, and 2021, revealing a great expansion in small pelagic fish harvesting (Fig. 16). The maps of the fishing areas show a large concentration of fishing units around Kelibia during the first years of the implementation of the NSDSPF (2003). Between 2010 and 2017, we see that there was an increase in this activity along with a move towards more remote areas. From 2020 to 2021, in addition to the stagnation in activity, we noticed a focalization of the fishing areas in the offshore areas. This is mainly due to the development of motorization and the modernization of fishing units targeting small pelagic fish in the eastern region of Tunisia (detection and conservation equipment).

The seasonal monitoring of small pelagic fish harvesting activities in the eastern region of Tunisia using the maps from the VMS showed a clear variability between seasons in 2021 (Fig. 17)

The greatest concentration of small pelagic fishing activity is observed during the summer, the time of year when the weather is good in this region, in contrast to the winter, when the maximum number of bad weather days are recorded [33, 42]. With the remoteness of the new fishing areas, the sea state factor becomes

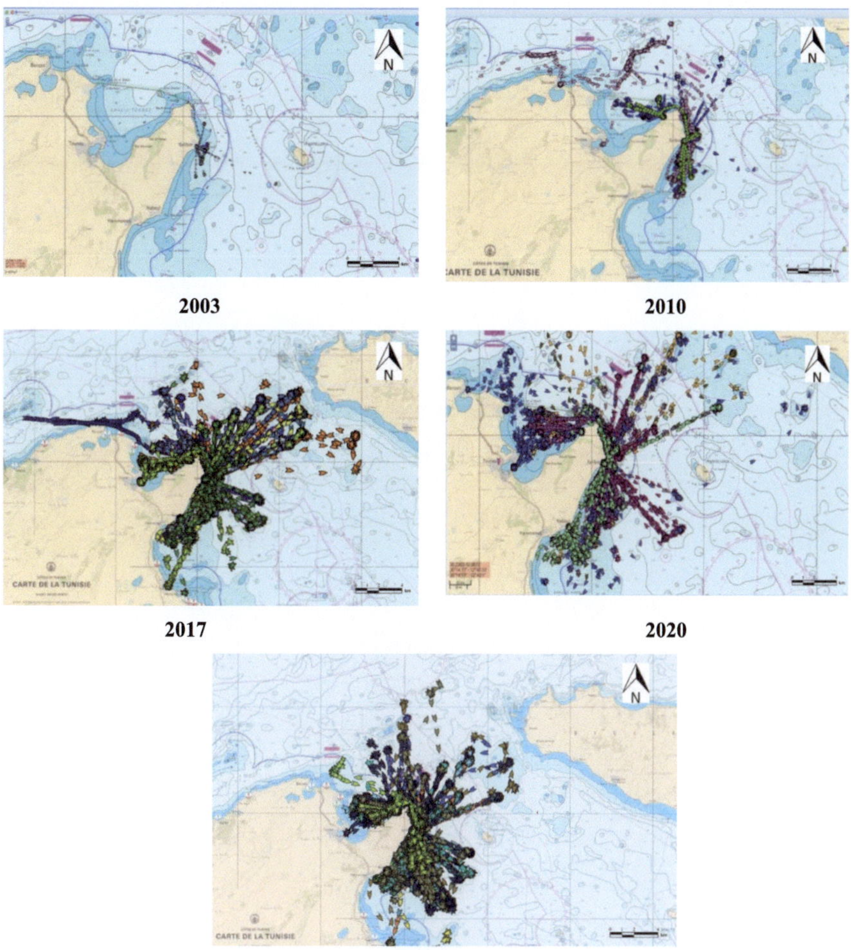

Fig. 16 Variation of the small pelagic fishing areas between 2003 and 2021 in the Northeast of Tunisia

Fig. 17 Seasonal variation of the small pelagic fishing areas in the Northeast region of Tunisia (GFCM-GSA 12 & 13) in 2021

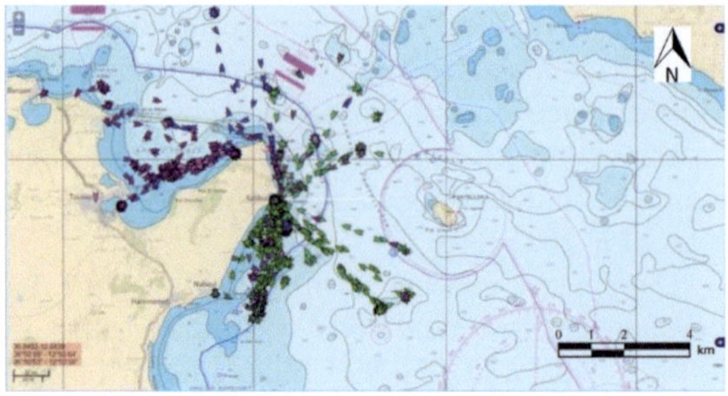

December 2020 - February 2021

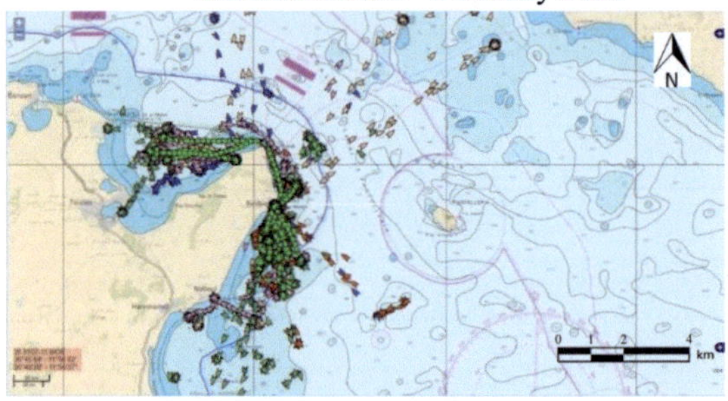

March-May 2021

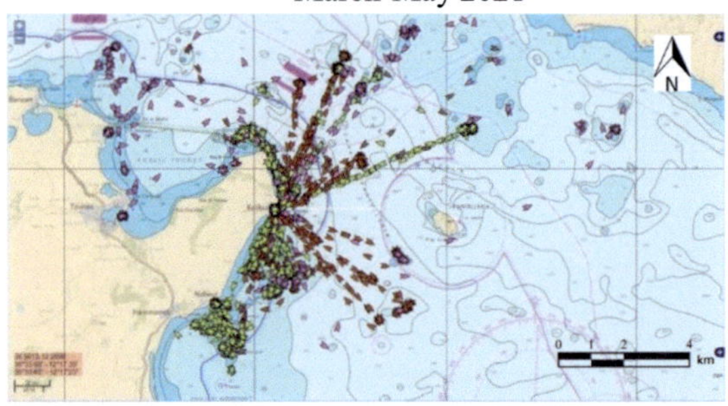

June - August 2021

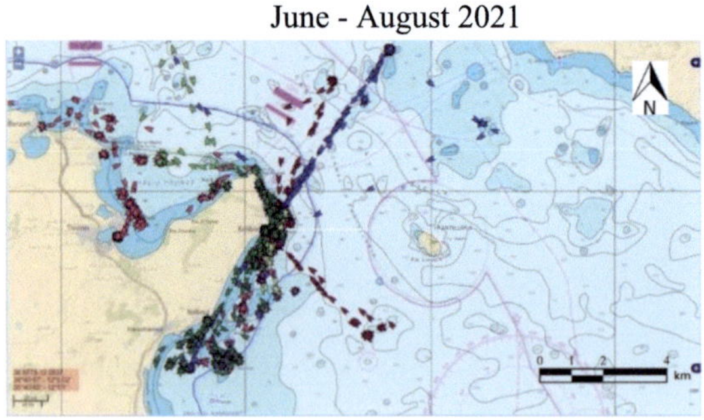

September - December 2021

primary for this fishing activity because the maneuvring of the seine requires good weather conditions. Despite the modernization of the equipment, the development of motorization, and the increase in the size of the vessels, the weather and sea state factors remain decisive for this fishing activity, especially in the offshore areas.

5 Discussion

Our findings suggest that the small pelagic fish production in Tunisia has experienced several fluctuations in the last 28 years (1991–2019). This is due to the implementation of the National Strategy for the Development of Small Pelagic Fisheries (NSDSPF). This strategy was based on the results of the research conducted by INSTM, which revealed the availability of an important exploitable potential of small pelagic species in Tunisian waters [23, 42].

When we compare small pelagic trawling to coastal small-scale fishing, we understand how important small pelagic is for satisfying national and international fish demand. The 2006 production peak accounts for 48% of Tunisia's total fisheries production. This increase was the result of the measures that were implemented, such as the modernization of the fishing fleet operating in offshore areas, the establishment of new marketing channels, and the promotion of small pelagic fish exports.

The last 10 years were marked by a decrease in the production of small pelagic fish in the Northeast of Tunisia (GFCM-GSA 12 & 13). This decline can be attributed to the overexploitation of these fish stocks and the impact of climate change on these resources [9, 42]. These phenomena generated variations in the specific composition of pelagic species over the years. The Northeast of Tunisia (Kelibia region) regrouped several types of fisheries with a remarkable abundance of small pelagic fish (63%) in 2019. The species caught are mainly mackerel (*Scomber scombrus* and *Scomber japonicus*), sardine (*Sardina pilchardus*), sardinella (*Sardinella aurita*), and horse mackerel (*Trachurus* sp.) [42].

The small pelagic landing in Kelibia port depends on several factors, such as the number of trips, fishing effort (number of seiners and their characteristics) and number of processing and conservation factories. Moreover, these fisheries are closely linked to the climatic conditions and the lunar cycle that determine the number of working days.

The remoteness of fishing areas can explain the loss in profitability for the fishing units, fishermen, and ship owners. However, it is worth noting that the shift to offshore areas was dictated by several factors, including the change in fish behavior and the overexploitation of these resources in the traditional fishing areas. As a consequence of the decrease in profitability of this sector, the direct workforce (13 jobs/fishing unit) shifted to coastal fishing and trawling. Despite the decrease in small pelagic fish production, the number of processing factories has been increasing due to the incentives for investment in this sector.

This study uncovered several issues, including a lack of scientific studies on the impact of small pelagic fisheries activities on the ecosystems; a low educational level and a lack of professional development for fishermen; a poor understanding of the rational exploitation of fish resources, and illegal fishing in shallow water or using small mesh-size purse seines. These non-regulatory behaviors associated with the capturing of juvenile species contribute to the decline of Tunisia's small pelagic fish stocks.

6 Conclusion

In this study, we reviewed the small pelagic fisheries state in the northeast of Tunisia before and after the implementation of the NSDSPF to determine the strengths and weaknesses of this sector and suggest recommendations for sustainable and rational exploitation. The Tunisian Northeast region contributes by 18% (8673 tons) to the national small pelagic landings, of which Kelibia port holds 80% [2].

The NSDSPF positively impacts Tunisia's small pelagic fisheries sector, increasing its potential. However, further support, supervision and accompaniment are required to enhance the exploitation of the small pelagic resources. The NSDSPF emphasizes the importance of the development and implementation of a coherent policy, which enhances the infrastructure, rationalizes, and optimizes the harvest of small pelagic fish. In addition, the modernization and development of pelagic fishing units should be a continuous process. The replacement of the old units with new powerful ones allows fishermen to operate in under-exploited marine areas. Besides, the establishment of a biological rest period for the small pelagic fisheries and a revision of the regulatory framework is strongly recommended. In addition, it is crucial to continue exploring new fishing areas for small pelagic fish stocks to ensure the durability and sustainability of this activity.

7 Recommendations

This chapter offers a comprehensive analysis to promote sustainable management of Tunisia's small-scale pelagic fisheries. Here, we outline several key recommendations to achieve this goal:

- Enhancing the monitoring of fish stocks by using advanced techniques to understand environmental impacts.
- Enhancing governance and enforcement capabilities at both national and regional levels to uphold compliance with sustainable fisheries and conservation measures.
- Modernization of fishing fleets through the integration of newer technologies
- Promoting responsible fishing practices by educating and training fishermen.

- Strengthening multi-stakeholder cooperation by involving partnerships between government agencies, research institutions, fishing communities and the private sector to simplify regulatory processes and improve market opportunities for small pelagic products.

By implementing these recommendations in Tunisia, it would sustainably manage small-scale pelagic fisheries, ensuring food security, promoting economic development, and conserving marine biodiversity.

Acknowledgements All our gratitude goes to Mrs. Hajer Zarrouk who considerably improved the manuscript. The authors also wish to extend their thanks to the General Direction for Fisheries and Aquaculture for the provision of statistical data on small pelagic fisheries and especially the VMS data, which helped realize the fishing maps. We are also very grateful to Sonia Nasraoui, the Head of the Fishing District of Kelibia, for her support. We would also like to thank the fishermen and the owners of the small pelagic fishing units in the region of Kelibia.

References

1. Weigel JY (1999) Exploitation and development dynamics of small marine pelagics in West Africa (Dynamiques d'exploitation et de valorisation des petits pélagiques marins en Afrique de l'Ouest). FAO Fisheries Technical Paper No. 390, 73p
2. GDFA (General Directorate of Fisheries and Aquaculture) (2019) Yearbook of fisheries statistics (Annuaire des statistiques de pêche). General directorate of fisheries and aquaculture. Ministry of Agriculture, 135p. (in Arabic)
3. Ben Messaoud R, Cherif M, Jaziri S, Koched W, Zaara K (2018) Mitigation of interactions between endangered species (delphinids and seabirds) and small pelagic fishing activities in the Kelibia region (Tunisia) (Atténuation des interactions entre les espèces menacées (delphinidés et oiseaux marins) et les activités de pêche des petits pélagiques dans la région de Kélibia (Tunisie)). Progress report MoU ACCOBAMS 5:57p
4. Fréon P (1988) Responses and adaptations of West African clupeid stocks to environmental and farming variability: Analysis and reflection based on the example of Senegal (Réponses et adaptations des stocks de clupéidés d'Afrique de l'ouest à la variabilité du milieu et de l'exploitation: Analyse et réflexion à partir de l'exemple du Sénégal). Study Collection and Thesis, ORSTOM, 287p
5. Cury P (1995) Comparison of the Ivoiro-Ghanaian fishery with other exploited upwelling systems in the world. In: Dynamique et usage des ressources en sardinelles de l'upwelling côtier du Ghana et de la Côte d'Ivoire (Dynamics and use of sardinella resources in the coastal upwelling of Ghana and Côte d'Ivoire). Bard FX, Koranteg KA (Eds) Conferences and seminars ORSTOM Paris, pp 169–193
6. Gascuel D (1995) Fishing effort and power: redefining concepts and application examples (Efforts et puissances de pêche: redéfinition des concepts et exemple d'application). In: Gascuel D, Durand JL, Fonteneau A (eds) french research in halieumetry (les recherches françaises en halieumétrique), Rennes Paris Orstom Edition, pp 159–181
7. Pauly DP (1997) Methods for assessing fishery resources (Méthodes pour l'évaluation des ressources halieutiques). In: Moreau J (ed) French adaptation (Adaptation française) Cepadues Toulouse, 288p
8. Ben Abdallah L, Gaamour A (2004) Geographical distribution and biomass estimation of small pelagic fish on the Tunisian coast (Répartition géographique et estimation de la biomasse des petits pélagiques des côtes tunisiennes). Med Sud Med Technical Documents 5:132

9. Ben Abdallah L, Barra M, Gaamour A, Khemiri S, Genovese S, Mifsud R, Basilone G, Fontana I, Giacalone G, Aronica S, Mazzola S, Jarboui O, Bonanno A (2018) Small pelagic fish assemblages in relation to environmental regimes in the Central Mediterranean. Hydrobiologia 821(1):113
10. Gaamour A, Ben Abdallah L, Khemiri S, Mili S (2004) Studies on the biology and exploitation of small pelagics in Tunisia (Etudes de la biologie et de l'exploitation des petits pélagiques en Tunisie). MedSudMed Tech Doc 5:132
11. Froese R, Pauly D (2021) FishBase: World Wide Web electronic publication.www.fishbase.org. version (08/2021)
12. FAO (2011) Fisheries management: marine protected areas and fisheries. FAO technical guidelines for responsible fisheries Rome 4, 198p
13. Monot J (2011) "Story 5. Fire fishing: from the torch of yesteryear to today's lamparo", Mediterranean fisheries. A journey into tradition ("Récit 5. La pêche au feu: de la torche d'antan au lamparo d'aujourd'hui", Les pêches méditerranéennes. Voyage dans les traditions), Éditions Quæ, pp 52–61
14. CGPO (Conseil General Pyrénées-Orientales) (2018) The big Site of Anse de Paulilles: Maritime Heritage, thematic sheet (Grand Site de l'Anse de Paulilles : Patrimoine Maritime, fiche thématique), 20p
15. He P, Chopin F, Suuronen P, Ferro RST, Lansley J (2021) Classification and illustrated definition of fishing gears. FAO fisheries and aquaculture technical paper no. 672. Rome, FAO, 110 p
16. Abdesselem F, M'Timet M, Jarboui O (1988) Benthic trawling on the northern coasts of Tunisia (La pêche au chalut benthique sur les côtes nord de la Tunisie). In: Lleonart J (eds) Dynamics of marine populations (Dynamique des populations marines). Cahiers Options Méditerranéennes Zaragoza CIHEAM 35:301–308
17. Ben Salem S, Franquesa R (2004) Socioeconomic indicators for fisheries in the northern and eastern regions of Tunisia. Study case. (Indicateurs socioéconomiques pour la pêche dans les régions Nord et Est de la Tunisie. Etude de cas) INSTM FAO-COPMED 36p
18. Ben Abdallah L, Gaamour A, Jarboui O, Missaoui H (2019) Geographical distribution and biomass estimation of small pelagic fish from Tunisian coasts during autumn 2018 (Répartition géographique et estimation de la biomasse des petits pélagiques des côtes tunisiennes pendant l'automne 2018). Bull Inst Natn Sci Tech Salammbô Sea, vol 46
19. Guennegan Y, Hattour A, Ben Abdallah L (1998) Assessment of pelagic resources in Tunisian waters (Evaluation des ressources pélagiques des eaux tunisiennes). IFREMER Direction des Ressources Vivantes: Halieutic Resources Laboratory of Sète, 26 p
20. Ben Abdallah L, Ben Salem S, Gaamour A, El Abed A (2000) Estimation of biomass and distribution of small pelagics in Tunisian waters (Estimation de la biomasse et distribution des petits pélagiques dans les eaux tunisiennes). National workshop organized by the Tunisian government and JICA on sustainable fisheries development in Tunisia. Mahdia 23–25 October
21. Hattour A, Ben Abdallah L, Guennegan Y (2004) Relative abundance and biomass estimation of small pelagics in Tunisian waters (Abondance relative et estimation de la biomasse des petits pélagiques des eaux tunisiennes). Bull Inst Natn Sci Tech Mer de Salammbô, vol 31
22. FAO (2020) Strengthening governance and development of the fisheries sector in Tunisia (Renforcement de la gouvernance et du développement du secteur de la pêche en Tunisie). T C P/TU N/3602, 13p
23. ANPE (Agence Nationale de Protection de l'Environnement) (2005) Sustainable fishing: Tunisian Observatory for the Environment and Sustainable Development (Pêche durable: Observatoire Tunisien de l'Environnement et du Développement Durable). French Ministry of the Environment and Sustainable Development, 61p
24. Tounsi S (2020) Profile of the small pelagic fishing industry in the North-East region of Tunisia (Profil de la filière pêche aux petits pélagiques dans la région Nord-Est de la Tunisie). Master thesis, Higher Institute of Fishing and Aquaculture of Bizerte, Tunisia, 61p
25. Cox BG, Binder DA, Chinnappa NB, Christianson A, Colledge MJ, Kott PS (1995) Business survey methods (Wiley series in probability and mathematical statistics). Wiley, New York, 762p

26. Bradai MN (2000) Diversity of the ichthyic population and contribution to the knowledge of sparids in the Gulf of Gabès (Diversité du peuplement ichtyque et contribution à la connaissance des sparidés du golfe de Gabès). PhD thesis, University of Sfax, Tunisia, 600p
27. JORT (Journal Officiel de la République Tunisienne) (1995) Order of the Minister of Agriculture of September 28, 1995 regulating fishing activities (Arrêté du Ministre de l'agriculture du 28 septembre 1995 réglementant l'exercice de la pêche). JORT 6(80):1896–1899
28. INM (Institut national de météorologie) (2021). web site www.meteo.tn
29. Kouki A (1984) Contribution to the study of sediment dynamics in the small Gulf of Tunis (Contribution à l'étude de la dynamique sédimentaire dans le petit golfe de Tunis). PhD thesis; University of Sciences of Tunis, Tunisia, 167p
30. Paskoff R (1992) Coasts in danger (Côtes en danger). Collection Pratiques de la Géographie, Masson, Paris, 250p
31. Oueslati A (2004) Coastal development in Tunisia (Littoral et aménagement en Tunisie). ORBIS, 534p
32. Halaouani N, Sabatier F, Gueddari M, Fleury J (2007) Evolution of the Tabarka Bouterfess coastline, northwest Tunisia (Évolution du trait de côte de Tabarka Bouterfess, nord-ouest de la Tunisie). Méditerranée 108:131–137p
33. Ben Ismail S, Sammari C, Béranger K, Lellouche JM (2010) Atlas of hydrological data for the Tunisian coastline (Atlas des données hydrologiques des côtes tunisiennes). Document technique de l'Institut National des Sciences et Technologies et de la Mer (INSTM), pp 1–169
34. Benmessaoud R (2008) Status of Delphinidae and study of the interaction between dolphins and fishing nets in the Kélibia region (Statut des Delphinidés et étude de l'interaction entre dauphins filets de pêche dans la région de Kélibia). Master thesis, INAT Tunisia, 163p
35. Romdhane MS (1993) Fishing techniques and gear used on Tunisia's northern coast. Fishing in Tunisia: coastal fishing and the environment (Techniques et engins de pêche utilises sur la côte septentrionale de la Tunisie. La pêche en Tunisie: pêche côtière et environnement). Cahier du CERES Série Géographique 8:195–210
36. Lopez-Jurado JL (1996) Evolution of a bottlenose dolphins population in the north-eastern waters of Sardinia (Italy). In: Evans PGH (ed) European research on cetaceans. Proceedings of the 15th European cetacean society conference, Roma, Italy, pp 70–72
37. GDFA (General Directorate of Fisheries and Aquaculture) (2020) Yearbook of fisheries statistics (Annuaire des statistiques de pêche). General directorate of fisheries and aquaculture. Ministry of Agriculture
38. Caloz R, Collet C (2011) Spatial analysis of geographic information (Analyse spatiale de l'information géographique). EPFL Press, 400p
39. GDFA (General Directorate of Fisheries and Aquaculture) (1991–2019) Statistical yearbooks of the Direction Générale de la Pêche et de l'Aquaculture from 1991 to 2019 (Annuaires statistiques de la Direction Générale de la Pêche et de l'Aquaculture Durant la période 1991 à 2019)
40. Gaamour A (1999) The round sardinella (Sardinella aurita Valenciennes, 1847) in Tunisian waters: reproduction, growth and fishing in the Cap Bon region (La sardinelle ronde (*Sardinella aurita* Valenciennes, 1847) dans les eaux tunisiennes: Reproduction, croissance et pêche dans la région du Cap Bon). PhD thesis, University of Occidental Bretagne, 246 p
41. Fezzani S, Rjeibi O, Cherif M, Saoudi A, Zaara K, Benmoumen Y, Jarboui O, Missaoui H, Mili S (2023) Abundance and operating area by benthic trawl of Atlantic horse mackerel in the central Mediterranean. Euro-Mediterranean J Environl Integr. https://doi.org/10.1007/s41207-023-00412-z
42. Fezzani S (2023) Ecobiological and dynamic study of horse mackerel Trachurus trachurus on the north and east coasts of Tunisia (Etude écobiologique et dynamique du chinchard *Trachurus trachurus* des côtes Nord et Est de la Tunisie). PhD thesis, National Agronomic Institut of Tunisia, 248 p

The Use of Geographic Information System (GIS) Data to Assess and Rebuild Stocks in Tunisia's Largest Freshwater Reservoirs

S. Mili, R. Ennouri, M. Fatnassi, S. Agrebi, T. Chargui, and H. Laouar

Abstract In Tunisia, freshwater fishery resources have just recently been exploited in comparison to sea fishing. Nonetheless, despite its recent history, the sector has garnered considerable attention. To ensure its durability and development, reservoirs have been filled with various types of fish fry gathered from coastal and estuary zones. Fish populations' stock assessment in reservoirs is also crucial for the sustainable management and conservation of ichthyologic freshwater resources. To assess the evolutionary trends of the fish resources available in the Siliana and Kasseb reservoirs, located in the Siliana and Beja governorates respectively, the Higher Institute of Fisheries and Aquaculture of Bizerte, in collaboration with the Aquaculture Technical Centre and the General Directorate of Fisheries and Aquaculture, carried out the present study. For sampling the researchers used multi-mesh nets regarding the European Standard CEN prEN 14,757, which are designed to capture most fish sizes in the reservoir water column. All of the outcomes of the fishing operations were converted into yields (number of fish) per unit of effort and were then entered into a Geographic Information System to map the results, allowing for the proposal of scenarios for improving stock status. The inventory settlement included six species: mullet, rudd, roach, pikeperch, and common carp, which are variably distributed. Our findings showed that the yield of these species is moderate in Siliana when

S. Mili (✉) · R. Ennouri · M. Fatnassi · S. Agrebi · T. Chargui
University of Carthage, Higher Institute of Marine Sciences of Bizerte, Bizerte, Tunisia
e-mail: sami.mili@issmb.ucar.tn

M. Fatnassi
e-mail: manel.fatnassi@issmb.ucar.tn

S. Mili
Laboratory of Fisheries Sciences, National Institute of Marine Sciences and Technologies, Salammbô, Tunis, Tunisia

R. Ennouri
Laboratory of Ecology, Biology and Physiology of Aquatic Organisms, Faculty of Sciences of Tunis, University of Tunis El Manar, Tunis, Tunisia

H. Laouar
Technical Center of Aquaculture, Tunis, Tunisia

F. Khebour Allouche (ed.), *Environmental Remote Sensing and GIS in MENA Region*, Earth and Environmental Sciences Library,
https://doi.org/10.1007/978-3-031-86574-9_4

compared to prior studies in terms of weight and on the digital map (23.72 ind/1000 m^2 and 4.7 kg/1000 m^2), while yields in Kasseb Reservoir are found to be quite low (123 ind/1000 m^2 and 18 kg/1000 m^2). The statistical analysis demonstrates that Siliana Reservoir possesses a high abundance of mullet population. Indeed, this reservoir is endowed with more effective habitat conditions than Kasseb, favouring the growth and survival of rudd and roach. In addition, the mapping of the distribution of different species and catch rates in these two reservoirs shows that the studied populations are diverse. The study also revealed that some ecological niches in the dams are underexploited, which requires certain measures to be undertaken, such as boosting fishing operations, seeding Mugilidae fry and introducing new species. The authors thus recommend that the various actors in the fishing sector take concerted action in this regard.

Keywords Freshwater Fisheries · Reservoirs · Stock Assessment · Multi-mesh Gill Nets · GIS · Tunisia

1 Introduction

Water is a fundamental component of life; no living creature can exist without it. It is used to drive socioeconomic development across a variety of sectors. Our society is no exception, since the water resources in the country, especially at a time of socioeconomic growth, affect numerous fields including agriculture and fishing [1].

Water resources in Tunisia, though, are unequally distributed in time and space due to the country's arid climate. A situation that imposed the deployment of all the water resources at hand, either through the construction of dams or the creation of hill lakes. There was also a proposal to exploit such water resources for aquacultural, economic, and ecological purposes. The reason is that certain types of fish can maintain an ecosystem's balance in these water bodies thanks to their feeding behavior [1, 4].

However, freshwater fishery resources in Tunisia have just been tapped compared to sea fishing activities. The initial implementations began in the 1960s and only carp *Cyprinus carpio*, roach *Rutilus rubilio*, rudd *Scardinus erythrophtalmus*, pikeperch *Sander lucioperca*, black-bass *Micropterus salmoides* and Tilapia *Oreochromis niloticus*, were able to acclimatize [1–4].

The preservation of fishery resources, particularly fishing in reservoirs in Tunisia, is based on the duty and need to ensure the sustainable exploitation of these resources. However, the low frequency of monitoring by the institutions involved, a lack of accurate quantitative information, and fishermen's data misreporting would jeopardize any attempt to evaluate the efforts of stocking these reservoirs. The rational and effective management of fisheries in these water bodies necessitates an understanding of the distribution of various species there.

In response to this urgent need, the Higher Institute of Fishery and Aquaculture of Bizerte (ISPAB) has conducted the current study in partnership with the Aquaculture Technical Center (CTA) and the General Directorate of Fishery and Aquaculture

(DGPA) to assess the evolutionary trends of available fish resources in Siliana and Kasseb reservoirs. This assessment is twofold. It aims to develop a strategy for adequate fisheries management and an array of technical support to be communicated to fishing industry decision-makers so that they can make the most informed choices about the way to operate these reservoirs.

2 Overview

Before starting analyses of data to assess and rebuild stocks in Tunisia's largest freshwater reservoirs, all aspects related to freshwater fishing activity in Tunisia need to be described in detail.

2.1 In-Land Fisheries

In this section, the freshwater fishing activity and regulations in dam reservoirs are described in detail. Additionally, we present fish stock assessment methods involving fish capture and without capture.

2.1.1 Water Potential in Tunisia

The surface water potential of Tunisia was projected to be 2,700 million m^3/year, ranging from 780 million m^3/year in 1993–1994 (Anonymous, 1994) to 11 billion m^3/year in 1969–1970. The maximum-to-minimum intake ratio in Tunisia ranges from 9 in the north to 180 in the south [5]. The contributions vary greatly from north to south; the average in the north consists of 81% of the total intake, the centre represents 11%, and the south accounts for 8% [6, 7].

Dam reservoirs, hill dams, and hill lakes can mobilize an average of 2100 million m^3/year. Water and soil conservation strategies could be used to mobilize the rest of the resources. Currently, the mobilization has reached around 1405 million m^3/year, with dams accounting for 1340 million m^3/year and hill dams accounting for 65 million m^3/year [DG/GTH, 1996 in 5].

Freshwater bodies cover 105,200 hectares in Tunisia. This is a huge area that can be categorized into three different types: large dam reservoirs, small dams, and hill lakes [1, 4]. Tunisia's dams are spread throughout various governorates in the country's north and centre (Fig. 1). There are 37 major dams covering an area of more than 100 acres. These enormous reservoirs can be classified into three types:

- Drinking water reservoirs, where the water is immediately used for the manufacturing of potable water. This group includes four reservoirs: Ben Metir, Kasseb, Joumine, and Lkebir, which occupy a total area of around 1,440 ha [1];

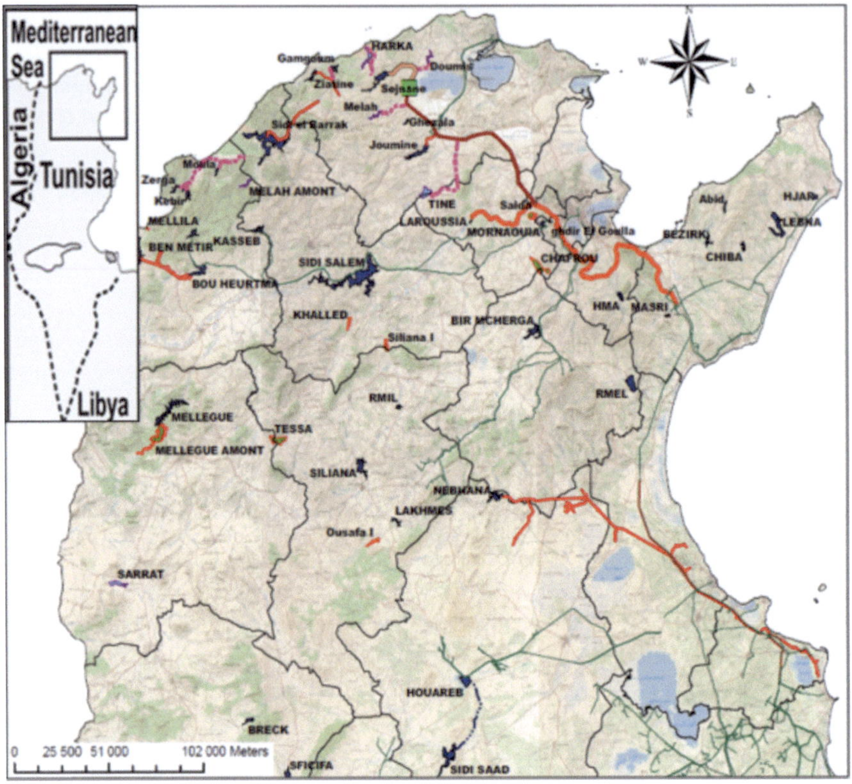

Fig. 1 Hydrographic map of Tunisia [4]. (Mili et al. 2024)

- Mixed-use reservoirs, where water is used for both human consumption and irrigation. This group includes the dam reservoirs of Mellegue, Bou Heurtma, Siliana, and Sidi Salem, making a notional area of 6850 ha [1]; and.
- The flood protection and irrigation dams, which include Sidi Saad and Bir Mcherga reservoirs, with an overall nominal area of about 4,200 ha [1].

In addition to these dams, there are 18 small reservoirs in the north of the country with an area ranging between 10 and 100 ha. The last category presents small collinear lakes that cover less than 10 ha and are fed by runoff during the rainy season [1, 4].

The operational dams irrigate roughly 130 thousand hectares of agricultural land out of a total of 370 thousand hectares, representing a 35% irrigation rate [8]. Among these dam reservoirs, sixteen are found in the northern part of Tunisia: Mellegue, Beni M’tir, Kasseb, Barbra, Sidi Salem, Bouhertma, Joumine, Ghezala, Sejnene, Sidi Barrak, Siliana, Lakhmes, Lhma, Bir Mcherga, and Rmel. Five other reservoirs are located in the centre, namely, Nebhana, Sidi Saad, Lhouareb, Sidi Aïch, and El Barrek. In the Cap Bon area, there are six other reservoirs: Bzirekh, Chiba, Masri, Lebna, Lahjar, Mlaabi, and Laabid [1, 4].

2.1.2 Freshwater Fishing Activity

Freshwater fishing activity is currently being conducted in nine governorates: Beja, Ben Arous, Bizerte, Jendouba, Kef, Nabeul, Zaghouan, Kairouan, and Siliana. The fishermen are generally peasants from the country's interior area with a restricted source of income. Extensive fish farming enables them to make a profit while producing cheap fish for sale or consumption [4, 9, 10].

The size of the dam reservoir determines the number of boats assigned to each impoundment. The overall number of boats inspected in this work was 232, equating to 464 fishermen at a rate of two fishermen per boat [15]. The length of the boat is limited to 6 m, and it is often made from wood or resin [10, 11].

Except for March and April, which coincide with the biological rest season, fish harvesting is allowed in the reservoirs all year round. The most common species caught during winter are carp, pikeperch, mullet, eel, catfish, and barbell [4, 9, 10, 12].

Monitoring the total productivity of Tunisian inland fish farming has revealed a moderate improvement in all reservoirs. This minor increase in overall species production is confirmed by an analysis of fishing data published by the General Directorate of Fisheries and Aquaculture from 2001 to 2021. Indeed, freshwater fish production rose from 831.5 tons in 2000 to 1073.194 tons in 2012 [4, 13–15]. This rise is attributable to many stocking operations, the transfer of broodstock of multiple species, and an increase in the number of fishing licenses provided to fishermen. Furthermore, improvements in the process of collecting fishing statistics have resulted in more reliable data on landings.

Freshwater fish found in reservoirs have a fair commercial value. Some of them are for human consumption, like mullet and carp, while others serve as feed fish for carnivorous species. The inventory of freshwater species in Tunisia is summarized in Table 1.

Table 1 Systematic list of freshwater fish species in Tunisia

Family	Species	Common name
Mugilidae	*Mugil cephalus*	Bighead mullet
	Chelon ramada	Hog mullet
Percidae	*Sander lucioperca*	Pikeperch
Siluridae	*Silurus glanis*	Catfish
Cyprinidae	*Barbus callensis*	Barbell
	Cyprinu scarpio carpio	Common carp
	Hypophthalmichthys molitrix	Silver carp
	Ctenopharyngodon idella	Herbivorous carp
	Scardinius erythrophthalmus	Rudd
	Rutilus rubilio	Roach
Anguillade	*Anguilla anguilla*	Eel

Carp production was 341 tons in 2021, accounting for around 35.2% of the total yield. Mullet came in second with around 275 tons or 28.4% of total production. Pikeperch ranked third with 212 tons of catch. In terms of landings, the rudd was the fourth most abundant species, accounting for 7.5% of the total yield. Barbells represented a modest portion of total production, followed by catfish and eel.

2.1.3 Fishing Regulations in Dam Reservoirs

The Minister of Agriculture's decree, dated September 20, 1994, establishes the regulations for reservoirs' fishing activities. Below is a translation of some excerpts from this decree.

Article 1: Fishing in the dam reservoirs must not in any case alter the watercourses or interfere with the actions of exploitation and maintenance of the hydraulic works. It cannot be conducted within less than 100 m of the catch tower of the dam in normal periods and within 200 m in periods of high water.

Article 2: Fishing in dam reservoirs, rivers, and freshwater bodies is forbidden from the 1st of March to the 30th of April. However, the competent authority may shorten or extend this period by one month according to the biological characteristics of the species.

Article 3: It is prohibited to fish in dam reservoirs, rivers, and freshwater bodies using gill nets, trammel nets, lines with hooks, and simple traps with spinnerets. Nets must be placed within 100 m of one another. The set nets must be carried on large floats or buoys that indicate the name of the fishing unit.

Article 4: A fishing unit may not operate more than eight nets at one time, and the nets cannot exceed 50 m in length. The drop of the nets used must be less than or equal to 1 m, the average depth of the water body where fishing is practiced.

Article 9: Article 9 of the law of September 28, 1995, issued in the Official Journal of the Republic of Tunisia on October 6, 1995, regulates the minimum size of fish that can be caught. It is, therefore, forbidden to capture the fish listed hereafter that are below the following sizes (the length from the extremity of the nose to the beginning of the tail): mullet 20 cm; barbell 15 cm; eel 30 cm; carp 20 cm; pikeperch 28 cm; and roach 12 cm.

However, in exemption from the rules of Article 9 above, Article 10 of the above-mentioned decree states that the landing of a quantity of undersized fish in a proportion not exceeding 10% for each of the landed species is tolerated.

2.1.4 Used Fish Stock Study Methods

Direct Methods Without Fish Capture

These methods consist of detecting the presence of fish by direct observation or by the immersion of detection instruments.

Echosounder was first used for surveying fish biomass in the open sea rather than in lakes after the Second World War [16]. Sonar was first deployed in the United States for fish counting in lakes and rivers [17], particularly for the quantification of salmon during migrations. The first attempt in Tunisia dates back to 2005 for the fish survey in Sidi Salem Dam Reservoir [3]. The downside of this method is that it only leads to an overall estimate; there is no way to identify the species of individuals that send an echo [9].

Direct Methods Involving Fish Capture

Electric Fishing

Electric fishing consists of exposing fish to a weak electronic field that attracts and temporarily stuns them. This time allows the identification and measurement of the captured fish. This technique has several benefits. It is not harmful to fish; it offers real guarantees of effectiveness; and presents a reproducible protocol [18].

Devices that capture fish electrically are classified as active gear. Spot electrofishing, using a reduced base unit, dictates a split in the sampling effort with defined transects [19]. The operation is carried out from a boat. The anode holder, which is located at the front of the boat, suddenly immerses the anode and holds it at the sampling point. This technique, while not very selective, allows us to survey all types of habitats. However, excessive depths ($\leq$1 m) and overly warm, highly mineralized, turbid waters can undermine its efficiency [19].

Multi-Mesh Net Sampling

This method involves a fishing technique using multi-mesh nets conforming to European Standard 14,757 [20], which was designed to evaluate biodiversity, species supply, and age distribution in freshwater bodies. The method of selection depends on stratified random sampling. The multi-mesh net sampling technique was first used in 2013 for the survey of Lahjar and Ghezala Dam Reservoirs in Tunisia [14].

Multi-mesh gill nets are straight nets known as stationary nets, generally set on the bottom and floating freely on the surface or between two layers. They are made up of several layers of different mesh sizes. This diversity of meshes allows us to catch all size classes and most types of species. Multi-mesh gill nets are mounted vertically between a top rope equipped with floats and a bottom rope sealed with lead (CEN 14,757). These nets can be either benthic or pelagic. This sampling technique was used to obtain specific abundances, numerical and weighted fishing yields, as well as length distributions of fish populations such a technique is commonly used both in natural lakes [21–23] and in artificial freshwater reservoirs [14, 24, 25].

3 Material and Methods

This section explains the methodology adopted to achieve the objectives of this study.

3.1 *Study Areas*

3.1.1 Siliana Reservoir (36°09′26″ N, 9°20′55″ E)

This reservoir, which covers a total area of 600 ha, was constructed in 1978. It is located to the north of Seliana City and is part of Medjerda's southern hydrological watershed. Seliana Reservoir, having a capacity of 23 mm^3/year, is used for irrigation around Gaafour/Laaroussia city (Fig. 2) [8].

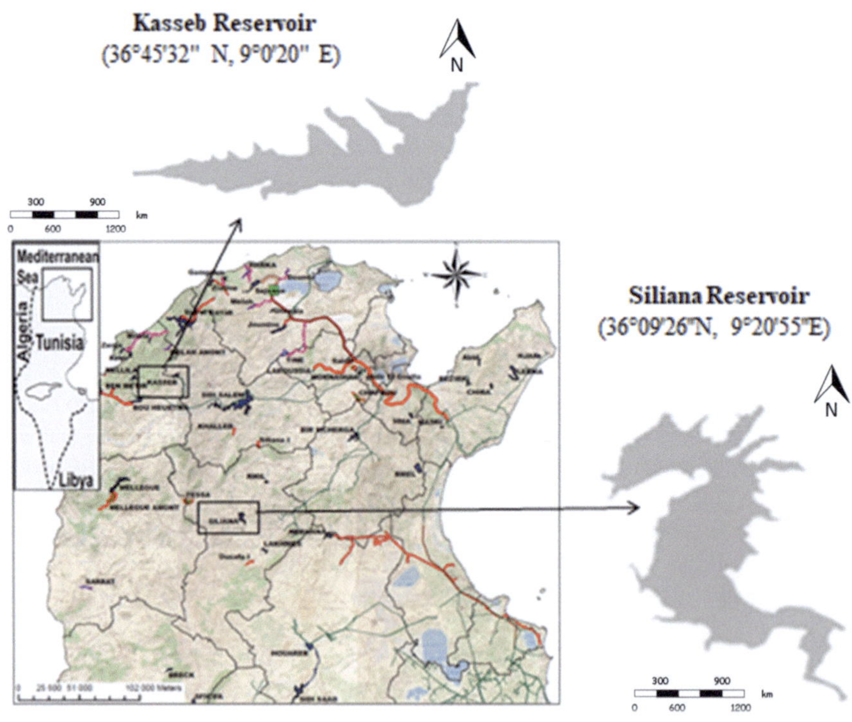

Fig. 2 Studies areas (Kasseb and Siliana Dam Reservoirs). (Mili et al. 2024)

3.1.2 Kasseb Reservoir (36°45′32″ N, 9°0′20″ E)

This reservoir was constructed in 1968 in Madjerda's north hydrological stream, about 20 km from Beja. This water body has an outer area of 430 ha and a capacity of 70 mm^3. It is primarily utilized for drinking water consumption in Tunis, with a capacity of 40 mm^3/year (Fig. 2) [8].

3.2 GIS Data Preparation

In our study, Geographic Information System (GIS) played a pivotal role in spatial data analysis and visualization. During the GIS data preparation phase, the reservoir boundaries were digitized and a foundational map was generated using GIS software (Arc GIS). Background data for vector layers were sourced from satellite imagery to ensure accuracy and relevance. GPS points were collected in the field using (GPS eTrex 10—GARMIN) to georeference and validate vector data. Bathymetric data were incorporated into the mapping process to depict the underwater topography comprehensively. Moving on to the spatial analysis, an assessment of the spatial distribution of fish species was conducted within the reservoir. GIS tools and spatial interpolation were utilized to estimate the distribution of species between distinct sampling points [26, 27]. The study deployed a spatial GIS analysis method known as Inverse Distance Weighting (IDW) to interpolate the obtained results. Measurements from various locations were organized into an Excel file, which was subsequently converted to CSV format, and read into ArcGIS. This CSV file was then superimposed onto the reservoir map, and the interpolation process was applied. The temporal analysis involved the creation of temporal layers in GIS that represent various sampling events. Using GIS functions, changes in fish distribution over time were visualized and analyzed. Subsequently, maps were generated to visually illustrate both the spatial and temporal distribution of fish species. To enhance clarity, species are differentiated by using color-coded symbols or layers. This systematic approach is designed to offer a structured and adaptable framework for mapping the spatial and temporal distribution of fish species within each reservoir, allowing for adjustments based on the specific requirements and available resources of the study.

3.3 Scientific Fishing Using Multi-Mesh Gill Nets

The study of fish populations requires a rigorous evaluation of their qualitative and quantitative composition. For this reason, the selection of a suitable multi-mesh net model is a crucial factor in studying the specific richness and abundance of freshwater fish in Tunisian reservoirs. This type of gear is easy to use, which is why they are popular among scientists. Fish sampling with multi-mesh nets is based on the

European standard CEN prEN 14,757 [20], which has been modified to be adapted to Tunisian species and dam reservoirs.

This fish population assessment method is based on the principle of a random and stratified fishing effort framing the peaks of fish activity. Fish sampling is carried out using benthic multi-mesh gillnets specifically designed for scientific use and composed of eight different mesh sizes ranging from 18 to 80 mm. This diversity of meshes aids in the capture of many species in the fishing space, regardless of their sorts or size classes. The CEN prEN 14,757 Standard also allows for fish sampling using pelagic nets with the same mesh size as benthic nets. Thus, additional sampling is intended for this study to Include Samples From The Pelagic Habitat.

3.3.1 Sampling Strategy

To determine the number of nets and their locations, the bathymetric maps of each reservoir were examined. The sampling period and fishing time frame were set according to the aforementioned standard [20]. The frequency of sample activities was periodic, and the fishing time was chosen based on suitable weather conditions. In the evening, all nets (benthic and pelagic) were set to be retrieved the next morning in the same sequence they were placed. This way, we guaranteed the same fishing effort for all the nets (total fishing effort means the number of benthic nets deployed, which is determined by the depth and surface of the reservoir).

The sampling protocol works by dividing the dam reservoir into depth strata where nets are randomly deployed. The depth stratum limits are established so that the volume of water within every depth stratum is roughly the same [20]. The depth strata are determined so that every stratum represents roughly an identical amount of water as follows: 0–3, 3–6, 6–12, 12–20, 20–35, 35–50, 50–75, and >75 m (Table 2) and (Figs. 3 and 4).

Pelagic multi-mesh gill nets were set in the deepest part of the dam impoundment. They were initially placed in the surface zone at intervals of 0–6 m covering the entire water column in a systematic way (Figs. 3 and 4).

Gill net catches were deployed to assess the number of fish per unit effort (NPUE/ 1000 m^2).

Table 2 The number of benthic and pelagic multi-mesh nets deployed in Siliana and Kasseb reservoirs (2016)

Dam Reservoir	Siliana	Kasseb
Benthic nets	32	48
Pelagic nets	0	6

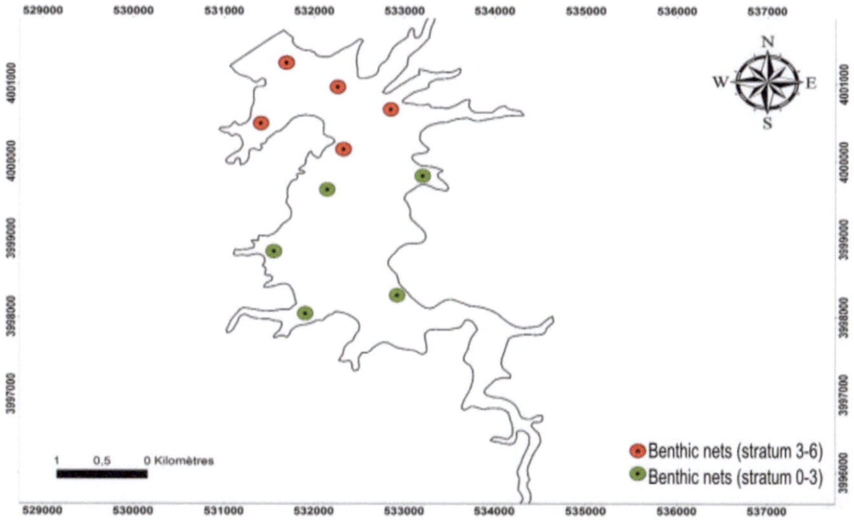

Fig. 3 Geographic localizations of the benthic multi-mesh nets in Siliana Reservoir 2016. (Mili et al. 2024)

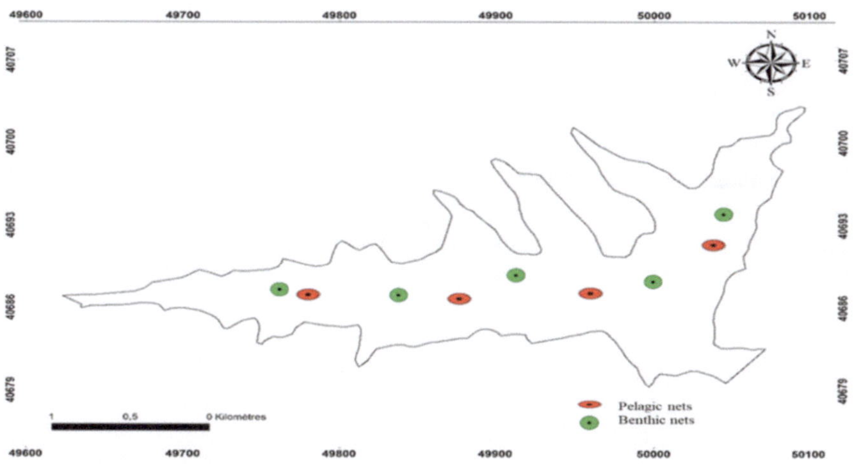

Fig. 4 Positions of benthic and pelagic multi-mesh nets in Kasseb Reservoir 2016. (Mili et al. 2024)

3.3.2 Laboratory Work

All species were identified based on their data sheets related to Fishbase. *Chelon ramada* and *Mugil cephalus*, two Mugilidae species, share striking physical resemblance, so we used the digestive system (number of *Pyloric Caecas*) to distinguish one from the other. *Mugil cephalus* has two *Pyloric Caecas* and *Chelon ramada* has seven or eight *Pyloric Caecas*. Roach and rudd are also morphologically similar, so we identified the location of the pelvic fins and the dorsal fins to tell them apart.

All fish caught were counted, measured, and weighed. The information was collected for each net and eventually for every mesh. Every single fish had its total length (TL in cm), standard length (SL in cm), fork length (FL in cm), total weight (TW in g), gutted weight (EW in g), gonad weight (GW in g), sex, and maturity stage measured.

4 Results and Discussion

During the four seasons of this study, we collected a huge database of species richness and abundance indicators thanks to the sampling survey that was conducted. All of the outcomes of the fishing operations were converted into yields (number) per unit of effort (NPUE) to make seasonal and interspecies comparisons. These yields were then entered into a GIS to map the results which helped improve the visibility of the yields. This analysis generated reliable and valid data on the status of fish populations in these impoundments, allowing for the proposal of scenarios for improving stock status.

4.1 Spatial and Temporal Species Distribution

4.1.1 Siliana Dam Reservoir

The sampling procedure enabled the catch of five fish species in Siliana Dam Reservoir: mullet *Chelon ramada*, pikeperch *Sander lucioperca*, barbell *Barbus callensis*, common carp *Cyprinus carpio communis* and rudd *Scardinius erythrophtalmus*.

The species were found to colonize the sampled strata with a quantitative predominance of the 0–3 m stratum, especially during the fall season. Rudd and mullet were found all over the reservoir during all the seasons and the presence of pikeperch depends on that of rudd. Only two specimens of rudd were sampled during March which is likely linked to the species' behavior throughout the breeding season (preparation and guarding of nests).

The common carp was sampled only during the summer season, while the barbell was present in the captures during both summer and spring.

During spring (March 2016), pikeperch was concentrated in areas along the dam searching for optimal reproductive conditions and shoal of rudd which represent the favored prey for this species. March seems to be the favorite month for mullet fishing, given that this species is abundant in all areas during this time of the year (Fig. 4d).

During summer (June 2016), mullet, rudd, carp, and barbell were present in the catches. Barbell was concentrated at the points of discharge of the rivers, while mullet and rudd were present in the entire reservoir (Fig. 4a). Mullets are the most important species at Siliana Reservoir which demonstrates the success of the CTA's stocking activities.

We recorded the presence of all fish species throughout the surveyed dam during the fall campaign (September 2016). Rudd were concentrated in the dam area, whilst mullets were abundant in the reservoir's center zone. Pikeperch had the same distribution as rudd and were dominant downstream of the reservoir where barbell was also profuse. Thus, we note that the distribution of this species is conditioned by the presence of its prey (Fig. 4b).

During the December 2016 sampling survey, mullet and rudd were abundant throughout the reservoir. Given the profusion of this resource at the dam upstream, the winter season appears to be the best time for mullet fishing. Similarly, rudd occupied all areas in Siliana Reservoir. However, its most significant biomass was recorded in the central zone (Fig. 4c).

Distribution maps during the sampling seasons showed that the dominant species in this reservoir were mainly mullet and rudd. Eventually, it would be wise to monitor carefully the status of the stocks of these two species, which appear to be in a state of optimal exploitation or overexploitation.

4.1.2 Kasseb Dam Reservoir

The sampling protocol permitted the catch of three fish types in Kasseb Dam Reservoir: mullet, pikeperch, and barbell. The majority of the catches (65–88%) were caught in the 0–3 and 3–6 m strata. The different fish species were present in the 6–12 m stratum. Barbell were caught in all strata, mainly in the 0–3 m stratum. Pikeperch were present only in the winter and autumn in strata above 3 m. Mullet were sampled in the 0–3 m stratum in spring and in the 6–12 m and 20–35 m strata in winter. This species was only captured with pelagic nets during the winter season. It was present during this season and in all strata, with a predominance of the 0–6 m pelagic stratum.

The sampling survey carried out in March showed that the fish resources in Kasseb Reservoir were well distributed. We noticed the presence of barbell in large quantities and the absence of their predator (pikeperch). The mullets were present in average quantities and concentrated in the deepest areas of the reservoir (Fig. 5d).

The fishing survey conducted in June showed the presence of one species in the catches (barbell). This indigenous ichthyic fish is concentrated in the mixing zones between rivers and reservoirs. The absence of mullet and pikeperch allows us to recommend not fishing these species in the Tunisian reservoirs during this period (Fig. 5a).

In September 2015, only barbell and pikeperch were caught in this reservoir. Barbell were present in the entire reservoir, unlike the previous season, when they were abundant at the level of the river's discharge area of the reservoir.

This distribution pattern is mainly due to the autumn rain, which induces an important runoff of water in the reservoir, mixing the water column contents (Fig. 5b).

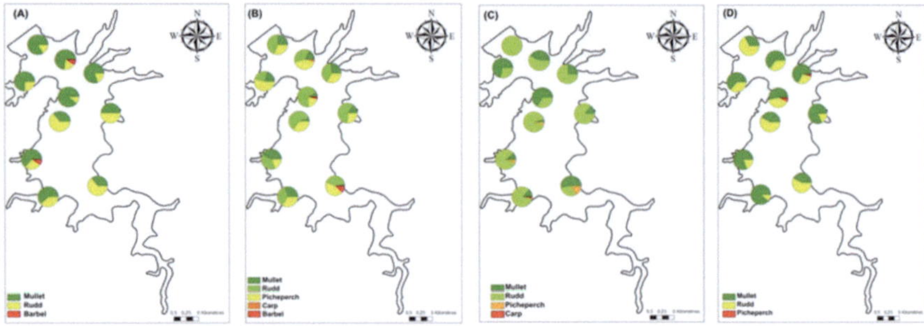

Fig. 5 Spatial and temporal species distribution in Siliana Dam Reservoir 2016 (**a** June; **b** September; **c** December; **d** March). (Mili et al. 2024)

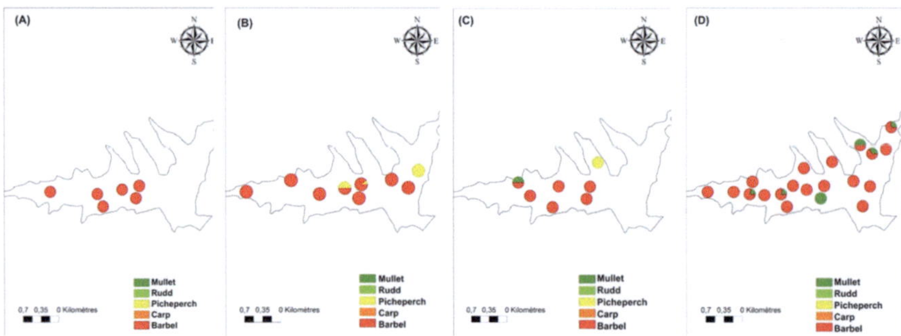

Fig. 6 Spatial and temporal species distribution in Kasseb Dam Reservoir (2016) (**a** June; **b** September; **c** December; **d** March). (Mili et al. 2024)

Barbell were the most abundant species caught during the winter season (December 2015). This result is similar to other sampling surveys. The distribution maps showed that barbell were present in the catches, which may reflect a state of optimal exploitation or even overexploitation of this ichthyic resource. Pikeperch and barbell were located in the same areas as in the fall season (Fig. 5c).

It would therefore be wise to review the regulations concerning the fishing season, which corresponds to March and April. The absence of pikeperch and the abundance of mullet, which do not reproduce naturally in freshwater reservoirs, during these two months, leads us to propose a shift in the rest period to allow an improvement in fishermen's income without affecting the state of the stocks.

The spatial distribution of fish in the depth layer greater than 3 m showed that 36.4 and 49.75% of specimens were located between 3 and 5 m depth during the day and night [9]. Large specimens were caught between 5 and 6 m in Kasseb impoundment during the day, and no such fish were detected at night at this depth [9]. Chargui et al. [28] demonstrated that the biomass of fishery resources at Tunisian reservoirs varies according to layers and strata. The surface layer contained higher fish biomass than the deep layer [28]. Longitudinally, the lowest biomass was recorded upstream of each reservoir [9].

4.2 *Numerical Yield in Siliana Reservoir*

Yields recorded in this reservoir were numerically important compared to other Tunisian dams and the highest yields were observed during the summer season (459 fish/1000 m^2). The numerical catches were mainly dominated by rudd (75.68–31.25%) and mullet (20–64.71%) followed by pikeperch (0.9–33.33%). Common carp and barbell seem to be in deficit abundance. Numerical yields were low not exceeding 1 fish/1000 m^2 for carp and 7 fish/1000 m^2 for barbell.

In Siliana Reservoir, each season was characterized by a specific relative abundance. Pikeperch was caught mainly in spring, rudd was captured in summer and autumn, and mullet was fished in winter and spring. The seasonal behavioral patterns of these species may be associated with mobility following the spawning seasons for pikeperch and rudd and to the spawning local migration for mullet. These species commonly migrate during the spawning season from freshwater to the sea, which inhibits female maturation. Spawning can occur only when all abiotic conditions such as salinity, and oxygen are favorable which is not the case in the freshwater reservoir [10, 29, 30].

4.2.1 Barbell

The NPUE of barbell ranged between 0 and 66 ind/1000 m^2 at stratum 0–3, and between 0 and 16 ind/1000 m^2 at stratum 3–6. The discharge area of the two rivers at Siliana Reservoir presents a high concentration of barbell. This observation is in congruence with previous research showing that barbells are common in Tunisian watersheds [31–33] (Fig. 7).

In Siliana Reservoir, barbell sizes ranged from 15 to 31 cm (average 23 cm). The size of fish caught reveals that the population of this species is composed of juveniles and that the stock is overexploited. This is mainly because this native species has low fertility and a slow growth rate, which is associated with an important fishing effort affecting this stock.

4.2.2 Mullet

Seeding reservoirs with Mugilidae fry dates back to the 1960s; the first operation took place in 1964. The Tunisian Technical Center of Aquaculture (CTA) ensures mullet stocking yearly. Mullet fry are first collected from the rivers, and then they are seeded in most of the reservoirs.

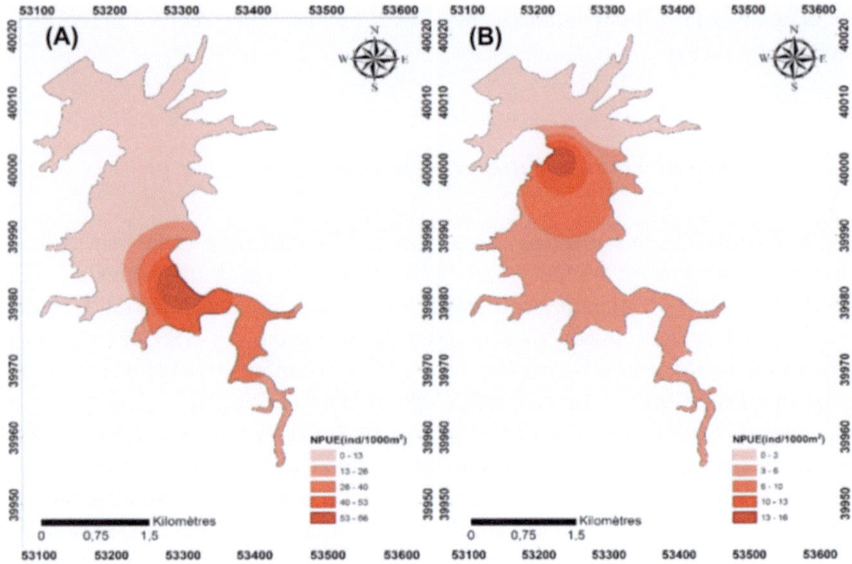

Fig. 7 Numerical yield of Barbell in Siliana Reservoir (2016) (**a** September 2016 (0–3); **b** September 2016 (3–6)). (Mili et al. 2024)

Mullets are concentrated in the uppermost layers of the most profound zones in Siliana Reservoir. Mugilidae represents the major target fish type in the Tunisian reservoir. The numerical yields show high values in this reservoir, ranging from 17 to 232 ind/1000 m^2 at the level of stratum 0–3 and from 33 to 146 ind/1000 m^2 in stratum 3–6 (Fig. 8).

During winter (December 2016), mullets make some local migrations. The numerical yields of this species ranged from 17 to 149 ind/1000 m^2 at stratum 0–3 and from 0 to 82 ind/1000 m^2 at 3–6 m. Mullet populations are affected by the fishing effort, which increases during the fall season.

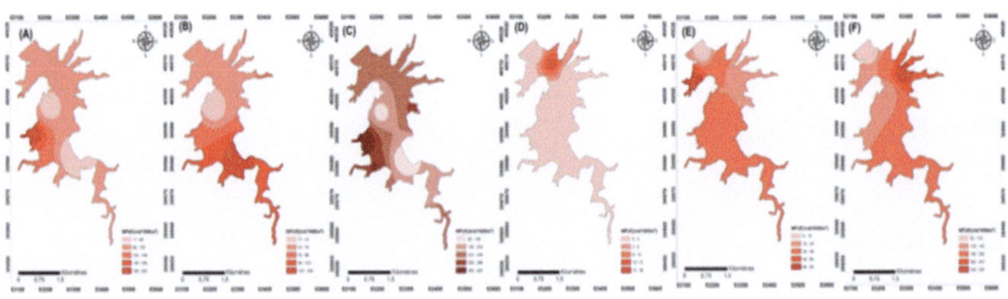

Fig. 8 Numerical yield of Mullet in Siliana Reservoir (**a** September 2016 (0–3); **b** December 2016 (0–3); **c** March 2016 (0–3); **d** September 2016 (3–6); **e** December 2016 (3–6); **f** March 2016 (3–6)). (Mili et al. 2024)

During the spring survey, conducted in March 2016, a large number of Mugilidae (*Chelon ramada*) were caught. The recorded numerical yields ranged from 50 to 482 ind/1000 m^2 at the 0-3 m layer and from 83 to 282 ind/1000 m^2 at the 3–6 stratum. This result confirms that mullets were concentrated in the upper layers of the deep areas of the reservoir. The 3–6 stratum had lower yields than the 0–3 layer.

This distribution appears to be controlled by both diet and oxygen levels. Spawning operations can take place only at sea, which is associated with thalassotopic or spawning migrations for this species. Mullets breed in autumn and early spring and they reach sexual maturity in the third year for males and the fourth year for females [10, 34–36].

At Siliana Reservoir, *Chelon ramada* is the dominant Mugilidae species. The sizes of specimens caught varied from 18 to 46 cm, which corresponds to ages from 2 to 4 years. The mean size increased from 22.64 ± 4.02 cm in summer to 28.98 ± 1.84 cm in spring. The population was composed of three age classes. Because sexual maturity only occurs when an individual is between 26.5 and 34 cm in length, we can affirm that the population in the spring was primarily composed of juveniles [37]. The majority of mullets were caught during the winter, a season which is characterized by a migration of the spawners to the sea. These findings are similar to those of Mili et al. in all Tunisian dam reservoirs [55].

4.2.3 Rudd

Rudd distribution by depth shows that numerical yields are important in Siliana Reservoir, varying from 150 to 300 ind/1000 m^2 in the layers 0–3 and from 100 to 282 ind/1000 m^2 at the stratum 3–6. Rudd is a forage species living in the center of the reservoir. Yields show that this species found favorable conditions for the development of its biomass in this reservoir (Fig. 9).

In March 2016, the numerical yields of rudd ranged from 50 to 133 ind/1000 m^2 at the 0–3 stratum and from 83 to 166 ind/1000 m^2 at the deepest layer (3–6). Once

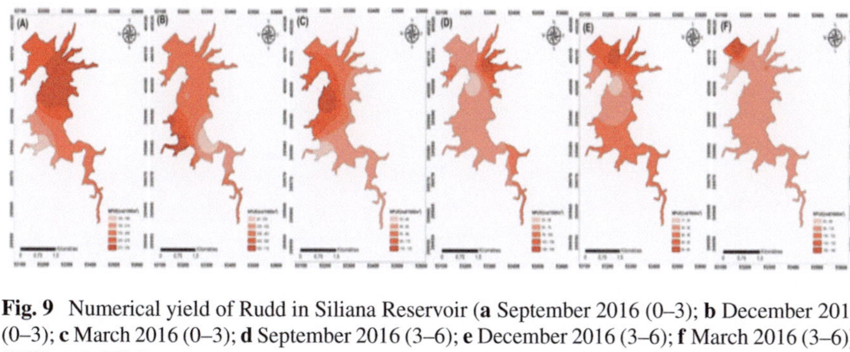

Fig. 9 Numerical yield of Rudd in Siliana Reservoir (**a** September 2016 (0–3); **b** December 2016 (0–3); **c** March 2016 (0–3); **d** September 2016 (3–6); **e** December 2016 (3–6); **f** March 2016 (3–6)). (Mili et al. 2024)

the spawning season was over, these forage fishes moved for feeding, while they opted for the upper layers of the deeper areas of the reservoirs during the spring.

During the winter season (December 2016), rudd fishing yields ranged from 83 to 716 ind/1000 m^2 at stratum 0–3 and from 17 to 66 ind/1000 m^2 between 3 and 6 m. This species' behavior is conditioned mainly by reproduction. Indeed, during the maturation and spawning period, the abundance was greater than that recorded during other periods indicating that the species prefers sheltered areas to spawn. Rudd is known to spawn from April to September in the vegetation. The fecundity varies from 1203 to 53,609 oocytes, and maturity is reached at one year of age [3]. The fry feed on plankton, while the adults feed on macrophytes and invertebrates (molluscs, insect larvae…) [38, 39].

Rudd sizes ranged from 9 to 31 cm, corresponding to four age classes (0+, I+, II+, and III+). Weights varied from 71.81 to 106.12 g. The average length increased from 17.45 cm in summer to 20.62 cm in spring. In this season, the rudd population was represented by specimens with sizes ranging from 17.1 to 25.6 cm. This finding demonstrates the capture of adults. Sexual maturity of this species occurs at 12.48 cm and 11.99 cm for females and males, respectively [40]. No fry or juvenile stages are caught because the minimal mesh size used in multi-mesh fishing nets is 18 mm.

4.2.4 Pikeperch

Pikeperch had a significant numerical abundance in Siliana Reservoir. Fishing yields ranged between 100 and 249 ind/1000 m^2 in the epipelagic area and from 83 to 216 ind/1000 m^2 in the stratum 3–6. This species feeds on rudd in the 3–6 stratum and changes their diet to barbell in 0–3 m (Fig. 10).

Belaifa et al. [41], affirmed that the diet of pikeperch varies seasonally and according to age. *S. lucioperca* is omnivorous with a piscivorous tendency. Rudd represents the major portion of the adult sander diet. Rudd morphology associated

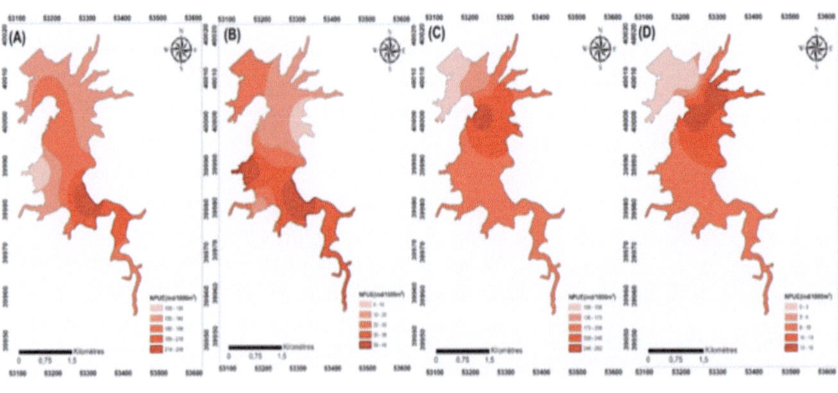

Fig. 10 Numerical yield of Pickeperch in Siliana Reservoir (**a** September 2016 (0–3); **b** December 2016 (0–3); September 2016 (3–6); March 2016 (3–6). (Mili et al. 2024)

with the small mouth opening of pikeperch [42–44], explains the variability in diet. Rudd is vulnerable to pikeperch predation their whole lives [41].

The analysis of the survey conducted in December 2016 shows that pikeperch was found far from the dam. The numerical yields ranged from 0 to 49 ind/1000 m^2 and all specimens of pikeperch were caught in the 0–3 stratum. The seasonal and bathymetric variability of the abundance of this species in Siliana Reservoir is linked to its reproductive and feeding behaviors. For instance, pikeperch has a significant feeding activity in the winter. According to Belaifa et al. [41], in this period food is plentiful and available, which explains the low vacuity indexes. While in the summer, due to the extended digestion time and the prey spawning season, stomach vacuity increases. The species' feeding activity decreases with age [41, 44].

In spring, fishing yields of pikeperch in Siliana Reservoir were poor and they ranged from 0 to 16 ind/1000 m^2. This low productivity is due to the reproductive activity of this species at this time of year. In the spawning period, females reduce their mobility, while males watch over the laid eggs and ensure their oxygenation. This lack of activity makes them less vulnerable to capture [45–47]. Another factor that interferes with this species' mobility and feeding behaviors is temperature as depicted by Belaifa and collaborators [41].

The pikeperch population showed a high abundance in this reservoir with sizes ranging from 8 to 50 cm (mean: 29 cm). The specimens caught were at their maturity stage considering that this fish attains sexual maturity when it is one year old measuring 24 cm and weighing 90 g [48]. Their weights varied from 44 to 1330 g (mean: 364 g). A total of four age classes were recorded (0+, I+, II+, and III+). Most of the fish were caught during the summer, indicating that they were moving and searching for food after the spawning season.

4.2.5 Carp

The common carp *Cyprinus carpio* [29] prefers two biotopes: shallow, calm environments rich in vegetation for growth and spawning, and deep water for wintering [49]. This omnivorous fish feeds on benthic invertebrates and plants [49]. Size ranges from 14 to 56 cm (mean size: 35 cm). Reproduction takes place between March and August in vegetation and shallow water. The eggs are attached to aquatic vegetation and the fecundity is estimated to be 100,000 eggs/kg [36, 54].

In Siliana Reservoir, carp yields were very low ranging from 0 to 16 ind/1000 m^2. This observation confirms the presence of small stock in this reservoir (Fig. 11). Most of the specimens caught were juveniles if we consider that males attain sexual maturity when they are one-year-old (TL = 15.8 cm) and females do so when they turn 2 years old (TL = 22.5 cm) [14, 36, 54].

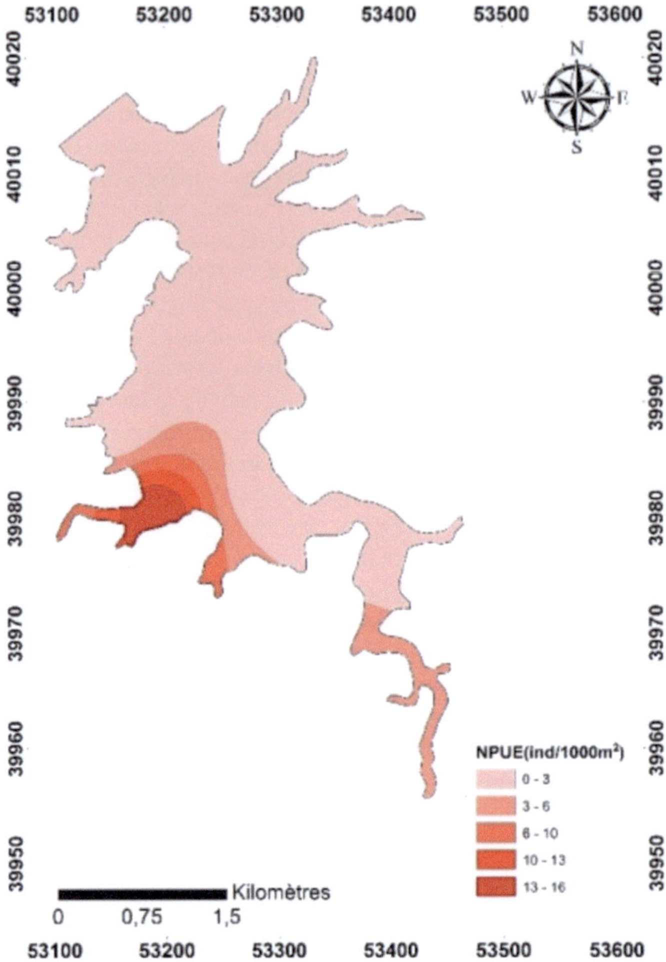

Fig. 11 Numerical yield of Carp in Siliana Reservoir (December 2016 (0–3)). (Mili et al. 2024)

4.3 *Numerical Yield in Kasseb Reservoir*

The yields in this reservoir were low numerically ranging from 10 fish/1000 m^2 to 38 fish/1000 m^2. Barbell, which is known to be a native species, displayed the most significant numerical yields (32 fish/1000 m^2 in winter). Pikeperch presented an overall deficit of abundance with low numerical yields. The maximum yield was recorded in winter (2 fish/1000 m^2). Since mullet is regularly harvested, the status showed a low numerical yield. The maximum yield of this species was 6 fish/1000 m^2 during the winter season.

4.3.1 Barbell

Barbell *Barbus callensis* is an endemic species quite widespread in the Tunisian watersheds, especially in rivers and reservoirs [50].

During spring (March 2016), yields oscillated between 0 and 10 ind/1000 m^2. The center of the reservoir had the largest abundance of the barbell. Two rivers supply the dam with native fish (barbell). This concentration is also related to the reproduction of this species (Fig. 12d).

The NPUE of the barbell in Kasseb (June 2016) varied from 0 to 12 ind/1000m^2 and was important in the deepest area of the reservoir. The spillway area of the two rivers in Kasseb Reservoir showed a high concentration of barbell (Fig. 12a).

The numerical yields recorded during the autumn season (September 2016) were low ($\leq$6 ind/1000 m^2). Barbell concentration areas were located at the river discharge zone in the reservoir. The low yields are attributed to the dry fall season in 2016. In this year, there was not enough rainfall to allow the mix and transfer of barbell from the river to the reservoir (Fig. 12b).

Numerical yields were low and varied from 0 to 5 ind/1000 m^2 in winter (December 2016). The low densities are related to overfishing occurring during winter when fish mobility is reduced because of reproduction (Fig. 12c).

The introduction of carnivorous species in Tunisian reservoirs caused the collapse of native fish populations, particularly barbell [32, 33]. This observation may be attributed to the species' low growth rate and fecundity. As a result, *B. callensis* has a limited stock renewal, especially with the significant fishing effort.

In Kasseb Reservoir, barbell sizes varied between 22 cm and 45.5 cm (weight: 126–858 g) with a mean size of 33.7 cm. According to Kraiem, the median total length of the barbell in Tunisia is 25 cm and it can reach 40 cm [50]. The size frequencies showed the presence of adults in this reservoir. Kraiem states that barbell spawned 3000–8000 eggs from May to July and that maturity is reached between 4 and 5 years [50]. Chaouachi reports that males attain sexual maturity at 19.6 cm while females reach this stage at 27.7 cm [51].

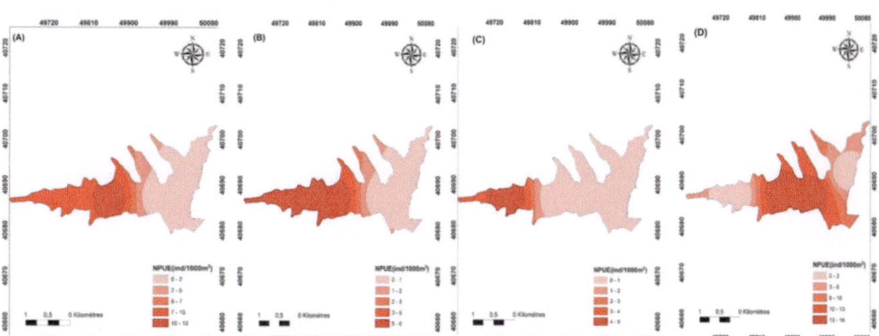

Fig. 12 Numerical yield of Barbell in Kasseb Reservoir (**a** June 2016, **b** September 2016, **c** December 2016, **d** March 2016). (Mili et al. 2024)

The mean size of barbell increased from 28.81 ± 3.12 cm in summer to 30.99 ± 4.1 cm in winter, and the average weight increased from 133 ± 49 g to 237.68 ± 79.5 g. The large size of the fish protected them against predation by pikeperch. A total of five size classes were caught during the sampling surveys which were conducted in winter (ages from 1 to 5 years). The results demonstrate that the conditions in this reservoir are favorable for the reproduction, growth, and survival of this species.

Gill net sampling revealed the importance of the barbell population in Kasseb Reservoir in all seasons. The presence of this species during the winter season can be attributed to an external contribution from the reservoir's watershed rivers. Significant rainfall during 2016, which was a year of water distress, was recorded by the General Directorate of Dams and Large Hydraulic Works (DG/BGTH) while we were conducting the sampling survey (March 2016). A water supply of 7.847 Mm^3, representing more than 10% of the existing stock, was recorded in March 2016 [9].

4.3.2 Pikeperch

Pikeperch lay their eggs (average: 200,000) between February and March ($T > 14$ °C) on sand or gravel covered with vegetation. The breeders guard the eggs and keep them off the ground until they incubate. The fry sticks together and move in schools but they tend to be solitary when adults are present [48]. Pikeperch attains sexual maturity when it is 1 year old.

A total of three age classes were caught: one juvenile (19 cm) and three adults (26, 27, and 87 cm). This is an alarming situation needing urgent intervention to boost the population's status. These findings contrast with those obtained in Sidi Salem Reservoir, where the Pikeperch population is composed of juveniles thanks to reproduction and good conditions [16].

4.3.3 Mullet

Mullets present a low yield, not exceeding 24 ind/1000 m^2. This is due to the recent introduction of mullet fry in the reservoir; specimens are still juveniles. The mullets migrate from downstream, where they are stocked, to the dam area where they congregate in the surface layers. These strata provide them with high oxygenation and abundance of prey.

The size structures of the mullet caught in this reservoir were composed of four age classes during the winter season. The size varied from 19.6 to 43 cm and the weight was between 71 and 688 g. The median size increased from 23.6 ± 0 cm in winter to 30.84 ± 8.38 cm in spring. These results are similar to those found by Mili et al. in the Tunisian dam reservoirs [55].

5 Conclusion

The current research seeks to identify the species richness, spatial and seasonal spread, along population status in Siliana and Kasseb reservoirs. Taking into account the non-regular regional spread of fish species in reservoirs, a random stratified sampling was adopted. This technique was adopted from the European Standard CEN prEN 14,757 and it allows the use of benthic and pelagic nets to cover the whole water column. The nets used for sampling are multi-mesh nets designed to capture most of the fish sizes in the reservoir.

The NPUE in Seliana and Kasseb reservoirs ranged from 10 to 495 fish/1000 m^2 of gill nets. Kasseb Reservoir has the lowest NPUE compared to other reservoirs in Tunisia. Yields in Siliana Reservoir are relatively optimal.

A differentiation was highlighted between the reservoirs of the northern sub-basins of Medjerda (Kasseb) and those of the southern sub-basins (Siliana). This disparity can be explained by the difference in water quality, specifically salinity, between these two basins. The salinity of the water in the southern sub-basins is relatively higher than in the northern sub-basins. This excess of salinity is due to the abundance of Triassic sulfurous soils, the leaching of salty soils, and the numerous salty sources emerging from the sulfurous basin [52]. This salinity has been proven to have an important impact on fish fauna's growth as demonstrated by several studies [53].

Our findings revealed that in Siliana Reservoir, mullets were abundant with small and medium-sized (immature) individuals, signing a notable fishing effort targeting large specimens. Forage fish, particularly rudd, were similarly numerous in this reservoir. This species seems to be the most common, with 54% of the overall numerical abundance, followed by mullet and pikeperch (26 and 16%). On the other hand, the carp and barbell populations were poorly represented.

The fishing activities conducted at Kasseb Reservoir highlighted the preponderance of barbell, which accounts for 90% of the catches, compared to a percentage of 7% for mullet and 3% for pikeperch. This is due to the presence of physicochemical and trophic factors that promote the growth of this indigenous species. The exploitation of barbells is, therefore, close to optimal. However, in the case of a substantial fishing effort, there will be overexploitation and an imbalance in the ecosystem.

Therefore, developing an effective and sustainable strategy for preserving the ichthyofauna diversity in these reservoirs is highly recommended. Siliana Reservoir requires pikeperch population replenishment (low capture rate). On the other hand, Kasseb Reservoir necessitates immediate and urgent intervention for the improvement of fish populations, such as the transfer of forage fish spawners, pikeperch, carp, etc. The stocking of these reservoirs with mullet fry must be reinforced, monitored, and assessed.

The improvement of freshwater fish production in reservoirs requires certain measures, such as boosting fishing operations and seeding Mugilidae fry (*M. cephalus* and *C. ramada*). The species richness in the reservoirs must be improved by optimizing the stocking operations of Chinese carp and transferring broodstock (pikeperch, roach, rudd, and tilapia). Attempts to introduce new species other than fish, such as the Turkish crayfish (*Astacus leptodactylus*) should be encouraged since some ecological niches in the dams are underexploited.

Fisheries management must be rigorous to protect the fish stock while making the most of it. The various actors in the fishing sector must take coordinated action in this regard.

6 Recommendations

Based on this comprehensive study several recommendations can be proposed to enhance the management and sustainability of fish stocks in Tunisian reservoirs.

-The implementation of a regular and systematic fish population monitoring programme in these reservoirs is essential. This should include the use of Geographic Information System (GIS) data to map the distribution and abundance of different species over different seasons. Such data-driven insights are essential for understanding population dynamics and identifying areas where certain species are thriving or declining.

- Improving the accuracy and reliability of fisheries recording, through better collaboration between fisheries management institutions and training of fishers to report catch accurately, would enhance longer-term assessment of fishing effort and resource status.
- Proactive measures should be taken to replenish and maintain fish populations, especially in reservoirs where species indicate potential risks of overexploitation.
- Promoting sustainable fishing practices can help prevent overfishing and ensure the long-term viability of freshwater fisheries in these reservoirs.

Through the implementation of these recommendations, stakeholders can work towards the conservation of biodiversity, the support of local livelihoods and the sustainable use of freshwater fisheries resources in Tunisia.

Acknowledgements The authors would like to thank the IRESA which financially supported the study for its added value to the research project (AMBISEPT). This work is the outcome of collaborative efforts between the Technical Center of Aquaculture (CTA), the General Directorate of Fisheries and Aquaculture (DGPA—Tunisia) and the Higher Institute of Fishery and Aquaculture, Bizerte (ISPAB—Tunisia); We extend our thanks to all the staff members (CTA, DGPA, ISPAB) for the efforts deployed in this work. Many thanks to Wissem Chawali for his implication in data collection and analyses. Our gratitude also goes to Hajer Zarrouk for proofreading the manuscript and editing the English language.

References

1. Loss GF, Nau W, Winter M (1991) Development of freshwater fishing in northern Tunisia (Le développement de la pêche en eau douce dans le nord de la Tunisie). Tunisian-German technical cooperation GTZ Gmbh : Tunisie, 403 p
2. Zaouali J (1981) Problems of aquaculture: brackish waters and aquaculture potential (Problèmes d'Aquaculture: eaux saumâtres et potentiel aquacole). Pasteur Institut of Tunis 58:93–103
3. Djemali I (2005) Assessment of fish biomass in Tunisian freshwater bodies: analytical and acoustic approach (Evaluation de la biomasse piscicole dans les plans d'eau douce tunisiens: approche analytique et acoustique). PhD thesis, National Agronomic Institut of Tunis, Tunisia, 250 p
4. Mili S, Ennouri R, Laouar H, Khedhri I, Zarrouk H, Chargui T, Romdhane N (2021) Freshwater fish farming and fishery management in tunisian reservoirs: limitations and opportunities. In: Agriculture productivity in Tunisia under stressed environment Springer water, 34 p
5. Louati MH, Khanfir R, El Echi ML, Marzouk A, Frigui L, Alouini A (1998) EAU 21: Tunisia's long-term water sector strategy 2030 (EAU 21: Stratégie du secteur de l'eau en Tunisie à long terme 2030). Tunisian Ministry of Agriculture, 89p
6. KAS (Konrad-Adenauer-Stiftun) (2017) The Mediterranean faced with dwindling water resources (La méditerranée face à la raréfaction des ressources en eau). Simpact (eds) Tunisia, 103p
7. Ncib O (2021) Water resources management in Tunisia (Gestion des ressources en eau en Tunisie). GG2 ENIS Tunisia, 113 p
8. Soudoud (2010) Specific database of large dams in Tunisia (Base de données spécifique des grands barrages en Tunisie). Ministry of Agriculture and Hydraulic Resources, Tunisia
9. Laouar H (2019) Assessment of fishery resources in Tunisian dam reservoirs: Inter-calibration of acoustic and gillnet samplings (Evaluation des ressources halieutiques dans les retenues de barrages de Tunisie: Inter-calibration des échantillonnages acoustiques et aux filets maillants). PhD thesis, INAT Tunisia, 133 p
10. Mili S, Ennouri R, Laouar H, Missaoui H (2015) Fisheries in the Tunisians dams: diagnosis of the current situation and development opportunities. FAO Fish Aqua Proc 39:95–106
11. Mtimet N (2010) Socio-economic analysis of the continental fish farming sector in northwest Tunisia (Analyse socioéconomique de la filière de la pisciculture continentale au Nord-Ouest de la Tunisie). Report by the Groupement Interprofessionnel des Produits de la Pêche in collaboration with the Italian Cooperation
12. CTA (2019) Technical report on the activities of the Aquaculture Technical Center in Tunisia (Rapport technique des activités du Centre Technique d'Aquaculture en Tunisie). CTA Tunisia
13. Agrebi S, Ennouri R, Fatnassi M, Zarrouk H, Chargui T, Romdhane N, Laouar H, Mili S (2022) Freshwater fish farming in Tunisian reservoirs: zoo-sanitary registration status. J Aquac Mar Biol 11(1):21–26
14. Mili S, Ennouri R, Laouar H, Ben Romdhane N, Missaoui H (2016) Study of fish populations in the Sidi Salem dam reservoir (Etude des peuplements piscicoles de la retenue du barrage de Sidi Salem). J New Sci Agri Biotech 27:1454–1465
15. General Directorate of Fisheries and Aquaculture (2020) Statistics yearbook of fisheries and aquaculture in Tunisia, Ministry of Agriculture, 140 p
16. Marchal E (1985) Acoustic detection in the study of fish populations (La détection acoustique dans l'étude des peuplements pisciaires). In: Gerdeau XD, BIllard R (eds) Fish management in artificial lakes and reservoirs (Gestion piscicole des lacs et retenues artificielles) INRA, Paris, pp 107–124
17. Mulligan T (2000) Shallow water fisheries sonar: a personal view. Aquat Living Resour 13:269–273
18. Chahed J, Besbes M, Hamdane A (2010) Water scarcity and food security: A global assessment of water potentiality in Tunisia. In: Re-thinking water and food edition: Taylor & Francis Group,

London, UK. Chapter: Section I: Are global water resources a limitation for food production Chapter 3, 400 p
19. Rousseau B, Nelva A, Persat H, Chessel D (1985) Creation of an ichthyological database using point-abundance sampling: application to populations in the French Upper Rhône region (Constitution d'une base de données ichtyologiques par l'échantillonnage ponctuel d'abondance: application aux peuplements du Haut-Rhône français). Cybium 9 (Z):157–173
20. CEN (2005) Water quality-sampling of fish with multi-mesh gillnets. EN 14757
21. Encina, Rodrıguez-Ruiz (2003) Abundance and distribution of a brown trout (*Salmo trutta* L.) population in a remote high mountain lake. Hydrobiologia 493:35–42
22. Tátrai IA, Specziár A, György AI, Bíró P (2008) Comparison of fish size distribution and fish abundance estimates obtained with hydroacoustics and gill netting in the open water of a large shallow lake. Int J Lim 44(4):231–240
23. Emmrich M, Winfield IJ, Guillard J, Rustadbakken A, Vergès C, Volta P, Mehner T (2012) Strong correspondence between gillnet catch per unit effort and hydro acoustically derived fish biomass in stratified lakes: comparison of gillnet catches with fish biomass estimates derived from hydroacoustics. Freshw Biol 57:2436–2448
24. Vasek M, Kubecka J, Peterka J, Cech M, Drastik V, Hladik M, Prchalova M, Frouzova J (2004) Longitudinal and vertical spatial gradients in the distribution of fish within a canyon-shaped reservoir. Hydrobiology 89(4):352–362
25. Loures RC, Pompeu P (2015) Seasonal and diel changes in fish distribution in a tropical hydropower plant tailrace: evidence from hydroacoustic and gillnet sampling. Fish Manage Ecol 22(3):185–196
26. Ansoumane M (2019) The geographic information system in mining and mine site redevelopment (case: Kettara Mine, Morocco (Le système d'Information Géographique dans l'exploitation minière et le réaménagement des sites miniers (exemple : Mine de Kettara, Maroc)). Master thesis, Abdelmalek Essaadi University, Faculty of Science, Tétouan, 43 p
27. GISRSStudy (2023) GIS & remote sensing: ArcGIS (Online: https://gisrsstudy.com/arcgis/)
28. Chargui T, Fatnassi M, Ennouri R, Zarrouk H, Laouar H, Romdhane N, Mili S (2021) Exploring freshwater fish assemblages and population structure in three Tunisian reservoirs for better fishery management. Biodiversity Int J 5(2):37–45
29. Brusle J (1981) Sexuality and biology of reproduction in grey mullets. In: Oren OH (ed) Aquaculture of grey mullet. Cambridge University Press, 99–154
30. Chauvet C (1984) Fisheries in the Lake of Tunis: fishery biology and increasing production by means other than regulation. In: Kapetsky JM, Lasserre G (eds) Management of coastal lagoon fisheries. FA0 Stud. Rev. GFCM No. 61, vol 1, pp 615–694
31. Kraiem MM (1984) Freshwater fish from Tunisia. Commented inventory and geographical distribution. Bull Natl Inst Sci Technol Sea 10:107–124
32. Mili S, Ennouri R, Laouar H (2016) Population status of freshwater fish in Tunisian reservoirs. Saving Freshw Fishes Habitats 11:32–34
33. Mili S, Ennouri R, Laouar H (2017) An update on the status of autochthonous freshwater fishes in Tunisian reservoirs. Saving Freshw Fishes Habitats 13:19–22
34. Farrugio H (1975) The mullet (teleost fishes) of Tunisia, distribution and fishing. Contribution to their systematic and biological study (Les muges (poissons téléostéens) de Tunisie, répartition et pêche. Contribution à leur étude systématique et biologique). Master thesis, University of Tunis, 201 p
35. Bedoui RF, Gharbi H (2005) Age and growth of *Liza aurata* (Mugilidae) along Tunisian coasts. Cybium: Int J Ichthyol 29(2):119–126
36. Hajlaoui W (2016) Reproductive biology of the Common carp, *Cyprinus carpio communis*, in Sidi Saad reservoir (central Tunisia). Bulletin de la Societé Zoologique de France 141(1):25–39
37. Mili S, Ennouri R, Fatnassi M, Chargui T, Zarrouk H, Thabet R, Laouar H, Romdhane N (2022) Variability of biological features of a mugilidae chelon ramada in two tunisian reservoirs. J Aqauc Mar Biol 11(2):41–48
38. Niederholzer R, Hofer R (1980) The feeding of roach (*Rutilus rutilus* (L.)) and rudd (*Scardinius erythrophthalmus* (L.)) studies on natural populations. Ekol Pol 28(1):45–59

39. Djait H, Bahru-Sfar L, Laouar H, Missaoui N, Ben Hassine OK (2019) Dietary comparison of pikeperch, Sander lucioperca (Linnaeus, 1758) and catfish, *Silurus glanis* Linnaeus, 1758 in Sidi Salem dam reservoir (Tunisia). Cybium 43(1):61–69
40. Mili S, Ennouri R, Laouar H, Aleya L (2017) Optimization of fishing for stock enhancement *of Rutilis rutilus* and *Scardinius erythropththalmus* in forage fish deficient Tunisian reservoirs. Environ Monit Assess 189:61
41. Belaifa B, Bouamra A, Attou F, Arab YI, Arab A (2013) Comparison of the diet of the pike-perch *Sander lucioperca* (L., 1758) in two Algerian lake systems (Comparaison du régime alimentaire du sandre *Sander lucioperca* (L., 1758) dans deux systèmes lacustres algériens). USTHB-FBS 4th International Congress of the Populations & Animal Communities "Dynamics & Biodiversity of the terrestrial & aquatic Ecosystems "CIPCA 4" TAGHIT (Bechar) – ALGERIA, 19–21 November
42. Smith PA, Leath, RT, Eaton JW (1998) A review of the current knowledge on the introduction, ecology and management of zander, *Stizostedion lucioperca*, in the UK. In: Stocking and introduction of fish (Cowx IG (ed.) Fishing News Books, pp 209–224
43. Yağci MA, Alp A, Yağci A, Uysal R (2014) Diet and prey selection of Pikeperch (*Sander lucioperca* Linnaeus, 1758) population in lake Eğirdir Turkey. Arch Biol Sci 66(4):1515–1527
44. Bousseba M, Ferraj L, Ouahb S, Ouizgane A, El Moujtahid A, Droussi M, Hasnaoui M (2020) Food preferences of PikePerch, *Sander lucioperca* (Linnaeus, 1758) in Morocco. E3S web of conferences, 150 p
45. Poulet N (2004) The zander (*Sander lucioperca* (L.)): Biology, behavior and population dynamics in the Camargue (Bouches du Rhône, France) (Le sandre (*Sander lucioperca* (L.)): Biologie, comportement et dynamique des populations en Camargue (Bouches du Rhône, France)). PhD thesis, Université Paul Sabatier- Toulouse III, 204p
46. Khodja S, Ameri A, Driss D (2019) Study of the reproduction of two species of freshwater fish: zander (*Stizostedion Lucioperca*) and royal carp (*Cyprinus carpio*) (Etude de la reproduction de deux espèces de poissons d'eau douce: sandre (*Stizostedion Lucioperca*) et la carpe royale (*Cyprinus carpio*)). Master thesis, University of Blida, Algerie, 113p
47. Sioux-fishing (2023). Online: www.meteo.tn. https://www.sioux-fishing.fr/poisson-riviere/ichtyologie/124-la-reproduction-du-sandre
48. Toujani R, Missaoui H, Romdhane MS (2000) Sexual cycle of female zander (*Stizostedion lucioperca*) in the reservoir of the Sidi Salem dam (Cycle sexuel des femelles de sandre (*Stizostedion lucioperca*) dans la retenue du barrage de Sidi Salem). Bull Inst Nat Sci Tech Mer 27:85–96
49. Brusle J, Quignard JP (2001) European freshwater fish biology (Biologie des poissons d'eaux douces européens). Lavoisier Tec & Doc, Paris, p 625p
50. Kraiem MM (1983) The freshwater fishes of Tunisia. Commented inventory and geographical distribution (Les poissons d'eau douce de Tunisie. Inventaire commenté et répartition géographique). Bulletin de l'Institut National des Sciences et Technologies de la Mer 10:107–124
51. Chaouachi B, Ben Hassine OK (1998) Barbel *Barbus callensis* valenciennes, 1842 (Fish, Teleosteen) from the Ichkeul lagoon (Le barbeau *Barbus callensis valenciennes*, 1842 (Poisson, Teleosteen) de la lagune de l'ichkeul). Archs Inst Pasteur Tunis 75(3/4):211–218
52. Ben Mamou A, Louati A (2007) Temporal evolution of silting in Tunisian dam reservoirs (Évolution temporelle de l'envasement des retenues de barrages de Tunisie). Rev Sci Eau 20(2):201–210
53. Boeuf G, Payan P (2001) How should salinity influence fish growth? Comp Biochem Physiol-Part C: Toxicol Pharmacol 130:411–423
54. Hajlaoui W, Chalghaf M, Troudi D, Cherif M, Mili S, Missaoui H (2022) Investigating age and growth of *Cyprinus Carpio* (Linnaeus, 1758) in Sidi Saad Reservoir (central Tunisia). Vie et Milieu Life Environ 72(1–2):9–16
55. Mili S, Ennouri R, Fatnassi M, Agrebi S, Louiz I, Khattab Y, Hedfi A, Ben Ali M, Laouar H (2023) Investigating freshwater mullet fisheries in tunisian reservoirs: future development prospects. Water 15:2554

Remote Sensing Applications with Google Earth Engine: Case from Algeria

Salah Eddine Ali Rahmani and Brahim Chibane

Abstract Landuse and land cover assessment is very important subject in earth sciences studies, the availability of data and computer resources are necessary to perform image processing and remote sensing manipulation. With the newweb based software for remote sensing developed by Google (Google earth engine), users around the world can work easily and without high cost and time efforts. This software does not need a high computer performance neither graphical or calculator processor unit; many applications can be realized using the GEE platform. This chapter aim to introduce some basic functions used to process data of remote sensing for some applications in earth sciences. Normalized difference vegetation index NDVI, Normalized difference salinity index NDSI are calculated from Sentinel and Modis dataset for the North of Algeria.

Keywords Remote sensing · Google earth engine · Power calculation · Earth sciences

Highlights Applications of Remote sensing data · Terrain Modling · Normalized vegetation index · Soil salinity index

1 Introduction

Since the introduction of Remote sensing applications in the early 1980s, research in the world has undergone a revolution in all scientific field; especially with the combination of geographic information system (GIS). Google Earth Engine (GEE) is a powerful cloud-based platform that allows researchers to analyze massive amounts of geospatial data for various applications [1, 2]. Using (GEE) for Landuse assessment includes identifying and classifying different land cover types, such as urban

S. E. Ali Rahmani (✉) · B. Chibane
Faculty of Earth Sciences, Geography and Territory Planning, Geo-Environment Laboratory (LGE), University of Sciences and Technology Houari-Boumediene USTHB, Algiers, Algeria
e-mail: salirahmani@usthb.dz

F. Khebour Allouche (ed.), *Environmental Remote Sensing and GIS in MENA Region*, Earth and Environmental Sciences Library,
https://doi.org/10.1007/978-3-031-86574-9_5

areas, agricultural activities, forests, water bodies, and barren lands. GEE provides a range of tools and datasets that enable accurate landuse mapping.

For Supervised Classification, GEE offers a wide range of satellite imagery, such as Landsat, Sentinel and Modis, which can be used for supervised classification algorithms like Support Vector Machines (SVM) and Random Forest (RF). Researchers have utilized these algorithms in GEE to classify land cover types effectively [3]. Also GEE can be used in change detection, land use change detection can be performed by comparing multi-temporal satellite images. Normalised techniques (e.g., NDVI, NDSI,NDWI) can be employed to detect land cover changes, vegetation health, salinity, and water bodies change over time [4].

The calculation of Normalized difference vegetation index NDVI using Google Earth Engine, providing valuable understandings of vegetation cover. GEE simplifies the process of analyzing NDVI over large spatial and temporal extents. GEE allows also the analysis of NDVI time series data, which facilitates monitoring vegetation dynamics and detecting changes in vegetation patterns. This is especially useful in ecological studies, agriculture, and land management [5]. For Phenology Studies, GEE's vast repository of satellite imagery enables the extraction of phenological metrics, such as the start and end of growing seasons, which are mandatory for assessing vegetation stress under climate change [6]. The integration of land use estimation and NDVI analysis provides valuable informations for many applications, including environmental planning, resource management, and climate change studies. By combining land use data with NDVI information, researchers can assess the provision of ecosystem services, such as carbon sequestration, water regulation, and biodiversity conservation. This integrated approach helps us in identifying priority areas for conservation efforts [7].

Google Earth Engine has emerged as a valuable tool for land use estimation and NDVI analysis due to its vast collection of satellite imagery, cloud-based processing capabilities, and easy-to-use interface. By utilizing GEE's functionalities, researchers have made significant contributions in different fields, especially in environmental sciences, agriculture, land and ecosystem management. As the platform continues to evolve, it is expected to further enhance our understanding of land use dynamics and vegetation patterns, leading to more informed decision-making and sustainable practices.

In this work we aim to highlights some practical and application with GEE, examples are given from Algerian territory.

2 Material and Methods

The analysis of remote sensing images aims to interpret the nature and function of the positioned objects. The interpretation process, which is done visually (Supervised classification) or automatically via unsupervised process, consists in transforming the data of the satellite image into spatially localized information based on pixel values. Google earth engine is new software for remote sensing data processing based on advanced web technology [20]. The advantage of this software is that it uses the hardware resources of Google servers, thus giving a great power of calculation and processing of remote sensing data, also using a easy code (Java script or python).

Figures 1 and 2 show the web interface of google earth engine.

The Fig. 3 show the main tools in the user interface.

In Fig. 4 we report the general flow chart for analyzing satellite's data in GEE environment.

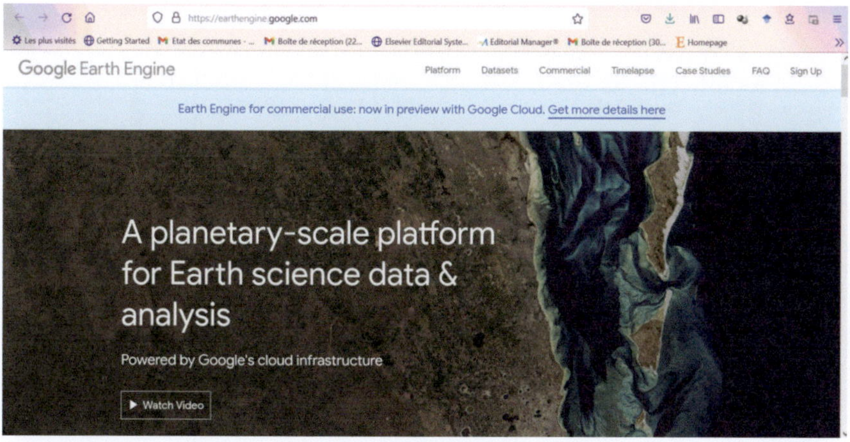

Fig. 1 Earth engine user interface: https://earthengine.google.com

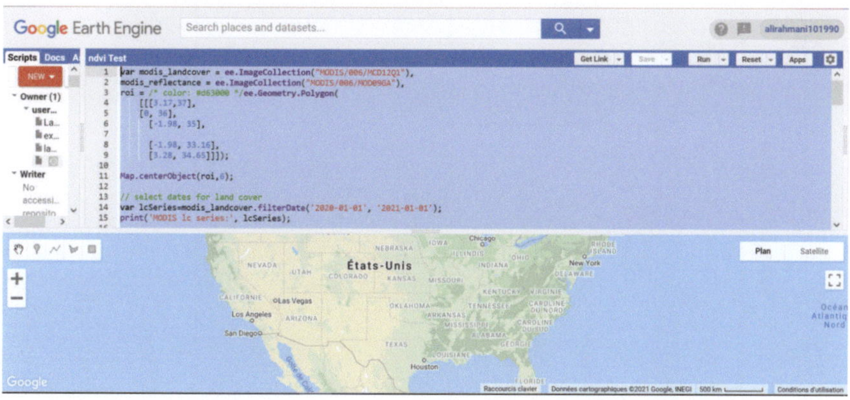

Fig. 2 User interface with code editor: https://code.earthengine.google.com/

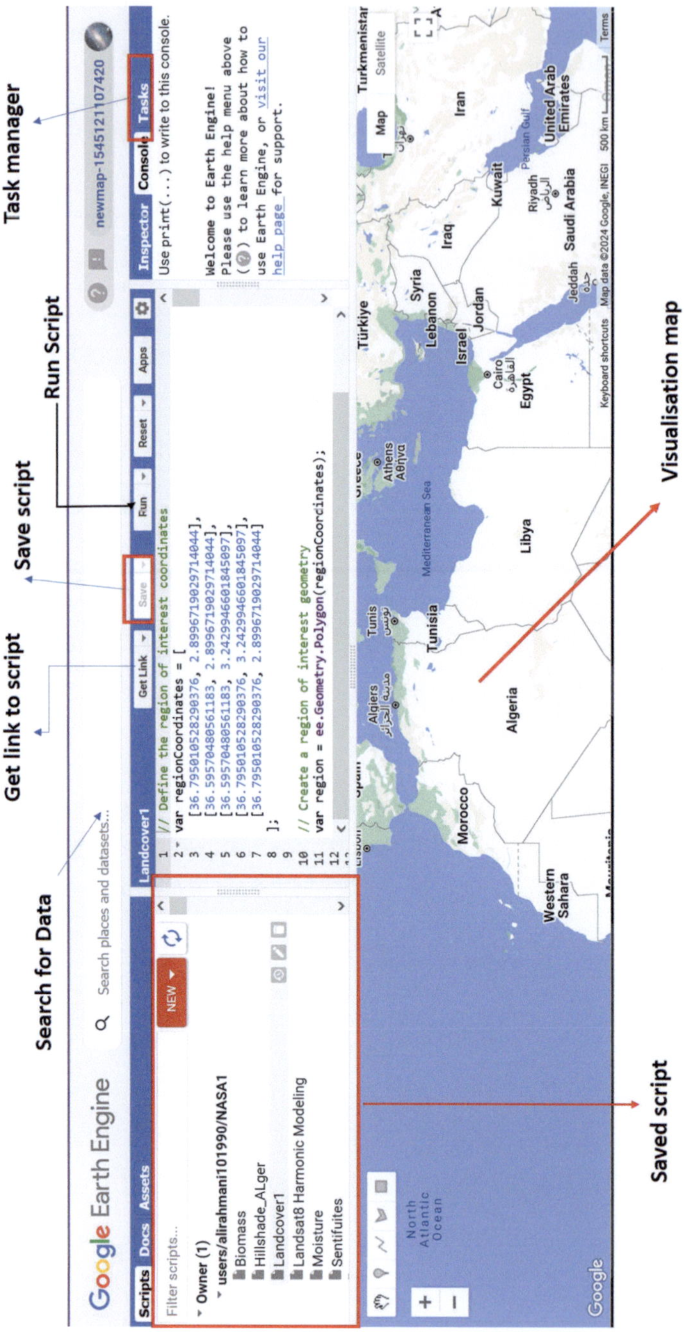

Fig. 3 User interface details (Google)

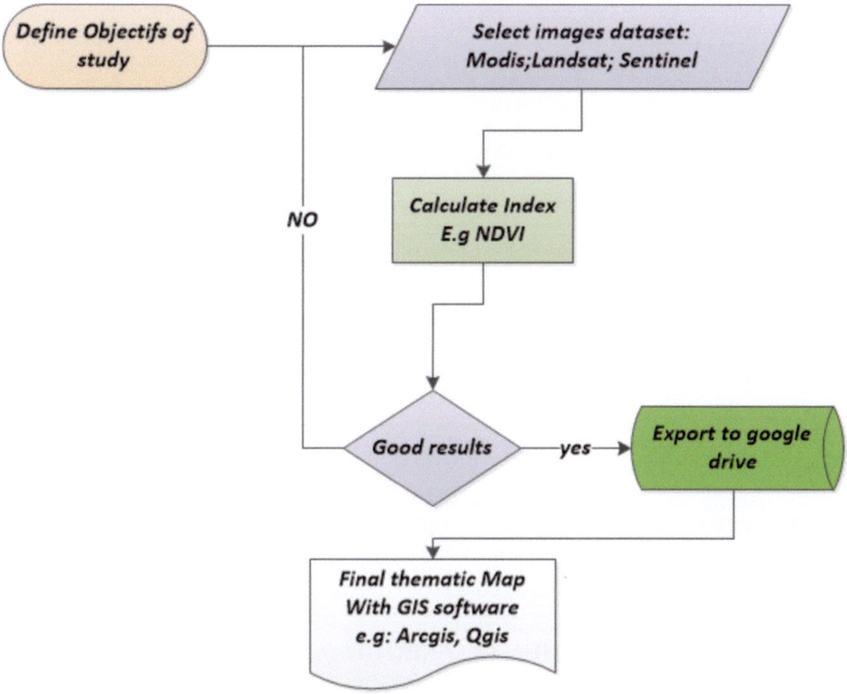

Fig. 4 Flowchart of GEE applications

2.1 Calculation of NDVI

The Normalized Difference Vegetation Index (NDVI) is a widely used remote sensing-derived vegetation index that provides valuable details about vegetation density and health [8, 9]. NDVI is calculated using near-infrared (NIR) and red (R) bands of satellite imagery [10–12]. The formula for NDVI calculation is given as follows:

$$\mathrm{NDVI} = (\mathrm{NIR} - \mathrm{R})/(\mathrm{NIR} + \mathrm{R})$$

Tucker [13] demonstrated the effectiveness of NDVI in monitoring vegetation changes over time, enabling the detection of drought stress and other environmental factors affecting vegetation health. The following code is used to calculate NDVI for different period in Algeria (Google Earth Engine [14]).

```
var img = ee.Image('MODIS/006/MOD09GA/2020_07_01');
var ndvi = img.normalizedDifference(['sur_refl_b02', 'sur_
refl_b01']);
```

```
var palette = ['FFFFFF', 'CE7E45', 'DF923D', 'F1B555',
'FCD163', '99B718',
'74A901', '66A000', '529400', '3E8601', '207401', '056201',
'004C00', '023B01', '012E01', '011D01', '011301'];
Map.setCenter(2, 36, 7);
Map.addLayer(img.select(['sur_refl_b01', 'sur_refl_b04',
'sur_refl_b03']),
{gain: [0.1, 0.1, 0.1]}, 'MODIS bands 1/4/3');
Map.addLayer(ndvi, {min: 0, max: 1, palette: palette},
'NDVI');
```

2.2 Calculation of Salinity Index

Salinity index is an important parameter to illustrate saline soils and identify saline depression (eg: sabkha, chott) [15].

The formula uses is given as follow [16]:

$$\mathrm{NDSI} = (\mathrm{R} - \mathrm{NIR})/(\mathrm{R} + \mathrm{NIR})$$

```
var dataset = ee.ImageCollection('MODIS/MOD09GA_006_NDSI')
.filter(ee.Filter.date('2018-04-01', '2018-06-01'));
var colorized = dataset.select('NDSI');
var colorizedVis = { palette: ['000,088', '0000ff', '8888ff',
'ffffff'],};
Map.setCenter(2, 36, 7);
Map.addLayer(colorized, colorizedVis, 'Colorized');
```

2.3 Hillshade Effect Calculation

Hillshade is a widely utilized technique in terrain analysis and cartography, offering a visual enhancement of the influence of illumination on digital elevation models (DEMs) or terrain data [17]. This method effectively simulates the shadows cast by the terrain when subjected to a light source, such as the sun. Hillshading allow users to better understand the characteristics of the land.

The calculation of hillshade includes the determination of the slope and aspect of each point within the DEM [18]. The slope signifies the steepness of the terrain, while the aspect presents the direction of the slope. Taking into account the position of the light source, usually the sun, the hillshade algorithm computes the level of shading or shadowing for each point based on its slope and aspect.

The hillshade image produced, generates a three-dimensional representation of the terrain, with shaded regions representing slopes facing away from the light source and illuminated areas depicting slopes facing towards the light source [19]. This visual output is valuable for terrain features analysis such as ridges, valleys, and slopes. Consequently, hillshade has many applications in diverse fields including geological studies, landform analysis, and cartographic representation in maps.

The following code allow to calculate hillshade and show results in multiples layers (Google Earth Engine [14]).

```
function radians(img) {return img.toFloat()
.multiply(Math.PI).divide(180);}
function hillshade(az, ze, slope, aspect) {
var azimuth = radians(ee.Image(az));
var zenith = radians(ee.Image(ze));
return azimuth.subtract(aspect).cos()
.multiply(slope.sin())
.multiply(zenith.sin())
.add(
zenith.cos().multiply(slope.cos()));}
var terrain = ee.Algorithms.Terrain(ee.Image('CGIAR/SRTM90_
V4'));
var slope = radians(terrain.select('slope'));
var aspect = radians(terrain.select('aspect'));
Map.SetCenter(2, 36, 7);
For (var i = 0; i < 180; i += 100) {
Map.AddLayer(Hillshade(I, 60, Slope, Aspect), {}, I + '
deg');}
```

2.4 Study Area

Algeria is located along the North African coast, bordered to the east by Tunisia and Libya, to the southeast by Niger, to the southwest by Mali, and to the west by Mauritania and Morocco. It is the largest country in Africa by total area, covering 2,381,741 km^2 with a coastline of 2.148 km [21], Algeria holds a special place in the Mediterranean basin with its geostrategic position and its demographic and economic weight.

Algeria is characterized by high climate variability across different regions and seasons [22]. The climate of Algeria is varied: the northern part has a Mediterranean climate; the rest of the country has mainly a semiarid to desert climate. However, between these two major types of climates, there is a transition zone with a semi-arid climate. The current study specifically carried out to the northen part of Algeria (Fig. 5).

3 Results and Discussion

The Fig. 6a–d, show the results of NDVI calculation for 2005, 2010, 2015 and 2020 consecutively.

NDVI calculation for huge area was performed without any expensive computer resources consumption, using google servers help us to avoid the problem of machine limit and time to download images. With GEE users can work about a large Area and long time's series.

Spatial and temporal variations in NDVI indicate that high values correspond to dense forest cover, particularly concentrated in the northeast and extreme west of Algeria. Detection of changes in NDVI values can serve as a valuable tool for monitoring fires and controlling deforestation. This capability, facilitated by Google Earth Engine (GEE), is made possible by its vast online datsets archives, which allow for quick comparison and computation. Figure 7 illustrates the loss of forest cover from 2000 to 2020.

The changes observed in NDVI can be attributed to variations in moisture levels and forest cover. The effects of drought have a significant impact on photosynthesis, consequently affecting the reflectance of infrared (IR) radiation. On the other hand, the reduction in NDVI in the study area is primarily due to an increase in forest fires, which have been increasingly prevalent in the North of Algeria in recent years.

The Fig. 8 illustrate the results of NDSI for the study area. The results demonstrate the existence of saline depression in color white.

The Normalized Difference salinity Index (NDSI) give the contrasting reflectance properties of red and near-infrared bands to estimate soil salinity. GEE makes calculating NDSI simple and quick. With a few lines of code, we can generate NDSI maps that reveal spatial patterns in soil salinity. This code can be easily adapted to specific study areas and user preferences, making GEE a versatile tool for soil scientists and environmental researchers.

The obtained results are compared to observed data. Circle 1 represents the Oran Sebkha, Circle 2 represents the Zahrez Chott, Circle 3 represents the Hodna Chott, and Circle 4 indicates the Melhrir Chott. This valuable tool enables users to identify and study salinity in the area.

The Fig. 9 show the Hillshaed effect for SRTM digital terrain model.

Fig. 5 Study area

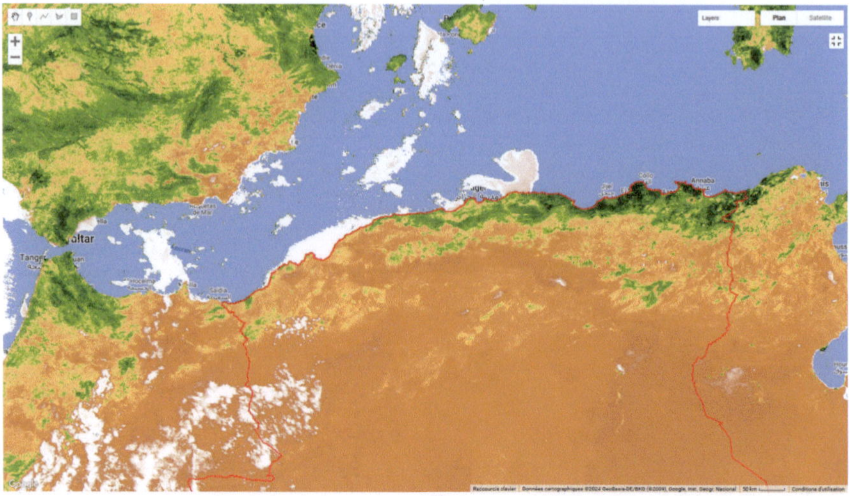

a

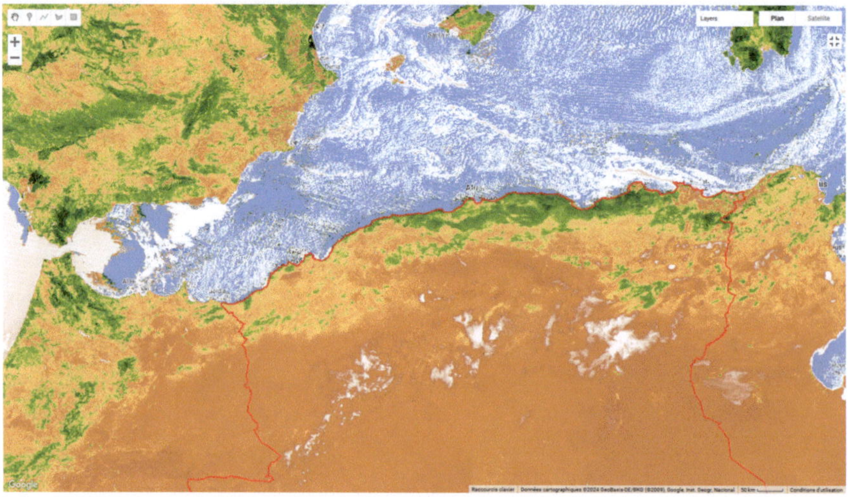

b

Fig. 6 **a** NDVI for the period of 2000_07_01. **b** NDVI for the period of 2005_07_01. **c** NDVI for the period of 2010_07_01. **d** NDVI for the period of 2015_07_01. **e** NDVI for the period of 2020_07_01

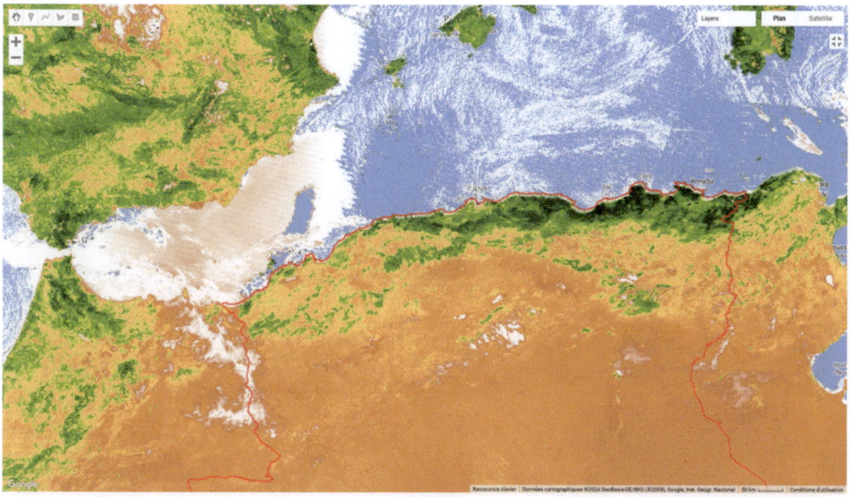

c

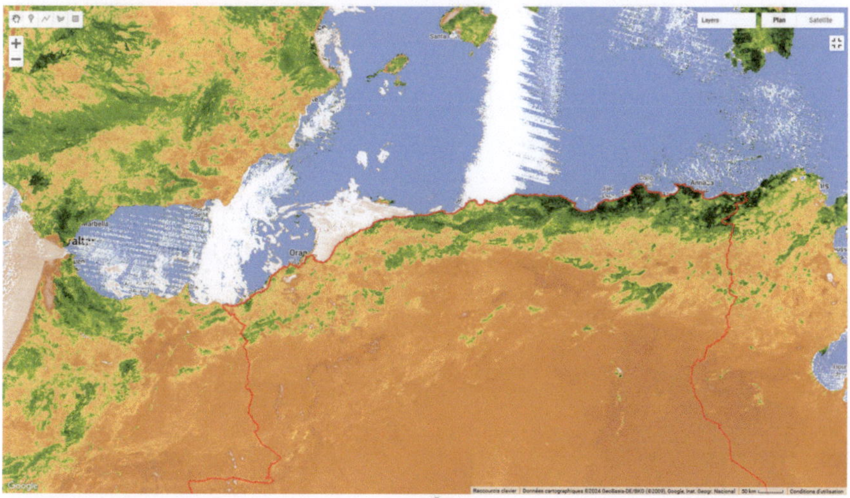

d

Fig. 6 (continued)

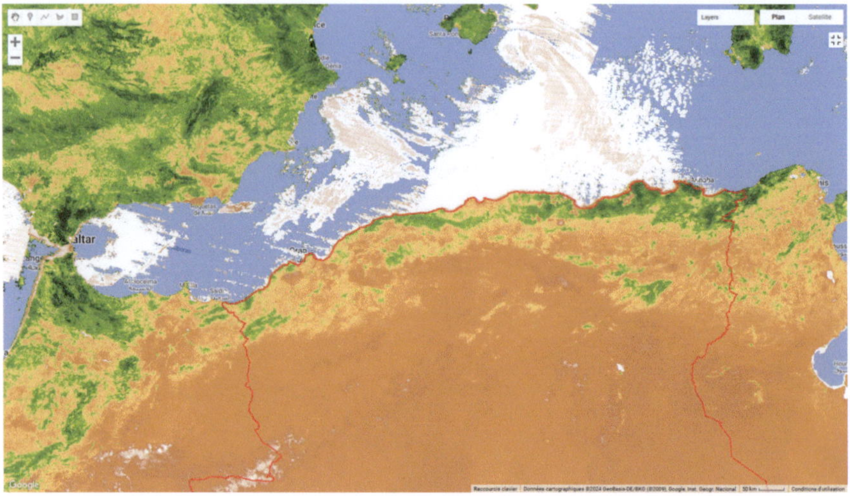

e

Fig. 6 (continued)

Fig. 7 Forest lost from 2000 and 2020

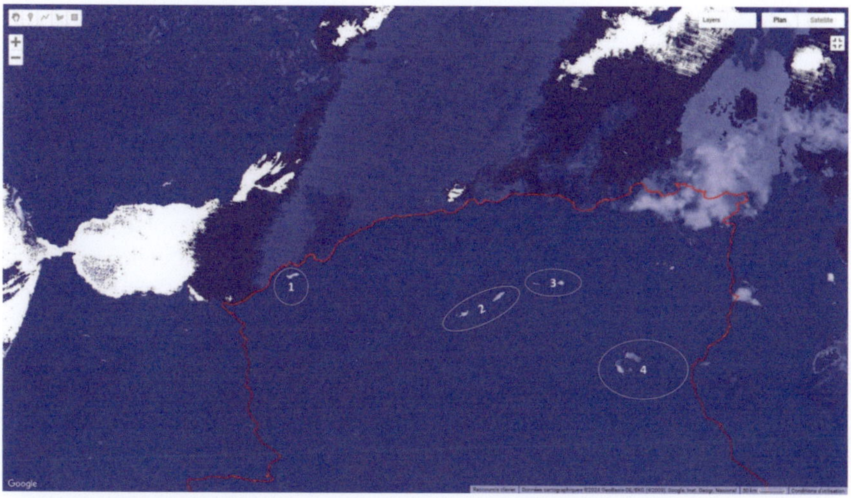

Fig. 8 NDSI results

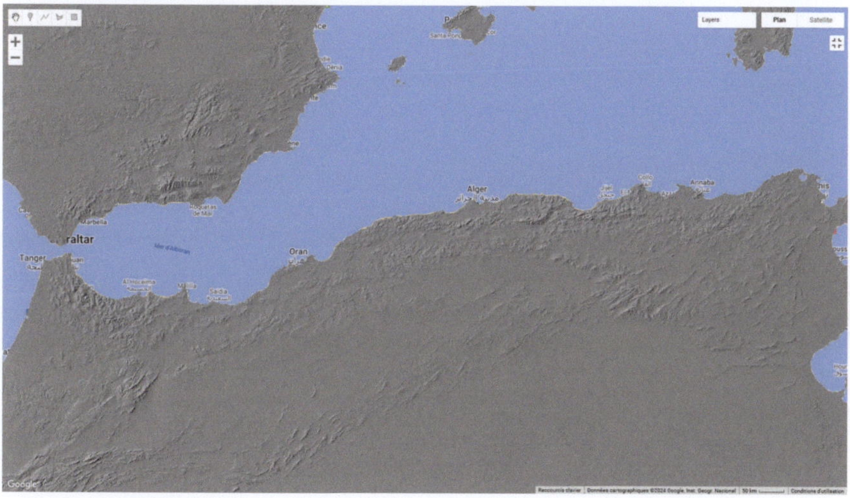

Fig. 9 Hillshade effect 60°

4 Conclusion

Google Earth Engine (GEE) is a powerful cloud-based platform for Earth observation data management and analysing. Its large data archive, combined with scalable processing power, an interactive geospatial analysis environment, and seamless integration with machine learning frameworks, make it an invaluable tool for revealing valuable insights from our planet. As GEE continues to evolve, its artificial intelligence capabilities hold the promise of further advancing our understanding of Earth and addressing complex environmental challenges. This makes GEE an even stronger tool for GIS and geomatics professionals around the world to solve environmental problems. The findings from the case study reveal that variations in NDVI reflect spatial and temporal changes attributed to drought and forest wildfires. Moreover, the NDSI calculations can effectively identify saline depressions in the North of Algeria, serving as a valuable tool for assessing soil salinity. By combining other capabilities of GEE, it can be used as a decision support tool to help stakeholders in different situations (e.g., wildfire assessment, flood sensitivity mapping, drought and water stress prediction, soil management).

References

1. Zhao Q, Yu L, Li X, Peng D, Zhang Y, Gong P (2021) Progress and trends in the application of Google Earth and Google Earth Engine. Remote Sens 13(18):3778
2. Mutanga O, Kumar L (2019) Google earth engine applications. Remote Sens 11(5):591
3. Chen J, Chen J, Liao A, Cao X, Chen L, Chen X (2017) A comparative analysis of classification algorithms for land-use and land-cover mapping using Google Earth Engine. Remote Sens 9(5):491
4. Ardila JP, Hernández DE, Pardo NA (2020) Land cover change detection using Google Earth Engine and earth observation data: a case study in the Colombian Amazon. Remote Sens 12(11):1780
5. Shi L, Wang J, Wang J, Wang F, Liang L, Shi T (2016) Monitoring vegetation phenology dynamics in response to climate change using MODIS NDVI time series. Int J Digital Earth 9(2):185–200
6. Zhao Y, Zhang J, Li Y, Cheng J, Yang H, Yang X (2019) Mapping the start of the growing season using NDVI time series from MODIS and Earth Observation Satellite (EOS) sensors in the Qilian Mountains, China. Remote Sens 11(20):2420
7. Gao Y, Li Y, Yang X, Jia Q, Liu W, Zhang Y, Ma W (2020) Mapping ecosystem services values and identifying priority areas for ecological conservation in the Qinghai-Tibet Plateau, China. Ecol Indic 109:105782
8. Huang S, Tang L, Hupy JP, Wang Y, Shao G (2021) A commentary review on the use of normalized difference vegetation index (NDVI) in the era of popular remote sensing. J For Res 32(1):1–6
9. Chen J, Jönsson P, Tamura M, Gu Z, Matsushita B, Eklundh L (2004) A simple method for reconstructing a high-quality NDVI time-series data set based on the Savitzky-Golay filter. Remote Sens Environ 91(3–4):332–344
10. Holben BN (1986) Characteristics of maximum-value composite images from temporal AVHRR data. Int J Remote Sens 7(11):1417–1434
11. Crippen RE (1990) Calculating the vegetation index faster. Remote Sens Environ 34(1):71–73

12. Nemani RR, Running SW (1989) Estimation of regional surface resistance to evapotranspiration from NDVI and thermal-IR AVHRR data. J Appl Meteorol Climatol 28(4):276–284
13. Tucker CJ (1979) Red and photographic infrared linear combinations for monitoring vegetation. Remote Sens Environ 8(2):127–150
14. Google Earth Engine Developers (2021) Google Earth Engine API documentation. https://developers.google.com/earth-engine
15. Li W, Liu J, Bao N, Mao X, Mao Y, Fu Y, Zhao Z (2021) Salinity monitoring at saline sites with visible–near-infrared spectral data. Minerals 11(10):1086
16. Venugopal V, Kumar DM, Selvaraj V, Prasanna Kumar SC (2023) Analysis of salinity indices using SVM based approach of Ballari Town, India. Math Model Eng Probl 10(5)
17. Evans IS (1980) An integrated system of terrain analysis and slope mapping. Z Geomorphol 29(1):15–42
18. Moore ID, Grayson RB, Ladson AR (1991) Digital terrain modelling: a review of hydrological, geomorphological, and biological applications. Hydrol Process 5(1):3–30
19. Wood J (1996) The geomorphological characterisation of digital elevation models. PhD thesis, University of Leicester
20. Gorelick N, Hancher M, Dixon M, Ilyushchenko S, Thau D, Moore R (2017) Google earth engine: planetary-scale geospatial analysis for everyone. Remote Sens Environ 202:18–27
21. https://www.agenzianova.com/en/news/a-study-reveals-the-real-length-of-the-Algerian-coast/
22. Sahnoune F, Belhamel M, Zelmat M, Kerbachi R (2013) Climate change in Algeria: vulnerability and strategy of mitigation and adaptation. Energy Procedia 36:1286–1294. https://doi.org/10.1016/j.egypro.2013.07.145

Coastal Eutrophication Detection and Assessement Using Sentinel-2 MSI Imagery

Nabila Allouche, Ghada Abdelkarim, Faiza Khebour Allouche, and Ali Harzallah

Abstract The Monastir region constitutes an invaluable reserve in terms of continental and marine biodiversity. However, the water bodies in Tunisia, such as bays, sebkhas and coasts have become the seat of recurrent appearance of extreme phenomena such as red tides or "Harmful Algal Blooms" (HABs), which are generally manifested by the eutrophication of the waters and the proliferation of algae or harmful bacteria. It is in this context we seek to identify and evaluate the phenomenon of eutrophication at the level of water bodies in the Monastir region. Our methodological approach uses new computer technologies, in particular data science, including sophisticated GIS and remote sensing tools. Results show the identification of four main water bodies using the NDWI index namely (i) the lake/sebkha of Monastir on the north-eastern side, (ii) the Sebkha of Moknine on the south-eastern side, (iii) the Sebkha of Sidi El Heni to the west and (iv) the body of water relating to the sea. Also, four degrees of eutrophication classified from very low to very high are determined based on the NDCI index. However, we found that the concentration of the eutrophication is more accentuated on the coast of Sayada and Teboulba, the bay of Dhkila and at the Sahline' Sebkha level. We have deduced that anthropogenic actions such as the realization of processes and the use of chemicals in agriculture have a considerable effect on this phenomenon.

N. Allouche (✉)
Water, Energy and Environment Laboratory (LR3E), National School of the Engineers, Sfax, Tunisia
e-mail: nabila.allouche@gmail.com

G. Abdelkarim · F. K. Allouche
Department of Engineering and Landscape Planning, High Institute of Agronomic Science of Chott Mariem, Sousse University, Chott Meriem Sousse, Tunisia

F. K. Allouche
Lr GREEN TEAM (LR17AGR01), Carthage University, National Agronomic Institute of Tunis, Tunis Mahrajène, Tunisia

A. Harzallah
INSTM Tunis, Marine, Environment Laboratory, University of Carthage, Carthage, Tunisia

F. Khebour Allouche (ed.), *Environmental Remote Sensing and GIS in MENA Region*, Earth and Environmental Sciences Library,
https://doi.org/10.1007/978-3-031-86574-9_6

Keywords Eutrophication · Environment · NDWI · NDCI · Anthropogenic · Monastir

1 Introduction

Although it constitutes an important reserve in terms of continental and marine biodiversity, the Monastir region is also one of the Tunisian regions that suffer from several environmental problems, and urgently require the implementation of protection and rehabilitation plans [3]. Indeed, this environmental problem is part of a larger framework that concerns the protection and management of biodiversity throughout the Mediterranean. For example, the bodies of water in Tunisia, such as bays, sebkhas and coasts have become the seat of recurrent appearance of extreme phenomena such as the red tides or "Harmful Algal Blooms" (HABs) which are generally manifested by the eutrophication of the waters and the proliferation of algae or harmful bacteria.

Environmental mapping is used to borrow some already mapped data (surface formations, geomorphology, land use … etc.) and to retain new elements including the dynamics of everything related to soil, air and water. It must, therefore, help to understand and judge the phenomena of correlation between man and nature and the potentialities of the areas studied for optimal use of the sites, but also be attentive to the trend of evolution of the natural components of the environment. The practical and specific objective of this study is to identify and evaluate the phenomenon of eutrophication by applying the concepts of mapping the framework and environmental data as well as their dynamics and evolution.

Tunisia is a North African country bordered to the north and east by the Mediterranean Sea which extends from the Atlantic Ocean in the west to the Asian continent in the east and separates Europe from the Maghreb region in Africa. The basin covers an area of up to 2.6 million km^2 with an average depth of 1460 m and a maximum depth of 5267 m, making the Mediterranean Sea the largest closed sea in the world [15].

The extension of the Atlas in Tunisia makes the western and northern regions mountainous. As an immediate consequence, the existence of several rivers called wadis which are mostly seasonal and periodically have low water levels or drying up after the rainfall. There are also some that flow all year round such as the wadis Medjerda, Meliane and R'Mel. The flow, the slope, the geological and morphological natures of the terrain, make these rivers flow into the Mediterranean Sea or into the depressions forming lakes.

A water body refers to anybody of running or stagnant water, whether created naturally (natural origin) or artificially (anthropogenic origin), and whether the flow or presence of water is continuous, intermittent or occurs only during a flood (Manitoba Department of Conservation 2009). Thus, there are several types of water bodies, namely fresh, brackish or salt water, which can be in motion or continuous. Generally, the size of these bodies of water distinguishes them from the others, as do their borders. In some cases, the vegetation that flourishes there differentiates them from each other [19].

According to [58], environmental risks including those affecting water bodies are numerous such as marine litter, overfishing and bycatch, distraction of habitats and decline of marine species, climate change, soil disturbance, and eutrophication.

The eutrophication of water bodies is a succession of biological processes triggered in response to an excessive supply of nutrients. It results in complex responses from all fresh, brackish or salty water bodies. It can be gradual or abrupt. The most notable effects are proliferations of primary producers, toxicity or anoxia phenomena (absence of oxygen) and/or biodiversity losses. It is a very localized environmental problem whose controlling factors are not only operate on a local scale and in a punctual basis. For example, nutrients arriving in coastal areas may come from upstream watersheds that are sometimes several hundred kilometers away, before being transported and diluted at sea for hundreds of kilometers. Eutrophication processes apply to all water bodies, although they have varied expressions and dynamics depending on the different types of fresh, brackish or salt water [52].

However, phytoplankton blooms are a common phenomenon that leads to eutrophication of water bodies, and there are several types: some are considered to be important contributors to ecosystem dynamics and marine primary production, while others are classified as harmful algal blooms (HABs) or 'red tides' or 'algal blooms' because of their production of powerful toxins. The word "bloom" must, therefore, be understood with caution because of the presence of species that present harmful or toxic effects even at low concentration levels [33].

So, as already mentioned, eutrophication is a complicated process caused by nutrient enrichment. It has been identified as a high-risk anthropogenic process in many coastal and marine waters for decades. The excess of nutrients, especially nitrogen and phosphorus, became a problem with the start of large-scale production and use of artificial fertilizers after the Second World War, and the effects on the aquatic environment became evident some two to three decades later [10, 53]. A major consequence of eutrophication is the development of algal blooms [28]. This can lead to harmful algal blooms (HABs) or even a regime changes from macrophyte-dominated to phytoplankton-dominated ecosystems in enclosed water bodies [26]. These harmful algae can develop asymmetrically as a result of eutrophication. Dinoflagellate species are the most abundant (HABs) species and produce the majority of toxins in the Mediterranean Sea [4, 24, 30]. Blooms of pelagic microalgae that discolor the water are traditionally known as red tides, which can be toxic or non-toxic. More specifically, in the Mediterranean, some blooms can be caused by algal species that produce toxins (e.g. Ostreopsis ovata) while others cause non-toxic blooms with high biomass e.g. *Alexandrium taylorii* [40]. For example, in the Gulf of Gabès, harmful microalgae were dominated by *Prorocentrum lima* (Ehrenberg) [21] and *Ostreopsis siamensis* [56], which are benthic or epiphytic species.

In coastal waters, this phenomenon has become more frequent over the last ten years on a global scale [29]. It can have serious effects on the marine ecosystem and human health, as well as negative economic consequences affecting fisheries and pleasure areas [41]. It can happen when the sea is luminescent or simply shiny. In the

Mediterranean, numerous cases have been reported for decades, particularly in the coastal waters of France, Italy, Yugoslavia, Greece and Tunisia. Depending on the case, the phenomenon is attributed to planktonic and/or bacterial manifestations [5]. In Tunisia, the "red water" phenomenon was detected in 1951 at Lake Tunis, which was the outlet for the surrounding urban areas [34].

The Organisation for Economic Co-operation and Development (OECD) defines the phenomenon of water eutrophication as the enrichment of water with nutrients, resulting in symptomatic changes such as increased production of macrophytes and algae, deterioration in water quality and other transmutations considered harmful and undesirable to the various uses of water [27]. This problem affects all types of water bodies, such as freshwater, which is most affected because it is so close to anthropogenic sources of phosphates and nitrates, brackish and saltwater (marine or inland water bodies) and deep or shallow marine environments. Eutrophication mainly affects standing water (such as ponds rich in dead leaves or collecting wastewater or water polluted by fertilisers), lakes, sebkhas and lagoons [17]. It also affects rivers with low flows or which receive large quantities of effluent from various agricultural, human or industrial operations, bays and other semi-enclosed areas. It is important to note that this environmental problem is largely due to the fact that these bodies of water are located downstream of watersheds that drain large quantities of nutrients, essentially when runoff is supported by human actions such as drainage, plowing, deforestation, waterproofing and many other activities (UNEP, 2000). Phosphorus, for example, is considered to be the limiting factor in the development of plants living in fresh continental surface waters. However, the concentration of nitrogen is less limiting in fluvial or lacustrine waters, and its concentration also remains higher in oligotrophic environments.

Eutrophication is considered synonymous with pollution, according to [62]. However, this phenomenon can have other aspects, such as biological, chemical (pesticides, metals, solvents, etc.) or physical contamination. For example, fishing in closed water bodies or very slow-moving watercourses (canals) is a cause of eutrophication or even water anoxia.

Recent developments in sensor technology have transformed remote sensing into an effective means of monitoring marine water bodies. Different remote sensing platforms and sensors have their own capacity to map and monitor water pollution of different types, characteristics and concentrations. There are remote sensors that capture the response of electromagnetic interaction with water [31].

This is because absorption and scattering are inherent optical properties (IOPs) of water, and variations in these properties change the reflectance of the water that is captured. O'Reilly et al. [50] pointed out that optical remote sensing makes it possible to observe water coloration due to algal pigments. In the deep sea, watercolor is mainly determined by phytoplankton; it is therefore relatively simple to develop algorithms using a bio-optical approach and remote sensing reflectance. Recently, ocean color remote sensing techniques have been widely used to monitor spatio-temporal variations in TSS (Total suspended solids) concentration and to map water turbidity. Thus, changes in ocean colour due to high sediment loads in the Yangtze River estuary [47] and the Pearl River estuary [16] have been detected (Fig. 25). It has been suggested that an algorithm using single bands provides a good estimate of TSS concentrations if an appropriate band is used [48]. In addition, [49] has demonstrated that a single-band approach can be adopted when the reflectance of water in the single band has a linear relationship with TSS concentrations. However, coastal waters are often a complex mixture of substances, resulting in wide variations in reflectance. In this case, several spectral bands must be adopted for the recovery of the TSS [22, 25, 48]. Karydis [37] noted that a good indicator for the evaluation of eutrophication can provide information on: (i) the detection of trends that cannot be easily observed from raw data, (ii) its use as an early warning system in decision-making when management practices are applied and (iii) the assessment of the degree of severity or sanitation in areas with eutrophication problems. The majority of recent assessments of eutrophication are based on traditional measurements based on punctual in-situ sampling and analysis of samples in the laboratory. As well as being more expensive and time-consuming than remote sensing, this method also lacks sufficient spatial and temporal resolution to adequately analyze and describe eutrophication phenomena. Research has therefore been carried out into the application of hyper-spectral sensors with high spectral and spatial resolution for monitoring eutrophication indicators in coastal and inland water bodies which, because of their biological and optical complexity compared with the open sea, represent a challenge for remote sensing. In addition, high and statistically significant empirical correlations were found between hyperspectral data and values of chlorophyll A (Chl-a), total suspended solids and turbidity measured in-situ.

Algorithms were developed to obtain concentrations of water quality parameters in water bodies from aerial images acquired (e.g. using the AISA Eagle sensor) and also from hyperspectral satellite images (e.g. using the Hyperion sensor). The best algorithm for Chl-a recovery was based on band ratios R498/R518 in coastal waters and R745/R418 in inland waters.

High correlations were detected between hyperspectral field data and nitrate and total phosphorus values in coastal water bodies, and nitrogen and total oxygen values in inland waters. This indicates the existence of optically visible constituents in the water that serve as indicators for the correlations established. The possibility of using remote sensing of invasive species as an additional indicator of eutrophication has also been examined [13, 35].

Caballero et al. [11] have shown that the Sentinel-2 twin satellite mission of the Copernicus program, in combination with in situ data, is a powerful tool that can offer valuable spatio-temporal information on a bloom of the dinoflagellate Lingulodinium polyedra that occurred in the southwest of the Iberian Peninsula. Using the robust ACOLITE atmospheric correction processor combined with the Normalized Difference Chlorophyll Index (NDCI), advanced mapping of small algal blooms can be performed at 10 m spatial resolution, revealing surface spots and heterogeneous distribution. Mishra et al. [45] propose a normalized chlorophyll difference index (NDCI) to predict the concentration of chlorophyll-a (chl-a) from remote sensing data in turbid estuarine and productive coastal waters. The results of the calibration and validation of the NDCI derived from simulated data sets and the Medium Resolution Imaging Spectrometer (MERIS) show its potential application to a wide variety of water types and climatic conditions. A quadratic function ($R^2 = 0.95$, $p < 0.0001$) accurately explained the variance in the simulated data for a chl-a range of 1 to 60 mg m^{-3}. In the case of remote coastal waters where ground truth data are not available, NDCI can be used to detect algal blooms and qualitatively deduce chl-a concentration ranges in a manner very similar to the application of NDVI (Normalized difference vegetation index) in terrestrial vegetation studies [45].

According to [8], these images are then spectrally and spatially scaled to simulate five current and future satellite image systems. The established and new Chl-a algorithm was then applied to the synthetic satellite image and compared to coherent Chl-a surface observations collected at 44 locations within one hour of image acquisition by the aircraft. They found several combinations of algorithm and satellite imager promising for routine estimation of Chl-a in small inland water bodies with satellite systems in operation and in the near future. The CI, MCI, FLH, NDCI, 2BDA and 3 BDA Chl-a algorithms work well in CASI imagery. The NDCI, 2BDA and 3BDA Chl-a algorithms work well on WorldView-2 and 3, Sentinel-2 and MERIS analogue imagery. NDCI is the most widely used Chl-a algorithm and performs well for CASI, WorldView 2 and 3, Sentinel-2 and MERIS imagery, but has limited performance for MODIS imagery. A new "green" fluorescent line height algorithm produces the best estimate of Chl-a using simulated Landsat-8 imagery.

2 Study Area

The northern part of Monastir Bay, located on Tunisia's eastern coastline, was chosen because it shows signs of eutrophication, with spectacular algal growth beginning during the winter season (green tide) and leading to chronic anoxia [55].

The Monastir region (Fig. 1) covers 1,024 km^2 and contains 13 delegations. More specifically, Monastir Bay (35.1° N–10.4° E and 36.6° N–11.8° E) belongs to the central part of the east coast of the Tunisian littoral [39]. It is bounded to the south-east by the Teboulba shoals, which continue as far as the Kuriat islands, and to the north-west by the Monastir peninsula [39]. The configuration of the coastline in the Monastir region is characterized by the existence of a cape, also known as the Monastir peninsula, which is sandwiched between two bays (Dkhila Bay to the east and Khniss Bay to the west). In the Monastir plain, there are large expanses of stagnant water with very high salt concentrations. Indeed, this plain is largely dominated by the sebkha and Sahline lake. It is separated from the Mediterranean Sea by the narrow coastline of Dkhila Bay and extends south-westwards into the Jammel corridor. The geography of the study area shows a slight variation in the topography. Elevation varies from 0 to 45 m above sea level [57].

According to [23], on a regional scale, rainfall is irregular and brutal with an average of 300 mm over six years over the period 2009–2014 and with strong inter-seasonal and inter-annual variability. The average annual temperature recorded in Monastir Bay is around 20 °C, with daily temperatures ranging from 10 to 13 °C in January, the coldest month, and 36 to 38 °C in August, the hottest month. Monastir

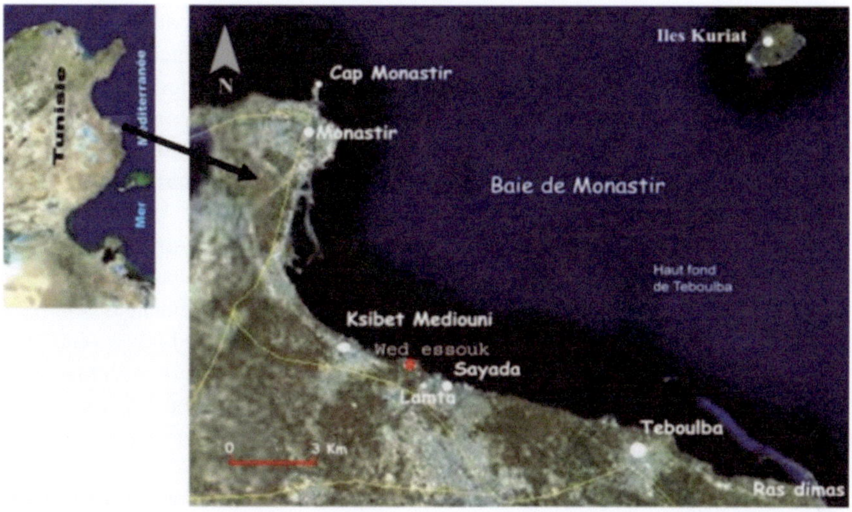

Fig. 1 Location of the study area, Tunisia

Bay is characterized by a wind speed between 1 and 5 ms^{-1} dominated in three main directions, namely, W, NNE, and E with an occurrence of 10.7%, 8.5%, and 7.6%, respectively [39, 57].

In the study area there are different types of soil: mineral soils, soils with little evolution, rendzina, calcareous brown soils, isohumic soils, brunified soils, hydromorphic soils and urban areas. However, the majority of the Monastir region is occupied by poorly evolved rendzina, followed by brunified soils, isohumic soils and calcareous brown soils, while hydromorphic soils characterize water bodies. According to [7], isohumic soils are the most widespread in irrigated areas (80%). Their depth and balanced texture make them ideal for irrigation. However, rendzina are very shallow, very calcareous and stony. Brown calcareous soils are not very stony, resting on a limestone slab or crust, and have an easily usable water reserve. Badraoui [6] added that poorly evolved soils are characterized by low weathering with the result of alluvial-colluvial erosion and formed on recent alluvium from the Atlas of Béni-Mellal, generally deep and suitable for agriculture.

Nevertheless, hydromorphic soils are deep, not very gravelly soils resting on a calcareous crust and located in basin or lowland areas, which results in hydromorphic and an accumulation of salts. Salinity is well represented in the upper horizons.

The study area includes eight hill lakes of 413,000 m^3 of water and dams of 170,000 m^3 of water as well as 54 recharge units. Most of the wadis are unable to connect with the sea. They flow into the sebkhas that multiply along the coast like Sebkhat Sidi Khalifa, Sebkhat Moknine, Sebkhat Halk Menjel, Sebkhat Sahline [3, 12]. The Monastir area contains the sebkhas of Moknine Est, Sidi El Hani to the south-west of Moknine and Garaat oued El Meleh in Monastir. These water bodies are the outlets for surface water inflows estimated at 13 million m^3. In addition, from north to south, there is the sebkha of Chérita, oriented N 50, and the sebkha of El Guerra, oriented N 60. The bottoms of these sebkhas are situated at altitudes of 50 and 65 m respectively. The Sahline sebkha is a natural hydrographic outlet. Its topographical position enables it to receive water. Streams born on the slope flow into the sebkha. The Khniss Bay coastline is low and has a lagoon. It is separated from the sea by a sand spit of coastline known as El Enf Island. It has been converted into an aquaculture station according to [2].

The Kheniss lagoon measures 3 km by around 500 m, with a total surface area of 170 ha and a depth of 1 m. It is separated from the sea by a largely artificial sandy-mud barrier. The macroflora is not very varied, and the macrofauna is relatively small, comprising around fifty species. Most of these species are highly tolerant of anthropogenic and environmental pressures [9]. This area is also the receptacle for urban wastewater discharged from neighboring towns, either via outfalls that empty directly into the sea along the shoreline, or by the Khnis drain and the aquaculture station.

The bay receives both wastewaters and rainwater. Wastewaters are discharged through two sea outfalls, from the treatment plants of Lamta-Sayada and El Frina [18, 57]. In addition, [23] found that it was at the mouth of Oued Essouk, at the bank of the town of Ksibet Mediouni and at the mouth of the Sidi Abdesslam rainwater drain in Sayada that the maximum values for heavy metal concentrations in the surface sediment and nutrient levels in the water in dissolved and organic form were recorded. These results may be due to the urban or industrial waste discharged into these areas, mainly from the ONAS purification and pumping stations, which treat wastewater from neighbouring towns. It is also important to note the high levels of nitrogen in the Sidi Abdesslam drain in Sayada, even though it is only designed for rainwater.

3 Methodology

The main objective is to assess the phenomenon of eutrophication in the Monastir region. The specific objectives include: (i) the study of water bodies; (ii) characterization of eutrophication using radiometric indexes and (iii) monitoring this phenomenon according to different factors. The methodological approach is based on the use of new technology in digital computing and remote sensing tools associated with GIS.

The dates on which eutrophication appeared on the Tunisian coast were collected at the INSTM in March 2021. This information is often accompanied by a description of the effects caused. However, some episodes were collected from a bibliographical review of private documents (INSTM Archive). The dates are validated on the basis of this archive. In total, we obtained a list containing 100 dates, from 1988 to 2021, covering the Tunisian coastline from Mahdia (Rejiche) to Djerba/Zarzis, and three zones were defined:

- The first zone includes the sites of Salakta, Chebba, Rejiche, Centre de Monastir, Sayada and Ksibet el Madiouni.
- The second zone covers the centre of Sfax, Sidi Mansour, the north of Sfax, Kerkennah, the commercial port of Sfax, Cercina, Ellouza, the south of Sfax, Mahres, Chaffar, Karboub, Casino 1 and 2, Hay Hamza and the coast of Sfax.
- The third zone includes the Centre of Gabes, Ghannouch, Boughrara, Gulf of Gabes, Centre of Zarzis, Djerba, Skhira, Ajim, North-West of Djerba, Elhicha and el akarit, Corniche of Gabes, Zarrat, Port of Gabes and Jaboussa.

The digital data were collected from the database of the agricultural map of the governorate of Monastir, an agricultural map produced in a GIS database. Only the vector layers useful for this study were collected, namely the boundary of the governorate, the soil, land use, soil depth and processes (Table 1).

Figure 2 shows a summary of the various GIS and remote sensing projects. The database created consists of vector layers of the agricultural map of Monastir and a time series of very high resolution images.

Table 1 Data collected from the Monastir agricultural map database

Layer name	Information
Adm_so32	Administrative divisions, areas and names of delegations
Tag_so32	Type of land use
Pédo	Soil types
Bv_so32	Sebkha
Profondeur	Depth of soil
Procédés	Types of process

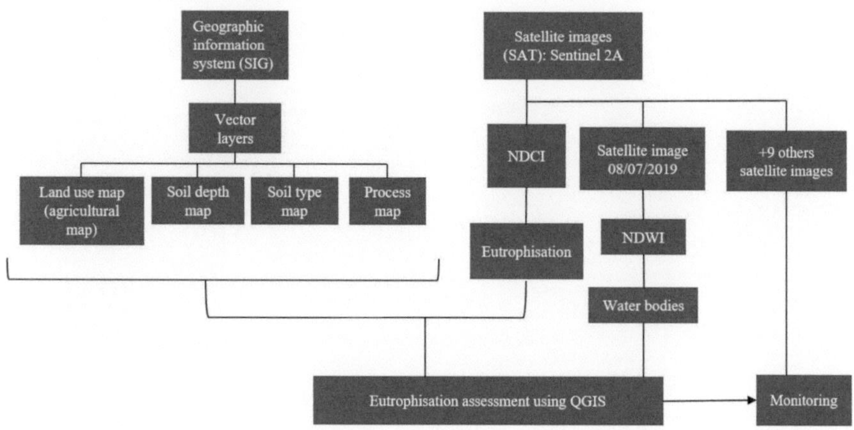

Fig. 2 Steps applied in the methodological approach using GIS and remote sensing

Satellite remote sensing, covering large and remote areas, is considered useful for detecting and monitoring marine pollution. Recent developments in sensor technology have transformed remote sensing into an effective means of monitoring marine areas. An initial Sentinel-2A (MSI) satellite image dated 08/07/2019 was downloaded free of charge from the site (Fig. 3). Then, to monitor the phenomenon, 10 Sentinel-2A corrected images were also downloaded at a rate of one image per month: 2019 (25/05/2019); (10/06/2019); (10/07/2019); (04/08/2019) and (09/08/2019) and 2020 (19/02/2020); (09/06/2020); (09/07/2020); (29/07/2020) and (28/08/2020). These images were filtered according to cloud cover (0 to 10%), acquisition duration and date (by year and month).

Sentinel-2 is a multispectral imager (MSI). It has 13 channels, the spectral range is from 0.4 to 2.4 μm, and the spatial resolutions are 10, 20 and 60 m [11]. The Sentinel-2 data product consists of level 1C (L1C) atmospheric top reflectance (TOA) data processed by geometric correction. The download address of the data source is https://scihub.copernicus.eu/dhus/#/home. The SENTINEL2A image contains 13 spectral bands in the visible spectrum, near-infrared (VNIR) and short-wave infrared (SWIR) wavelengths.

Fig. 3 Extract from a Sentinel 2A satellite image of 08/07/2019

Figure 4 shows the different stages in the GIS and remote sensing section. The normalized difference water index (NDWI) is used to monitor changes related to the water content in water bodies. As bodies of water absorb a lot of light in the visible-to-infrared electromagnetic spectrum, the NDWI uses green and near-infrared bands to highlight bodies of water. It is sensitive to built-up areas and can lead to an overestimation of water bodies. The NDWI was first proposed by McFeeters in 1996 to identify non-urban surface water associated with wetlands [43]. It has been successfully used with other sensor systems in applications where measurement of open water extent is required [14, 61].

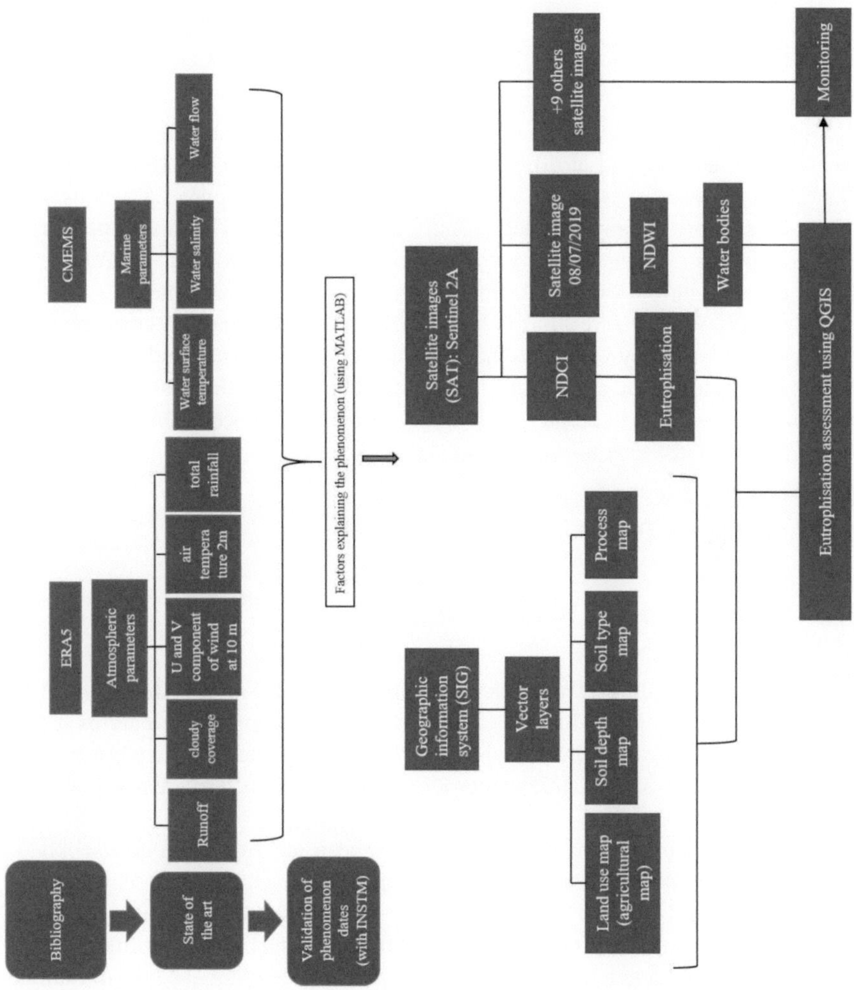

Fig. 4 Diagram explaining the methodological approach adopted

It can effectively highlight information about water bodies and eliminate soil and terrestrial vegetation features but misses built-up areas [44, 59].

The NDWI is calculated as follows:

$$NDWI = B08A - B04/B08A + B04 \quad (1)$$

According to [8], the normalized difference chlorophyll index (NDCI) is the most widely used algorithm and performs well for Sentinel-2 imagery. It aims to detect algal blooms and qualitatively deduce ranges of chlorophyll A concentrations and is very similar to the application of NDVI in terrestrial vegetation studies.

The NDCI index can be obtained according to the following equation [63]:

$$NDCI = B05 - B04/B05 + B04 \quad (2)$$

Once these indices have been calculated, the water bodies and the eutrophication phenomenon are mapped at different spatial scales, firstly in the Monastir region, then along the coast and in the Sebkhas. The thresholding approach was used for both indices. This is a simple image banarization technique. This image processing method consists of converting a greyscale image into an image in which the pixel values can only have a value of 1 or 0. This is known as a binary image.

For the NDWI, the threshold is 0.3; any value greater than 0.3 corresponds to bodies of water and any value less than or equal to 0.3 represents land. However, the NDCI' threshold is fixed at 0.25; any value greater than 0.25 indicates eutrophication, while a value less than or equal to 0.25 indicates a total absence of the phenomenon.

4 Results and Discussion

4.1 Identification of Water Bodies by NDWI

Figure 5 shows that the NDWI value varies from −1 to 1. Water bodies have a high NDWI, while zero or low values are easily associated with non-water features such as roofs and paved roads. According to [43] NDWI values greater than zero are assumed to represent aquatic areas, while values less than or equal to zero are assumed to be non-aquatic zones.

In addition, four main bodies of water have been identified: the lake/sebkha of Monastir on the north-east side, the Sebkha of Moknine on the south-east side and the Sebkha of Sidi El Heni on the west side. The fourth body of water is the sea. Nevertheless, the lake of Monastir contains different parts, from the north the bay of dkhila and from west to east the salt marshes, sebkha of Sahline, the lake of Sahline. To the east, the Khniss lagoon is located on the coastal side, further south it is the Khnis Bay. A few spots located in the study area may correspond to other surfaces, but their magnitude was less than the water surface. A threshold value of 0.3 was

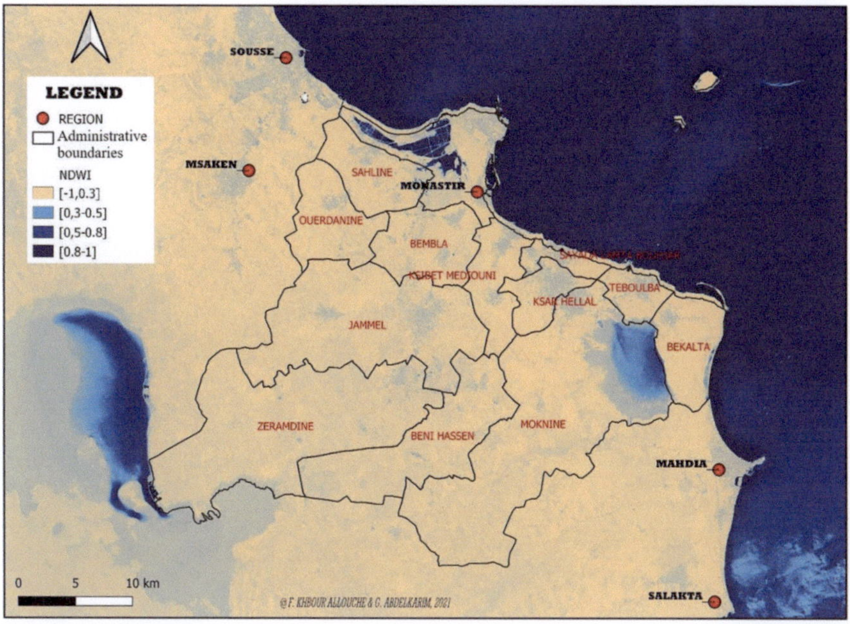

Fig. 5 Spatial delimitation of water bodies in the Monastir region using the NDWI radiometric index

therefore applied to the NDWI image to better isolate surfaces with no detectable water (NDWI less than 0.3) from those with detectable water surfaces (NDWI greater than or equal to 0.3). The increase in this value can be attributed to the turbidity of the water. On the other hand, the decrease in the index, presented by the light blue color around the dark one located at the level of the water bodies detected, is related to the presence of vegetation [36]. Figures 6 and 7 show the location of the main bodies of water in the study area.

4.2 *Identification of Eutrophication by the NDCI*

Figure 8 shows that values less than or equal to 0.25 indicate an absence of the phenomenon. The NDCI detected the proliferation of algae at the level of water bodies and qualitatively located ranges of concentrations of chlorophyll A. This location is located at the coast, in the bay and the bodies of water with different concentrations.

Figures 9 and 10 show that on the coast at Sayada and Teboulba, this concentration is more pronounced along a strip 1800 m long with two extremities with low eutrophication rates, but further inland, on the eastern side, it decreases along a strip 1500 m long. A small core is also observed on the west side. The first band

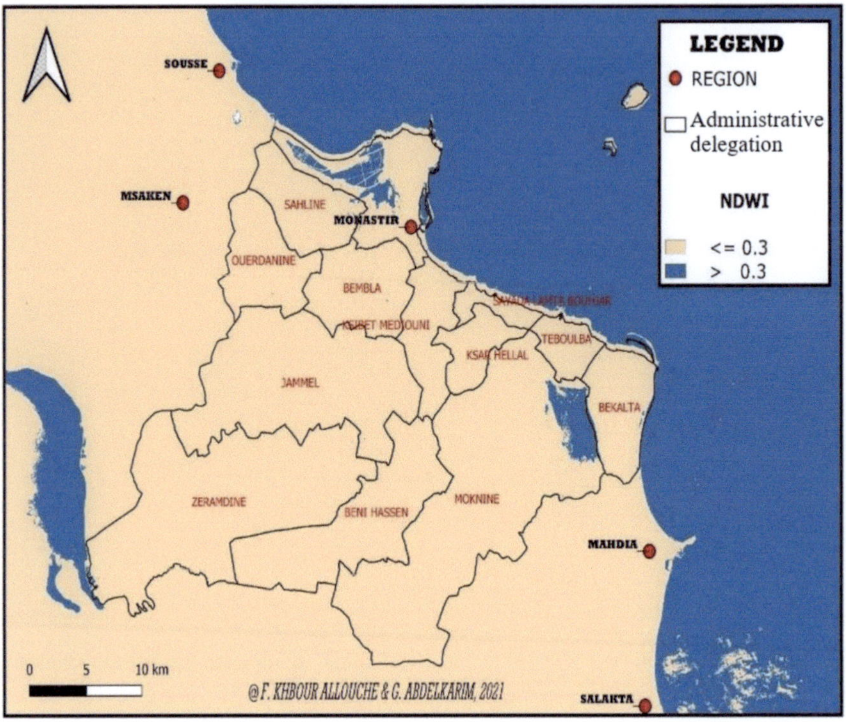

Fig. 6 Spatial delineation of water bodies in the Monastir region using NDWI thresholding

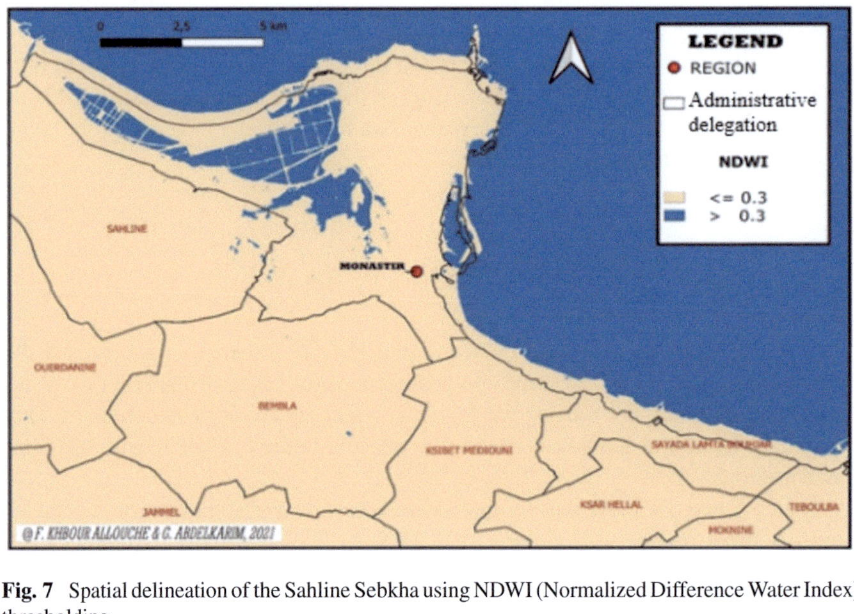

Fig. 7 Spatial delineation of the Sahline Sebkha using NDWI (Normalized Difference Water Index) thresholding

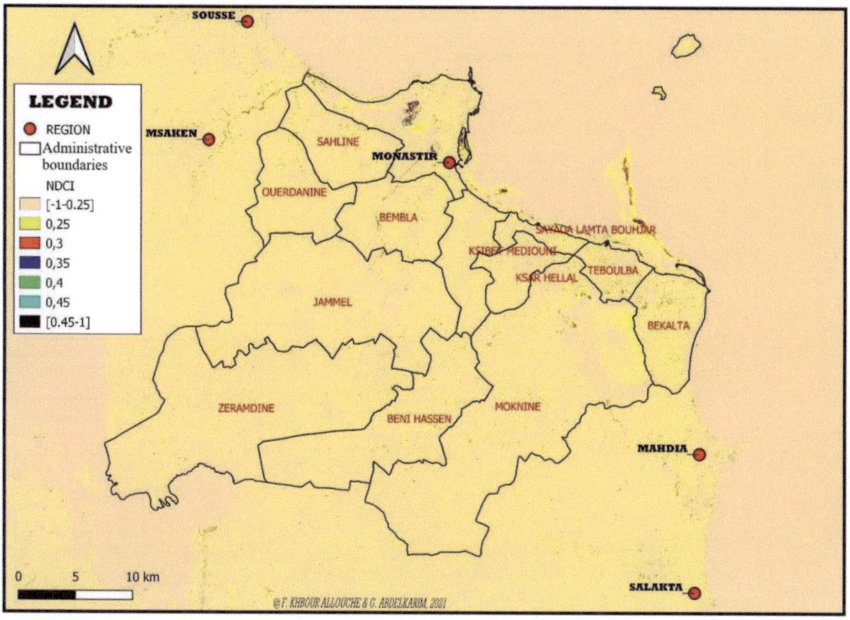

Fig. 8 Spatial distribution of eutrophication in the Monastir region, Tunisia

is characterized by a heterogeneous dispersion of eutrophication, with a gradient of increasing concentration towards the north-west and south-east. This band is subdivided into different sections with alternating concentrations. Different degrees of eutrophication can thus be classified; (i) NDCI $\geq$ 0.45 represents the class with a very high degree of eutrophication; NDCI = [0.4–0.45] relates to the class with a high degree of eutrophication; NDCI = [0.35–0.4] corresponds to the average degree of eutrophication.

NDCI = [0.3–0.35] represents the class with a low degree of eutrophication; NDCI = [0.3–0.25] refers to the class with a very low degree of eutrophication. Eutrophication is visible in the eastern lagoon of the "sebkha de sahline" water body, where the water has a brown color caused by phytoplankton bloom. The species responsible for this brown water is a diatom. In fact, the alternation of different degrees of eutrophication can be explained by the alternation of "dry water/flooded water". In addition, the high levels of silicon, nitrogen and phosphorus may be favorable to the development of the phytoplankton compartment, particularly for diatoms with siliceous tests.

The spatial distribution of eutrophication in the Sebkha/Sahline lake is very heterogeneous (Figs. 11 and 12). Three areas with a high degree of eutrophication mark the water body, the first oriented north-east and the other two south-west. This variation can be explained by the impact of several factors that contribute to the variation in chlorophyll A content. For example, Khiari [38] explained that the coastal fringe of Monastir is the receptacle for multiple wastewater outfalls from urban areas, in

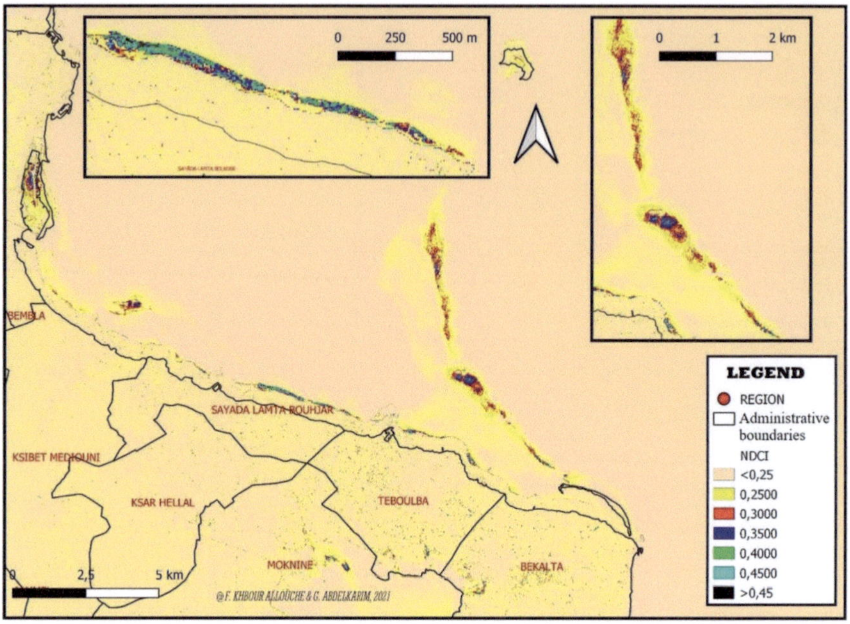

Fig. 9 Spatial distribution of eutrophication in Monastir Bay, Tunisia

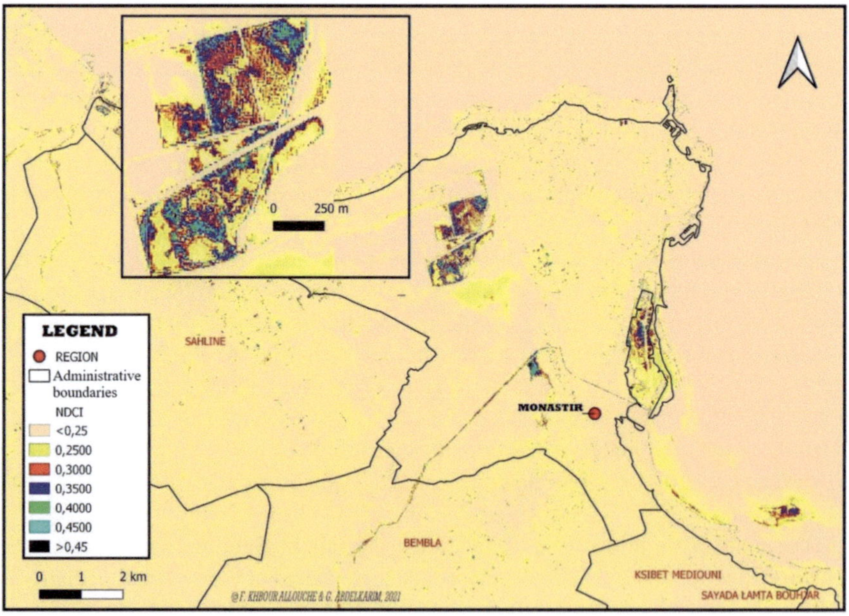

Fig. 10 Spatial distribution of eutrophication in the Sahline Sebkha, Tunisia

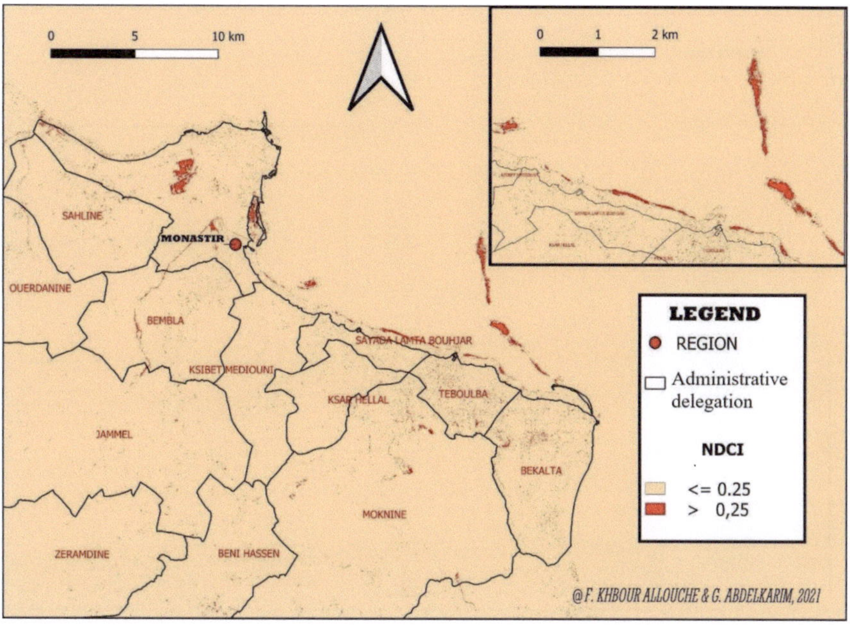

Fig. 11 Spatial distribution of the eutrophication phenomenon with thresholding at Monastir Bay, Tunisia

particular those from Khnis and Ksibet El Mediouni which discharge at numerous points along the shoreline as well as those from the city of Monastir which are discharged to the north of the port. Zaouali (2006) added that a lower level of salinity is due to the presence of numerous wastewater discharge points. So these parameters can contribute to the variation in the chlorophyll A rate in the Sebkha.

Thresholding was applied to better locate the eutrophication of the water bodies and a higher concentration was observed in the Sahline lake, the Khniss lagoon and the coastal fringe. These high concentrations can be explained by a high concentration of chlorophyll primary production. Indeed, a study carried out at Soliman's Sebkha in 2002 showed that chlorophyll (A) contents (such as nitrogen, phosphorus and turbidity) can be used to assess the degree of eutrophication of the water. Chlorophyll has two origins, phytobenthic and phytoplanktonic.

Furthermore, Sassi et al. [56] have shown that the north of Monastir Bay is affected by both organic and mineral pollution. Metal pollutants (Cu, Pb) are associated with dissolved or particulate organic matter brought in by urban wastewater discharged onto a coastline affected by low hydrodynamics. This can be explained by the high rate of eutrophication in Monastir Bay.

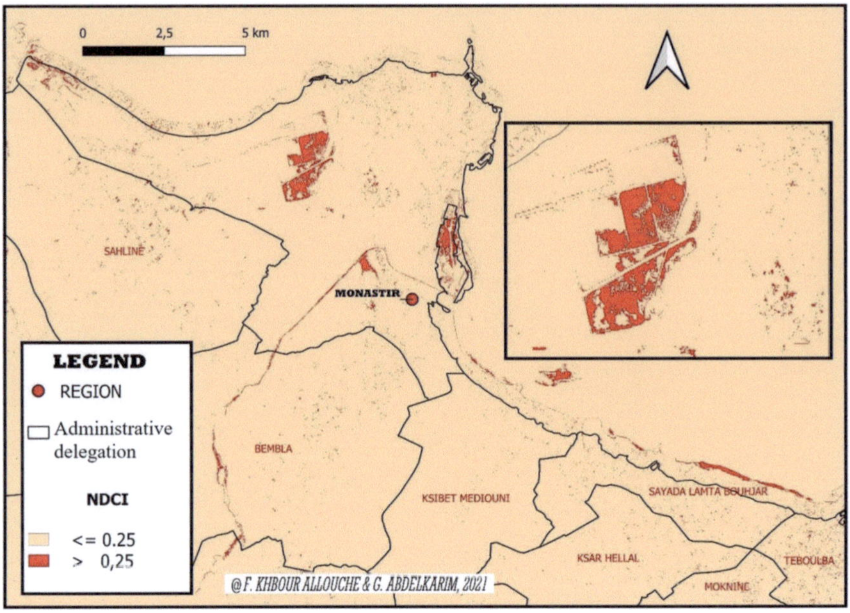

Fig. 12 Spatial distribution of the eutrophication phenomenon with thresholding at the Sahline Sabkhaa, Tunisia

4.3 Characterization of the Eutrophication Phenomenon According to Physical and Anthropogenic Factors

4.3.1 Eutrophication and Soil

Figures 13 and 14 show that the phenomenon of eutrophication is well developed in hydromorphic soils in water bodies, and is of minor importance in soils with little evolution and crude mineral soils. Hydromorphic soils are deep, not very gravelly that rest on a limestone crust and are located in troughs or low-lying areas, which leads to hydromorphic and an accumulation of salts. Salinity is clearly visible in the upper horizons. At sebkha level, we can deduce that eutrophication is more limited on the eastern side by the brown soils, while a few spots persist on the southern side of Monastir and to the west of Bebmbla on the soils with little evolution.

In addition, Figs. 15 and 16 show that eutrophication is well situated in shallow soils below 60 cm.

4.3.2 Eutrophication and Land Use

There has been a proliferation of eutrophication on the eastern side of the sebkha and Sahline lake, where annual crops and olive groves persist (Figs. 17 and 18). This area

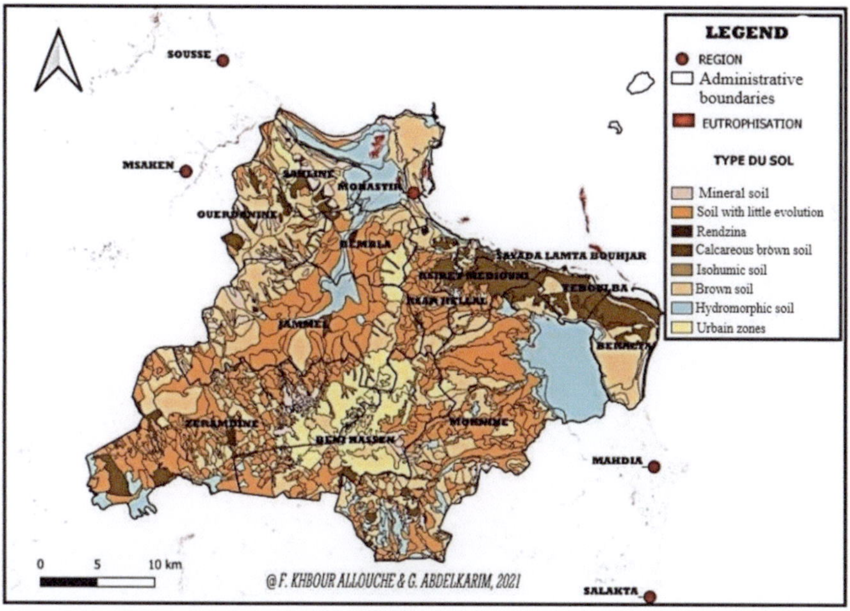

Fig. 13 Location of the eutrophication phenomenon with thresholding at the level of the different types of soil in the Monastir region, Tunisia

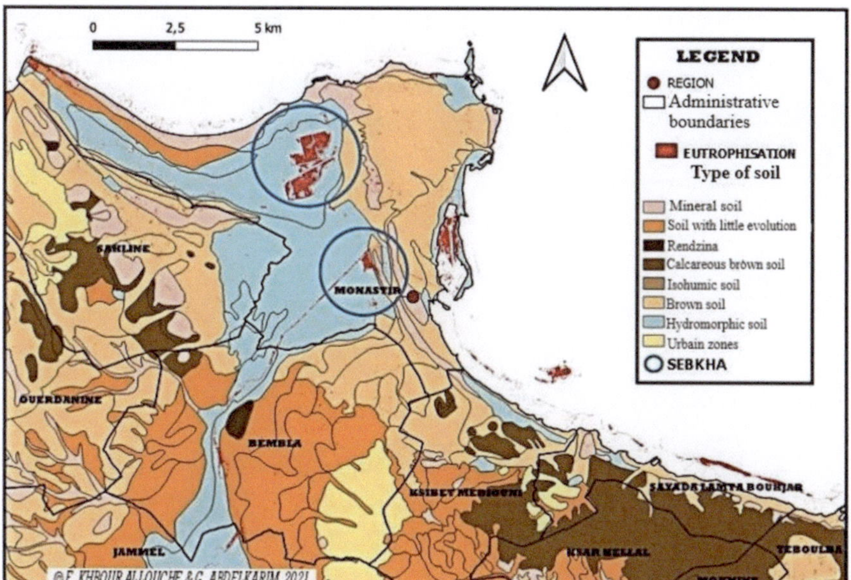

Fig. 14 Location of the eutrophication phenomenon with thresholding at the level of the different types of soil in the area around the Sahline sebkha, Tunisia

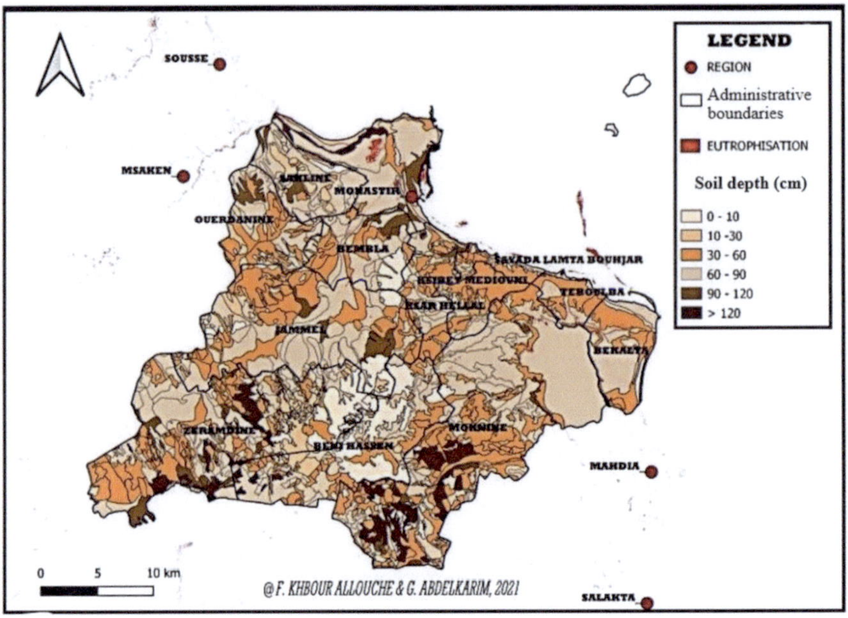

Fig. 15 Location of the eutrophication phenomenon with thresholding at different soil depths in the Monastir region, Tunisia

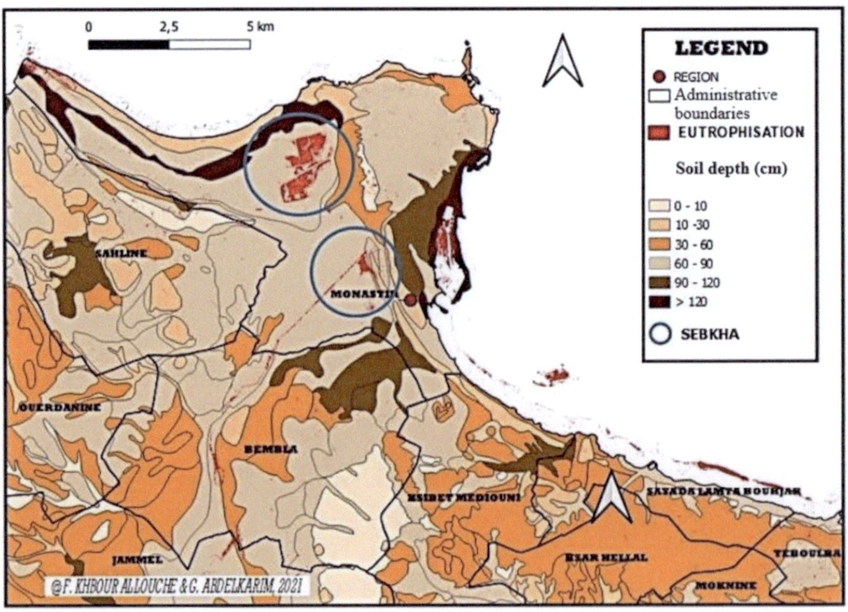

Fig. 16 Location of the eutrophication phenomenon with thresholding at different soil depths in the area around the Sahline sebkha, Tunisia

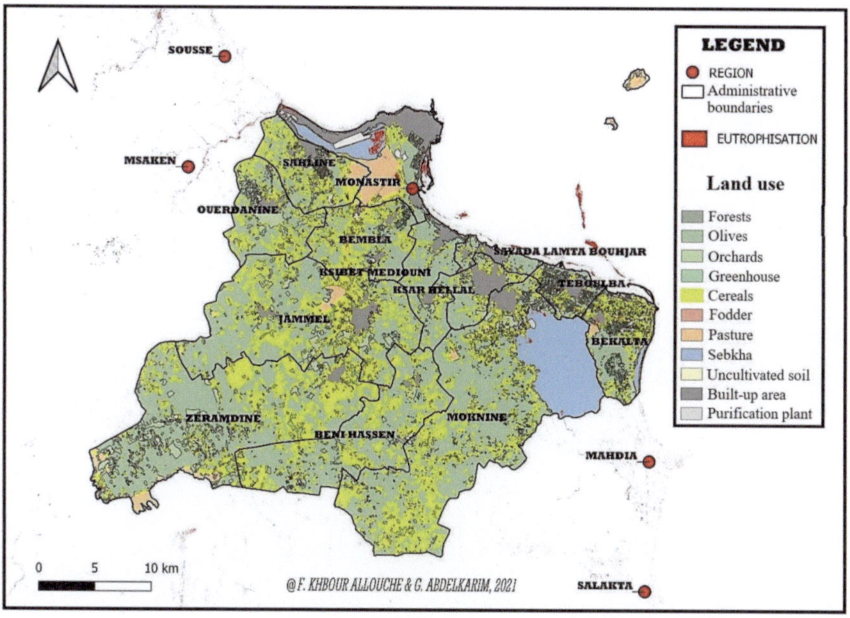

Fig. 17 Location of the eutrophication phenomenon with thresholding at the level of the different types of land use in the Monastir region, Tunisia

is characterized by irrigated cropping and livestock rearing, so we can conclude that this eutrophication may be intensified by the input of nitrogen, whether organic or mineral. In this context, Hagmann [32] highlighted that agricultural soils can transport nitrogen and phosphorus from urine, excreta, excrement and the decomposition of organic matter (OM) and necromass.

4.4 Eutrophication and Processes

Figures 19 and 20 illustrate that, to the south, there is a natural lagoon station with a treatment process that ensures sludge stabilization. Almost all urban wastewater purification processes, whether biological or physico-chemical, lead to the concentration of pollutants in the form of mud which constitute bulky waste. Lagooning is a natural water purification technique based on oligotrophication. In the case of natural lagooning, micro-organisms that develop in the environment using oxygen from photosynthesis provide biological purification. Lagooning is a natural water purification technique based on oligotrophication. In the case of natural lagoon, micro-organisms that develop in the environment using oxygen from photosynthesis, allow biological purification.

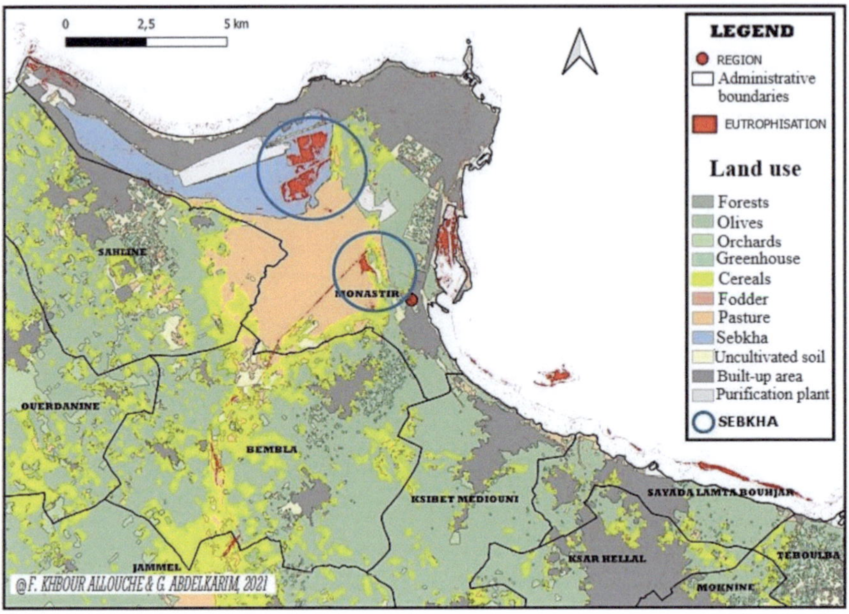

Fig. 18 Location of the eutrophication phenomenon with thresholding at the level of the different types of land use in the area around the Sehline Sebkha, Tunisia

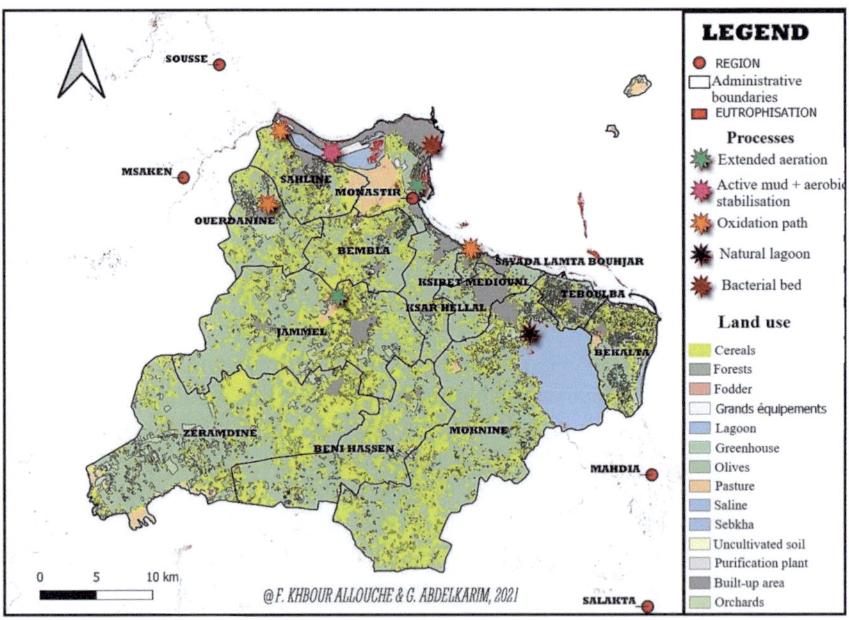

Fig. 19 Location of the eutrophication phenomenon with thresholding and processes at the level of the different types of land use in the Monastir region, Tunisia

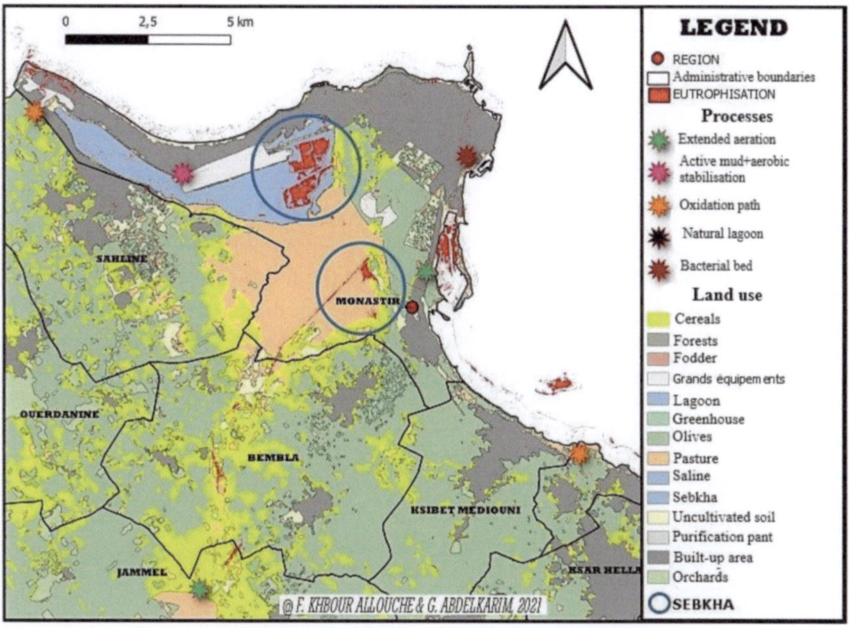

Fig. 20 Location of the eutrophication phenomenon with thresholding and processes at the level of the different types of land use in the area around the Sahline Sebkha, Tunisia

These processes help to form lagoon sludge at the bottom of the basins. This explains why there is no eutrophication in the Moknine sebkha. This process is located in the Moknine delegation. Going to the northeast, west of the Khniss lagoon, we find the oxidation channels, where there is prolonged aeration, which is a treatment allowing the degradation of carbon and nitrogen pollution. Since eutrophication is sensitive to nitrogen and phosphorus, which are catalysts for the phenomenon, the degree of seriousness of eutrophication in these bodies of water is high. A bacterial bed is located further north of the town of Monastir. According to [20], the liquid to be treated trickles over the support on which the purifying biomass gradually forms. This treatment is used in addition to the activated sludge process, enabling a wider range of pollutants to be eliminated Moulin et al. [46]. Indeed, its purpose is to concentrate the bacteria and thus localize their action, making it more effective, which explains the oligotrophication of the surroundings of this process. In addition, this technique has several advantages since it gives a good yield on the elimination of carbon pollution at moderate operating costs. It also contributes to the mineralization of sludge. In the north-west and within the sebkha of sahline, there are active sludge and aerobic stabilization.

Almost all urban wastewater purification processes, whether biological or physico-chemical, lead to the concentration of pollutants in the form of sludge, which constitute bulky waste since they contain between 95 and 99% of water and generate nuisances insofar as they consist of fermentable organic matter and may contain toxic substances. Fermentable waste is composed exclusively of biodegradable organic materials.

4.5 Monitoring Eutrophication

Figures 21, 22, 23, 24, 25, 26, 27, 28, 29, 30, 31 show temporal variations in the appearance of the eutrophication phenomenon in the north of the Monastir region. These variations can be detected on seasonal and multiannual scales. Figures 23 and 25 reveal the absence of eutrophication, which can be explained by the low temperatures and lack of precipitation on 04/08/2019 and the low wind speed on 10/06/2019.

Fig. 21 Appearance of eutrophication in Monastir on 09/06/2020, Tunisia

Fig. 22 Presence of eutrophication in the water body in the north of Monastir on 25/05/2019

Fig. 23 Absence of eutrophication in the water body in the north of Monastir on 10/06/2019

Fig. 24 Slight presence of eutrophication in the body of water in the north of Monastir on 10/07/2019

On the other hand, for the rest of the figures, the phenomenon is present to different degrees according to various atmospheric criteria:

- 25/05/2019 presence of precipitation on the day of the phenomenon, the day before and the day after (Fig. 22);
- 19/02/2020 increase in wind speed of around 35 km/h (Fig. 27);

Fig. 25 Absence of eutrophication in the water body in the north of Monastir on 04/08/2019

Fig. 26 Slight presence of eutrophication in the water body in the north of Monastir on 09/08/2019

- 29/07/2020 temperature rise of 4 °C compared with the average temperature for July (28 °C) with an increase in wind speed on the day of the event from 17 to 25 km/h (Fig. 30);
- 28/08/2020 increase in wind speed on the day of the event from 14 to 28 km/h, with temperatures rising from 31 to 35 °C on the same day (Fig. 31)
- 04/08/2019 increase in temperatures one day before the phenomenon (Fig. 25)
- 09/06/2020 Rainfall was recorded on days d−1, d and d+1 of the event. However, the temperature increased slightly and the wind speed was low. Consequently, the low degree of this phenomenon (Fig. 28);

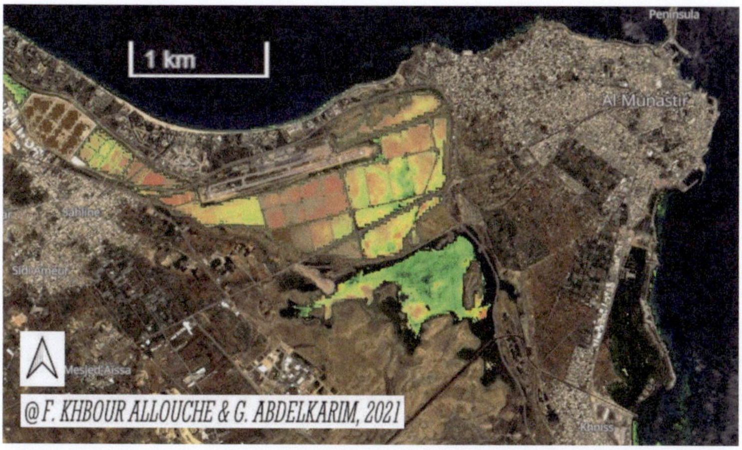

Fig. 27 Eutrophication accentuated in the body of water to the north of Monastir on 19/02/2020

Fig. 28 Presence of eutrophication in the water body to the north of Monastir on 09/06/2020

- On 09/07/2020, the wind increased compared to days d−1 and d+1 (24 km/h). But, the degree of this phenomenon is fairly low because of the low temperature (Fig. 29);

The starting of the phenomenon of eutrophication and the variation in its degree are strongly related to atmospheric factors. However, marine, physical and anthropogenic parameters contribute to the accentuation or weakening of the phenomenon. Thus, at the level of water bodies, eutrophated spaces can have daily, monthly, seasonal or multiannual variations.

Fig. 29 Slight presence of eutrophication in the water body to the north of Monastir on 09/07/2020

Fig. 30 Eutrophication accentuated in the water body to the north of Monastir on 29/07/2020

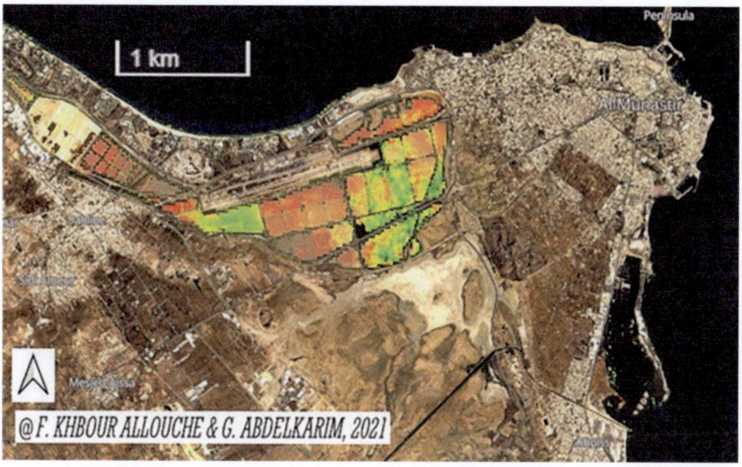

Fig. 31 Increased eutrophication in the water body north of Monastir on 28/08/2020

5 Conclusion

The use of remote sensing by applying two radiometric indices, namely the NDWI and the NDCI, made it possible to locate the water bodies and the phenomenon of eutrophication in the Monastir region. In addition, the thresholding of the two indices at 0.25 and 0.3 respectively helped to improve the accuracy of their locations. Four main bodies of water are identified, on the north side it is the lake/sebkha of Monastir, the Sebkha of Moknine on the south-east side and to the west it is the Sebkha of Sidi El Heni. The fourth body of water is relative to the sea.

Four degrees of eutrophication are established (i) NDCI $\geq$ 0.45 presents the class with a very high degree of eutrophication; NDCI = [0.4–0.45] relates to the class with a high degree of eutrophication; NDCI = [0.35–0.4] corresponds to the average degree of eutrophication;

NDCI = [0.3–0.35] represents the class with a low degree of eutrophication and NDCI = [0.3–0.25] refers to the class with a very low degree of eutrophication.

Analysis of the results according to physical criteria showed that this phenomenon persists in hydromorphic soils and in the proximity of areas occupied by irrigated annual crops. In addition, active sludge, aerobic stabilization and prolonged aeration contribute to increasing the process.

On the other hand, natural lagoons like Lake Moknine play an important role in de-eutrophication. Marine, physical and anthropogenic parameters contribute to accentuating or weakening the phenomenon. In this way, eutrophicated areas in bodies of water can be characterized by daily, monthly, seasonal or multiannual variations.

6 Recommendations

- Application of drone technology remote sensing in eutrophication process monitoring gives many advantages over the conservative monitoring techniques. The ability to view the indicators' overall spatial distribution instead of point-scale measurements is the main advantage of the UAV remote sensing on applications. It has been highlighted that the implication of additional tools in monitoring aquatic environments is particularly important for applying preventive and corrective measures that can be taken to minimize socioeconomic impacts.
- We recommend the setup of a coordinated target-focused monitoring program or experiment that utilizes in situ measurements in combination with remote sensing data. This may improve future monitoring and management of coastal ecosystem health as well as other ecosystems that are subject to regular monitoring.
- Consideration must be given to carefully timing in-situ sampling in association with the fly-over times of satellites. This will improve the validity and statistical power of comparisons.

References

1. Amari A, Bedir M (1988) The Quaternary basins of the central Sahel of Tunisia. Genesis and evolution of the sebkhas in a compressive and distensive strike-slip context. (Les bassins quaternaires du Sahel central de la Tunisie. Genèse et évolution des sebkhas en contexte décrochant compressif et distensif. Géodynamique 4(1):49–65. ISSN 0766-5105
2. APAL (2000) Sensitive area management study, Khniss Lagoon, Report prepared for the Coastal Protection and Development Agency, Monastir (Etude de gestion des zones sensibles, la lagune de Khniss, Rapport préparé pour l'Agence de Protection et d'Aménagement du Littoral, Monastir), pp 96
3. APAL (2009) Study of the coastal fringe of Monastir - Rehabilitation strategy - phase 1, Tunis, Ministry of the Environment and Sustainable Development (*Étude de la frange littorale de Monastir-Stratégie de réhabilitation-phase 1*, Tunis, ministère de l'Environnement et du Développement durable), p 432
4. Arff J, Miguez BM (2016) Marine microalgae and harmful algal bloms: a European prespective. In: Tsaloglou MN (eds) Microalgae: current research and applications). Caister Academic Press, pp 45–71. https://doi.org/10.21775/9781910190272.04
5. Aubert M, Aubert J (1986) Eutrophy and dystrophy in the marine environment, planktonic and bacterial phenomena. (Eutrophie et dystrophie en milieu marin, phénomènes planctoniques et bactériens). Rev Intern Océanogr Médicale 53–54:1–302
6. Badraoui, M (2003) Development of an eco-balance model for the environmental assessment of irrigated agriculture in Morocco, Case of the Tadla irrigated perimeter.(Élaboration d'un modèle d'écobilan pour l'évaluation environnementale de l'agriculture irriguée au Maroc, Cas du périmètre irrigué de Tadla)
7. Barbouchi M, Lhissou R, Chokmani K, Abdelfattah R, El Harti A, Ben Aissa N, INRS-ETE (2013) Characterization of soil salinity using satellite radar imagery: case of Tunisia and Morocco, Tunisia, Morocco (Caractérisation de la salinité des sols à l'aide de l'imagerie radar satellitaire : Cas de la Tunisie et du Maroc, TunisieMaroc), p 99

8. Beck R, Zhan S, Liu H, Tong S, Yang B, Xu M, Su H (2016) Comparison of satellite reflectance algorithms for estimating chlorophyll: a in a temperate reservoir using coincident hyperspectral aircraft imagery and dense coincident surface observations. Remote Sens Environ 178:15–30
9. Ben Cherifa B (2015) (Invertebrates of Tunisian lagoons) Les invertébrés des lagunes tunisiennes. https://fr.slideshare.net/BenCherifaBilel/les-invertbrs-des-lagunes-tunisiennes
10. Bouwman AL, Beusen A, Glibert PM, Overbeek C, Pawlowski M, Herrera J, Mulsow S, Yu R, Zhou M (2012) Mariculture: significant and expanding cause of coastal nutrient enrichment. Environ Res Lett 8(044026):5
11. Caballero I, Ruiz J, Navarro G (2019) Sentinel-2 satellites provide near-real time evaluation of catastrophic floods in the west mediterranean. Water 2019(11):2499
12. Chairi R (2004) Study of recent filling and associated organic matter in the sebkha of Moknine. Doctoral thesis, University of Tunis El Manar (Étude du remplissage récent et de la matière organique associée dans la sebkha de Moknine, Thèse de doctorat, UniversitéTunis El Manar), 249 p + 2 annexes
13. Chandra KBSP, Kumar A, Saikia P, Lele N, Srivastava P, Pulla S, Suresh H, Bhattarcharya BK, Khan ML, Sukumar R (2024) Mapping of understorey invasive plant species clusters of Lantana camara and Chromolaena odorata using airborne hyperspectral remote sensing. Adv Space Res 73(2):1379–1396. https://doi.org/10.1016/j.asr.2022.12.026
14. Chowdary VM, Vinu Chandran R, Neeti N, Bothale RV, Srivastava YK, Ingle P, Ramakrishnan D, Dutta D, Jeyaram A, Sharma JR (2008) Assessment of surface and sub-surface waterlogged areas in irrigation command areas of Bihar state using remote sensing and GIS. Agr Water Manag 2008(95):754–766
15. Coll M, Piroddi C, Steenbeek J, Kaschner K, Frida Ben Rais L, JacopoAguzzi J, Ballesteros E, Bianchi CN, Corbera J, Dailianis T, Roberto D, Estrada M, Froglia C, Galil SB, Gasol JM, Gertwagen R, Gil J, Guilhaumon F, Kesner-Reyes K, Miltiadis-Spyridon K, Athanasios K, Lampadariou N, Laxamana E, Lo´pez-Fe´ de la Cuadra CM, Heike KL, Martin D, Mouillot D, Orol D, Raicevich S, Rius-Barile J, Ignacio Saiz-Salinas J, San Vicente C, Somot S, Templado J, Turon X, Vafidis D, Villanueva R, Voultsiadou E (2010) The bioversity of the mediterranean sea : estimated, patterns, and threat. 36:136
16. Copernicus E (2018). Available from: https://scihub.copernicus.eu/dhus/#/home
17. Daloz A et al (2009) the Lagoon Monitoring Network: a diagnostic tool and action support for the fight against eutrophication of lagoons: an information and transfer strategy to stakeholders (le Réseau de Suivi lagunaire : un outil de diagnostic et d'aide à l'action pour la lutte contre l'eutrophisation des lagunes :une stratégie d'information et de transfert aux acteurs). European Conference on Coastal Lagoon Research. Poster 94
18. Dammak M, Fourati R, Ellech B, Kallel M (2019) Assessment of organic and metallic contamination in the surface sediment of Monastir Bay (Eastern Tunisia): spatial distribution, potential sources, and ecological risk assessment. Mar Pollut Bull 149:110500. https://doi.org/10.1016/j.marpolbul.2019.110500
19. Daniel C, Joël C, Yerubandi R (2013) water bodies. Based in part on the article "Rivers, canals and estuaries" by Steve Monismith, which appeared in the Encyclopedia of Environmetrics. https://doi.org/10.1002/9780470057339.var054.pub2
20. Dehbi FZ (2015) Comparative study of the performance of a bacterial bed with pozzolan packing from Beni Saf and a bacterial bed with plastic packing, Final dissertation for obtaining a Master's degree in Hydraulics, Abou bekr belkaid University (Etude comparative des performances d'un lit bactérien à garnissage en pouzzolane de Beni Saf et d'un lit bactérien à garnissage plastique, Mémoire de fin d'études pour l'obtention de diplôme du Master en Hydraulique, Université Abou bekr belkaid), Tlemcen, p 13
21. Dodge JD (1975) The prorocentrales (dinophyceae). II. Revision of taxonomy within the genus Prorocentrum. Bot J Linn Soc 71:103–125
22. Doxaran D, Froidefond J-M, Lavender S, Castaing P (2002) Spectral signature of highly turbid waters: application with SPOT data to quantify suspended particulate matter concentrations. Remote Sens Environ 81:149–161

23. El Khoja H (2014) Monastir Bay: State-NGO consultation, Monastir, Tunisian Association for Studies and Research on Cetaceans - HOUTIYAT (La baie de Monastir: concertation Etat-ONG, Monastir, Association Tunisienne des Etudes et Recherches sur les Cétacés – HOUTIYAT), p 15
24. Feki W, Hamza A, Frossard V, Abdennadher M, Hannachi I, Jacquot M, Belhassen M, Aleya L (2013) What are the potential drivers of blooms of the toxic dinoflagellate Karenia selliformis? A 10-year study in the Gulf of Gabes, Tunisia, southwestern Mediterranean Sea. Harmful Algae 23:8–18
25. Feng L, Hu C, Chen X, Song Q (2014) Influence of the Three Gorges Dam on total suspended matters in the Yangtze Estuary and its adjacent coastal waters: observations from MODIS. Remote Sens Environ 140:779–788
26. Friedland R, Meier HM, Edman MK, Eilola KJ, Placke M, Neumann T, Andersson HC, Brunnabend S-E, Dieterich C, Frauen C, Gröger M, Gustafsson BG, Gustafsson E, Isaev A, Kniebusch M, Kuznetsov I, Müller-Karulis B, Omstedt A, Ryabchenko V, Saraiva S, Savchuk OP (2019) Assessment of eutrophication abatement scenarios for the Baltic Sea by multi-model ensemble simulations. Front Mar Sci 5:1–24
27. Galvez-Cloutier R, Sylvaine I, Sylvain A (2002) Deterioration of water bodies: manifestations and means of combating eutrophication (La détérioration des plans d'eau: Manifestations et moyens de lutte contre l'eutrophisation). VECTEUR environnement 35(6):18
28. Glibert PM, Al-Azri A, Allen JI, Bouwman AF, Beusen AHW, Burford MA, Harrison PJ, Zhou M (2018) Key questions and recent research advances on harmful algal blooms in relation to nutrients and eutrophication. In: Glibert PM, Berdalet E, Burford MA, Pitcher GC, Zhou M (dir.) Global ecology and oceanography of harmful algal blooms. Springer International Publishing, Cham, pp 229–259
29. Glibert PM, Seitzinger S, Heil CA, Burkholder JM, Parrow MW, Codispoti LA, Kelly V (2005) The role of eutrophication in the global proliferation of harmful algal blooms. Oceanography 18(2):198–209. https://doi.org/10.5670/oceanog.2005.54
30. Gomez F, Roselli L, Zhang H, Lin S (2024) Misidentifications of the bloom-forming dinoflagellates Gymnodinium litorale and Margalefidinium polykrikoides in the Mediterranean Sea. Regl Stud Mar Sci 70:103376. https://doi.org/10.1016/j.rsma.2024.103376
31. Hafeez S, Wong MS, Abbas S, Kwok CYT, Nichol J, Lee KH, Tang D, Pun L (2018). https://doi.org/10.5772/intechopen.81657
32. Hagmann DF, Goodey NM, Mathieu C, Evans J, Aronson MFJ, Gallagher F, Krumins JA (2015) Effect of metal contamination on microbial enzymatic activity in soil. Soil Biol Biochem 91:291–297. https://doi.org/10.1016/j.soilbio.2015.09.012
33. Hallegraeff GM, Anderson D, Cembella A (2004) Manual on harmful marine microalgae. Edited by HO Enevoldsen
34. Heldt J (1952) Red waters. Bull Tunisian Soc Nat Sci (Eaux rouges. Bulletin de la Société des Sciences Naturelles de Tunisie) 5:103–106. http://hdl.handle.net/1834/15891
35. Hestir Erin L, Khanna S, Andrew ME, Santos MJ, Viers JH, Greenberg JA, Rajapakse SS, Ustin SL (2008) Identification of invasive vegetation using hyperspectral remote sensing in the California Delta ecosystem. Remote Sens Environ 112(11):4034–4047. https://doi.org/10.1016/j.rse.2008.01.022
36. Huang X, Xie C, Fang X, Zhang L (2015) Combining pixel- and object-based machine learning for identification of water-body types from urban high-resolution remote-sensing imagery. IEEE J Selected Top Appl Earth Observ Remote Sens 8(5):2097–2110. https://doi.org/10.1109/JSTARS.2015.2420713
37. Karydis M (2009) Eutrophication assessment of coastal waters based on indicators: a literature review. Global NEST J 11:373–390
38. Khiari N, Atoui A, Khalil N, Charef A, Aleya L (2017) Dynamics of sediments along with their core properties in the Monastir-Bekalta coastline (Tunisia, Central Mediterranean). J Afr Earth Sc 134:320–331
39. Khiari N, Charef A, Atoui A, Azouzi R, Khalil N, Khadhar S (2021) Southern Mediterranean coast pollution: long-term assessment and evolution of PAH pollutants in Monastir Bay (Tunisia). Mar Pollut Bull 167:112268. https://doi.org/10.1016/j.marpolbul.2021.112268

40. Kudela RM, Berdalet E, Bernard S, Burford M, Fernand L, Lu S, Roy S, Tester P, Usup G, Magnien R, Anderson DM, Cembella A, Chinain M, Hallegraeff G, Reguera B, Zingone A, Enevoldsen H, Ed Urban (SCOR) (2015) Harmful algal blooms. A Scientific Summary for Policy Makers. IOC/UNESCO, Paris (IOC/INF-1320)
41. Lopez-Rodas H, Rouco MV, Costas E (2008) Warming will affect phytoplankton differently: evidence through a mechanistic approach. Proc Roy Soc Biol Sci 278:3534–3543
42. Mawdsley JR, O'Malley R, Ojima DS (2009) A review of climate-change adaptation strategies for wildlife management and biodiversity conservation. Conserv Biol 23:1080–1089
43. Mcfeeters SK (1996) The use of the Normalized Difference Water Index (NDWI) in the delineation of open water features. Int J Remote Sens 17(7):1425–1432. https://doi.org/10.1080/01431169608948714
44. Meng H, Zhang J, Zheng Z, Lai Y, Geng H (2024) Risk assessment and spatio-temporal characteristics analysis of water bloom in three large-scale eutrophic reservoirs in Fujian Province, China. Ecol Ind 158:111539. https://doi.org/10.1016/j.ecolind.2023.111539
45. Mishra S, Mishra DR (2011) Normalized difference chlorophyll index: a novel model for remote estimation of chlorophyll-a concentration in turbid productive waters. Remote Sens Environ 117(2012):394–406
46. Moulin S, Rechels DR, Stankovic M (2013) Wastewater treatment. Workshop water quality vs quantity. CERES: Center for teaching and research on the environment and society (Traitement des eaux usées. Atelier Lõeau qualité vs quantité. CERES: Centre d'Enseignement et de Recherches sur l'Environnement et la Société. ERTI): Environmental Research and Teaching Institute
47. Nasa VE (2018) Modis: mouth of the Yangtze. Available from: https://visibleearth.nasa.gov/view.php?id=55219
48. Nechad B, Ruddick KG, Park Y (2010) Calibration and validation of a generic multisensor algorithm for mapping of total suspended matter in turbid waters. Remote Sens Environ 114(4):854–866
49. Novoa S, Doxaran D, Ody A, Vanhellemont Q, Lafon V, Lubac B, Gernez P (2017) Atmospheric corrections and multi-conditional algorithm for multi-sensor remote sensing of suspended particulate matter in low-to-high turbidity levels coastal waters. Remote Sens 9:61. https://doi.org/10.3390/rs9010061
50. O'Reilly JE, Maritorena S, O'brien MC, Siegel DA, Toole D, Menzies D, Chavez FP (2000) In: SeaWiFS postlaunch calibration and validation analyses. Part 3. NASA tech. Memo? 206892(11):3–8
51. Panagos P, Van Liedekerke M, Montanarella L (2011) Multi-scale European Soil Information System (MEUSIS): a multi-scale method to derive soil indicators. Comput Geosci 15:463–475
52. Pinay G, Gascuel C, Menesguen A, Souchon Y, Le Moal M (2017) Eutrophisation. Manifestations, causes, conséquences et prédictibilité. Rapport de l'Expertise scientifique collective. https://doi.org/10.35690/978-2-7592-2757-0
53. Reusch TBH, Dierking J, Andersson HC, Bonsdorff E, Carstensen J, Casini M, Czajkowski M et al (2018) The Baltic Sea as a time machine for the future coastal ocean. Sci Adv 4(5). https://doi.org/10.1126/sciadv.aar8195
54. Sassi R, Souissi F, Soussi N, Baccar F, Added A, Charef A, Abdeljaoued S (1998) Environmental diagnosis of the north of Monastir Bay (eastern Tunisia) by geochemical study of surface sediments (Diagnostic environnemental du nord de la baie de Monastir (Tunisie orientale) par l'´etude g´eochimique des s´ediments superficiels). Bull des laboratoires des ponts et chauss´ees 218:49–58
55. Sassi R, Souissi F, Soussi N, Boukaaba M, Belayouni H, (1998) Geochemistry of organic matter for the analysis of the degradation of the Monastir-Ksibet el Mediouni coastline (eastern Tunisia) (Géochimie de la matière organique pour l'analyse de la dégradation du littoral de Monastir-Ksibet el Mediouni (Tunisie orientale)). C.R. Acad. Sci. Paris. Sciences de la terre et des planetes, pp 303–308
56. Schmidt J (1901) Flora of Koh Chang. Contributions to the knowledge of the vegetation in the Gulf of Siam. Peridiniales. Botanisk Tidsskrift 24:212–221

57. Souissi R, Turki I, Souissi F (2014) Effect of submarine morphology on environment quality: case of Monastir Bay (eastern Tunisia). Carpathian J Earth Environ Sci 9:231–239
58. SPF (2021) Federal Public Service Public Health, Food Chain Safety and Environment (service public fédéral Santé publique, Sécurité de la Chaîne alimentaire et Environnement). https://www.health.belgium.be/fr
59. Stoops CA, Gionar YR, Shintaz PS, Sismadi P, Rachmat A, Elyazar IF, Sukowati S (2008) Remotely-sensed land use patterns and the presence of Anopheles larvae (Diptera: Culicidae) in Sukabumi, West Java, Indonesia. J Vector Ecol 33:30–39
60. UNEP/MAP (2012) State of the Mediterranean marine and coastal environment. In: United nations environment programme/mediterranean action plan (UNEP/MAP)—Barcelona Convention: Athens, p 96
61. US Geological Survey (USGS) and US Department of the Interior (2013) Normalized difference water index (NDWI). Available online: http://deltas.usgs.gov/fm/data/data_ndwi.aspx. Accessed on 3 May 2013
62. Vollenweider RA (1970) The scientific basis of eutrophication of lakes and running waters with particular regard to phosphorus and nitrogen as eutrophication factors (Les bases scientifiques de l'eutrophisation des lacs et des eaux courantes sous l'aspect particulier du phosphore et de l'azote comme facteurs d'eutrophisation, p 218
63. Xu M, Liu H, Beck R, Lekki J, Yang B, Shu S, Benko T (2018) A spectral space partition guided ensemble method for retrieving chlorophyll-a concentration in inland waters from Sentinel-2A satellite imagery. J Great Lakes Res

Performance of Modeling Based on Statistical Methods in Spatial Flood Susceptibility Mapping: Case of the Upper Oum Er-Rbia Watershed

Sonia Hajji, Abdenbi Elaloui, Abdelghani Boudhar, and Mustapha Namous

Abstract Floods, being the most prevalent natural disasters, result in significant economic losses and have a profound impact on people's lives as well as infrastructures. Flood susceptibility mapping can help enhance mitigating strategies. The primary objective of this study is to delimit the flood-prone area within the high Oum Er Rbia watershed. To achieve this objective, an extensive initial field survey was conducted, coupled with the interpretation of Google Earth images. These efforts allowed us to identify a total of 60 flood locations in the area under study. Out of these historical flood locations, 70% selected for training the models, 30% were reserved for the validation process. By partitioning the historical flood locations into training and validation datasets, we aim to develop robust models that can accurately delineate flood-prone areas. This approach ensures that the models are tested against independent data, thereby enhancing their reliability and applicability. Fifteen determining factors including, elevation, slope, aspect, curvature, Stream Position Index, Topographic Position Index, Topographic Wetness Index, Topographic Ruggedness Index, distance from the road, distance from the river, stream density, lithology, Rainfall, Normalized Difference Vegetation Index, land use/land cover, were employed to map the flood susceptibility in the study region. The relationship between each class of flood factors and frequency of floods was determined using the weighting factor, weight of evidence, and frequency ratio approaches. The natural break approach was used to classify all of the factors into nine classes after they were resamples with a pixel size of 30 by 30. Finally, a hazard map was produced and categorized into

S. Hajji (✉) · A. Elaloui · A. Boudhar · M. Namous
Faculty of Sciences and Techniques, Data Sciencefor Sustainable Earth Laboratory (Data4Earth), Sultan Moulay Slimane University, Beni-Mellal, Morocco
e-mail: sonia.hajj@usms.ma

A. Elaloui
e-mail: a.elaloui@usms.ma

A. Boudhar
e-mail: ab.boudhar@usms.ma

M. Namous
e-mail: mustapha.namous@usms.ma

F. Khebour Allouche (ed.), *Environmental Remote Sensing and GIS in MENA Region*, Earth and Environmental Sciences Library,
https://doi.org/10.1007/978-3-031-86574-9_7

five susceptibility levels: very low, moderate, high and very high. The area under the curve (AUC) of the receiver operating characteristic (ROC) was used to validate and assess the performance of models. The outcomes demonstrate that the Weight of Evidence on the training dataset had a high performance (Woe = 76%, FR = 71%, WF = 73%), and the performance of models, of 71%, 73%, and 75% respectively. The outcomes of this study could be very helpful in managing flood risk and can be applied by agencies that deal with flood risk management, planners, and local disaster management.

Keywords Floods · Frequency ratio · Weighting factor · The weight of evidence · High Oum Er-Rbia watershed · Morocco

1 Introduction

Flood is a natural disaster that harms human health. Moreover, flood present a significant risk to human life, infrastructures, agricultural areas, buildings, and ecosystems. According to the literature, floods cause more than 2000 deaths each year [17]. Many factors, contribute to floods including, natural and anthropogenic factors. Generally, floods occur due to heavy rain and melt snow overflows, or flood plains. Recent research indicates that climate change is a fundamental factor inducing floods. According to [8] changes in land use patterns can have a significant impact on flood disasters in region by creating an impermeable surface that may increase flow velocity. Several factors can participate in the occurrence of floods, including topographic factors (slope, elevation, curvature, …), geological factors (lithology…), climatic factors (rainfall, snow…), and anthropogenic factors (Land use, NDVI…). Several scientists have devoted significant effort to improve the mitigation, management and understanding of flood hazards, considering flood is a natural phenomenon that cannot be completely prevented. Several studies found that identifying flood hazard areas and implementing mitigation measures can effectively reduce flood damage to an acceptable level [5]. Furthermore, flood hazard mapping is important in flood planning, and the development of flood risk reduction measures [12]. Several research has been conducted to map flood and hazards using different statistical approaches: frequency ratio (FR) [30], the weight of evidence (WoE) [4], evidential belief function (EBF) [2], artificial neural networks [21], neuro-fuzzy systems [29] fuzzy logic [18], analytical hierarchy processes (AHPs) [3], Shannon entropy (SE) [1], logistic regression [9], and statistical index [24]. The statistical methods were used in this study, namely frequency ratio, weighting factor, and weight of evidence. These models are considered the most crucial method that is simple to implement and provides appropriate risk analysis and mapping [27].

2 Problematic

In many developing countries, especially in arid and semi-arid regions, these strategies do not yet address flood risk. One such instance is the Middle East and North Africa (MENA) region where countries have different climatic, hydrological, storms characteristics, extreme and flash floods recently have occurred in majority of MENA's arid region causing significant economic and property damage. Morocco is located in one of the driest regions of the world where the impacts of climate change would be marked by the amplification and increase in the frequency of extreme events (droughts and floods).

It is submitted, like all the countries around the Mediterranean Sea, to very serious and dramatic flooding events. The character of river discharges may be aggravated by the absence and degradation of vegetation cover, which can support runoff at the expense of infiltration. Flooding has affected a large number of Moroccan cities and regions over the past few years, for example, Beni Mellal [13], Ourika watershed (Marrakech) [23], and Guelmime [16]. The impact of these floods is aggravated by the accelerated degradation of the natural environment. The region of Beni Mellal-Khenifra is subject to flood risk. Also, the city of Khenifra is located within a lower basin, and is crossed by the river and several Chaabats. The flood risks are more serious as the city is located in the Middle Atlas Mountains, in a bowl position. And that it is under the combined influence of the cold continental climate humid in winter and warm in summer. The main objective of this chapter is to investigate flood susceptibility in the Oum Er Rbia watershed using statistical index and evaluate the accuracy of each model. The next section explains the methodology used to achieve these objectives.

3 Methodology

The methodology flowchart of this research is presented in Fig. 1. The first step is to prepare a flood inventory, which considered the most critical factor in predicting flood occurrences. In this present research, flood inventory map with 60 flood locations was recorded in the high Oum Rbia watershed using extensive field surveys (i.e., GPS points). The flood map was divided randomly into sets, 70% (42 flood points) and 30% (18 flood points), for training and validation, respectively.

The next step is the elaboration of condition factors for developing an assessment approach for flood susceptibility mapping. Fifteen conditioning factors commonly used in the flood susceptibility are covered in the following section Fig. 2. these factors are elevation, slope, curvature, aspect, topographic wetness index (TWI), stream power index (SPI), topographic position index (TPI), topographic ruggedness index (TRI), distance to river, distance to roads, stream density, rainfall, following parts of the study, the first step is the preparation of Digital Elevation Model (DEM). The slope, elevation, and curvature maps were created in ArcGIS software using DEM of 30 × 30 resolution (2944 row, 2726 column) download from Earthdata (the ASTER Global Digital Elevation Model version 3). Additionally, topographic wetness index (TWI), topographic position index (TPI), stream power index (SPI), were calculated. A Landsat OLI image was download on July 19, 2019 and was treated to edit NDVI and LULC maps. To create the final flood susceptibility map, we sum the model values (FR, WoE, WF) of factor using Eq. (1). This equation allows us the calculate the Floods Susceptibility Index (FSI), which represents the relative probability of flood occurrence based on historical flood data obtained through the frequency ratio analysis.

$$FSI = \sum FR(elevation) + \ldots FR(Rainfall) \quad (1)$$

To visualize and interpret the prone areas, it is necessary to generate susceptibility maps for each model. The susceptibility values are categorized: very low, low, moderate, high, and very high. This version provides a clear representation of the varying levels of susceptibility in different areas.

4 Study Area

The high Oum Er Rbia watershed is located in the South-West of the Central Middle Atlas, northern Morocco, and administratively it is located in the Khenifra province. The basin extends over an area of 3387.7 km^2, and is localized between longitudes 5°05′ and 5°50′ W and latitudes 32°35′ et 33° N. It is bordered by the Bouregreg Basin in the North, the Tensift and Draa Basin in the South, and the Rheris and Moulouya Basin in the East (Fig. 2). The basin is characterized by an altitude that varies from 661 m to the southeast to 2400 m to the northeast with an average of 1373 m. The basin of high Oum Er Rabia is devised in two topographic components: cause middle Atlasic which is characterized by high slope values from 19° to 31°. The annual average varies between 400 and 700 mm/year. Moreover, the region receives rain even in the dry season, due to the mountain effect, with thunderstorms which are quite frequent, causing serious damage. The study area is a forested region, particularly at higher elevations with gorgeous, huge forest dominated by green chains, cedars, and some conifers. Generally, the province's soil is generally distributed as Harch soils (crude minerals): representing 57% of cultivated land, Hamri soils: representing 20% of cultivated land, and Tirs soils: representing 14% of cultivated land (Fig. 3).

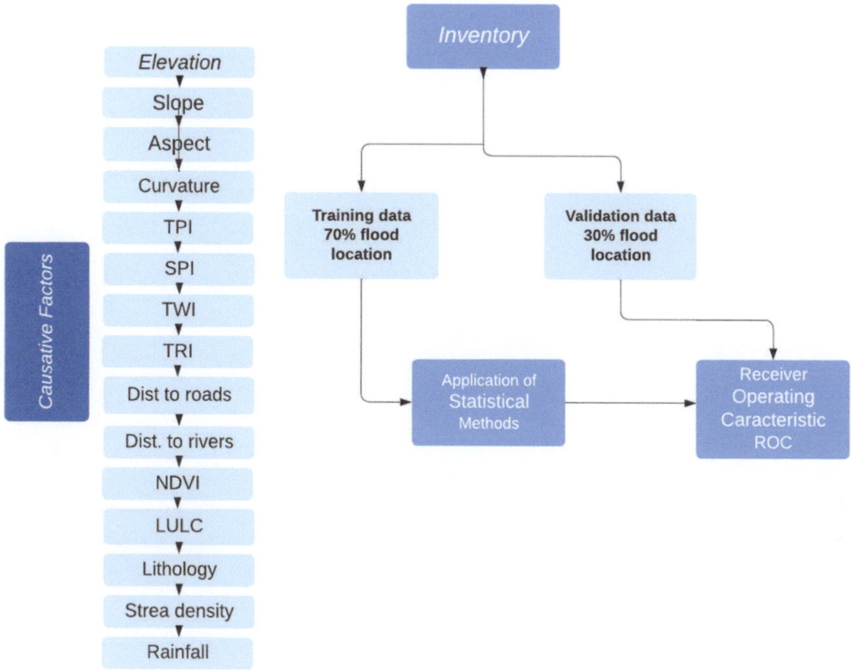

Fig. 1 Flow chart of the study by Sonia HAJJI

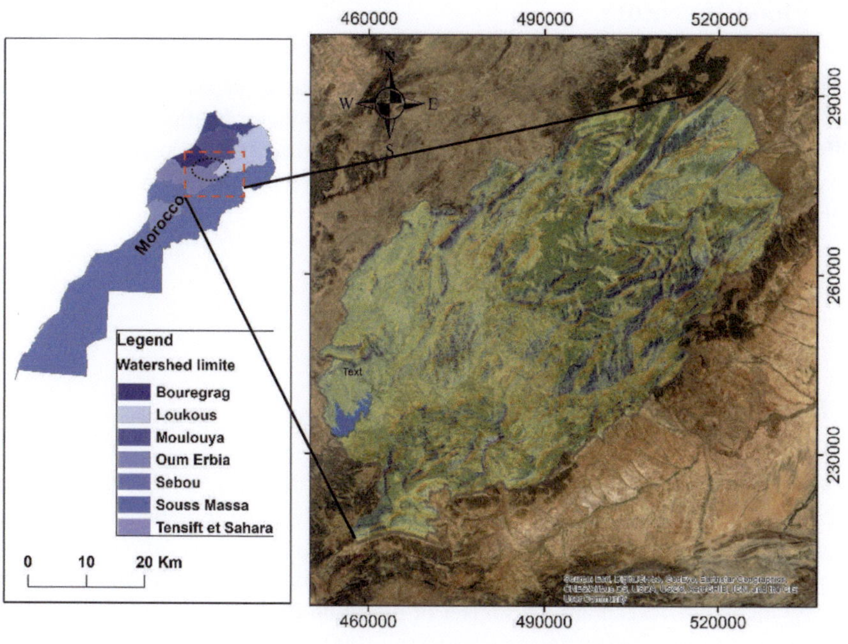

Fig. 2 Geographic situation of the high basin of Oum Er Rbia (Morocco)

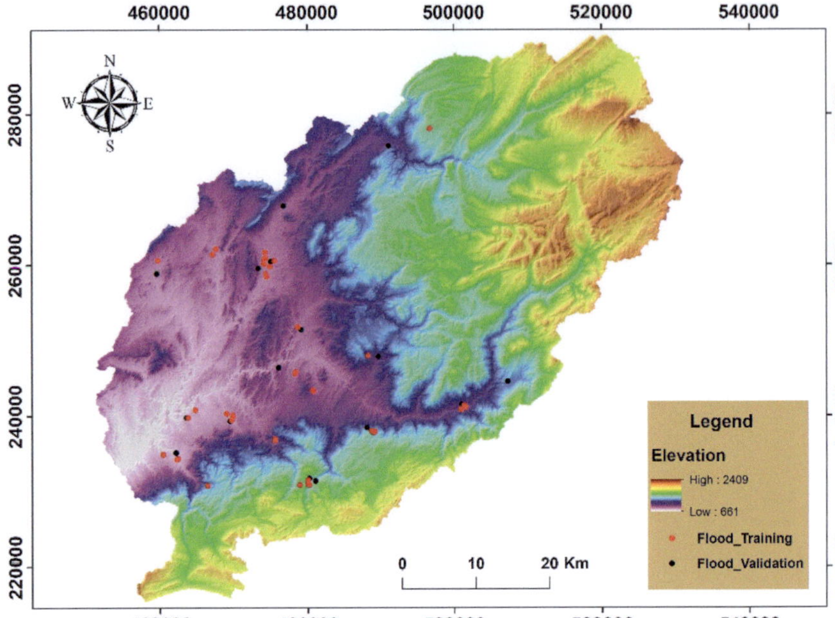

Fig. 3 Elevation map of the study area (Morocco)

5 Flood Inventory

The future flood could be predicted by analyzing previous records of its occurrence. As a result, the flood inventory is considered critical factor for flood prediction. In this research study, include a flood inventory map with 60 flood points. The flood inventory map was divided randomly [20] into two datasets such as training and validation of statistical models.

6 Flood Conditioning Factor

6.1 Elevation

The digital elevation is a primary data source for the extraction of topographic factors. Topography has a significant impact on the spatial distribution of hydrological conditions, including soil moisture and groundwater flow. Among the flood-causative factors, the elevation factor holds significant importance in flood susceptibility mapping. To capture the impact of elevation on flood susceptibility, it was classified into nine distinct classes: <800, 800–1000, 1000–1200, 1200–1400, 1400–1800, 1800–2000, 2000–2200, and >2200 m (Fig. 4). This classification allows for

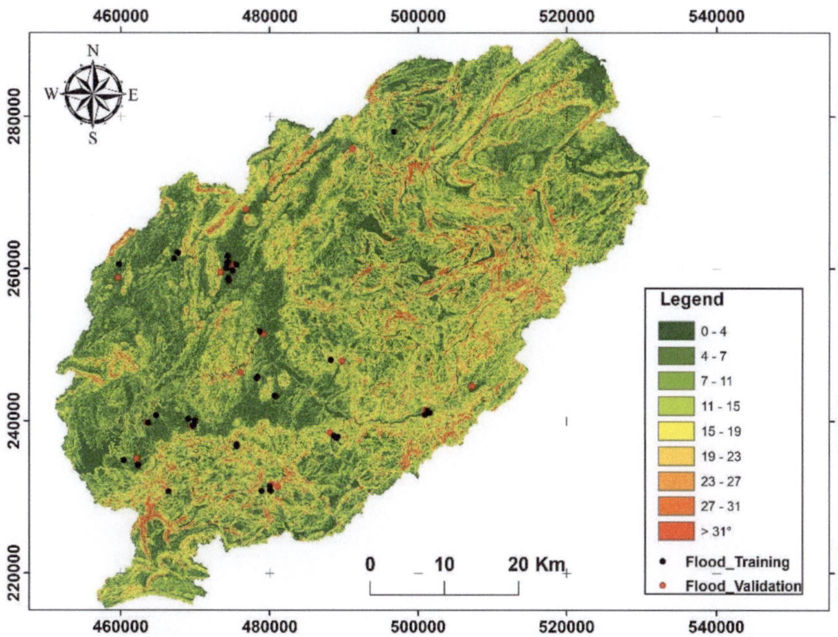

Fig. 4 Spatial distribution of slope angle (in degree) of the Oum Er Rbia watershed (Morocco)

a detailed representation of the elevation range and its influence on the likelihood of flood occurrence in different areas.).

6.2 *Slope Angle*

The slope angle regulates surface runoff infiltration and water flow velocity. In areas with gentle slopes, runoff occurs at a slower rate, giving more time for infiltration. However, as the slope angle increases, water infiltration decreases, while the velocity of water increases. This leads to a significant volume of runoff reaching rivers and flat areas, resulting in flooding. Interestingly, flood events commonly occur in regions with lower slopes. Although an increase in slope can provide in more runoff and higher velocity, areas with flatter terrain are typically more vulnerable to floods. Regions that have low slopes or regional flat surfaces are more vulnerable to flash floods, due to water accumulates and rises in such areas.

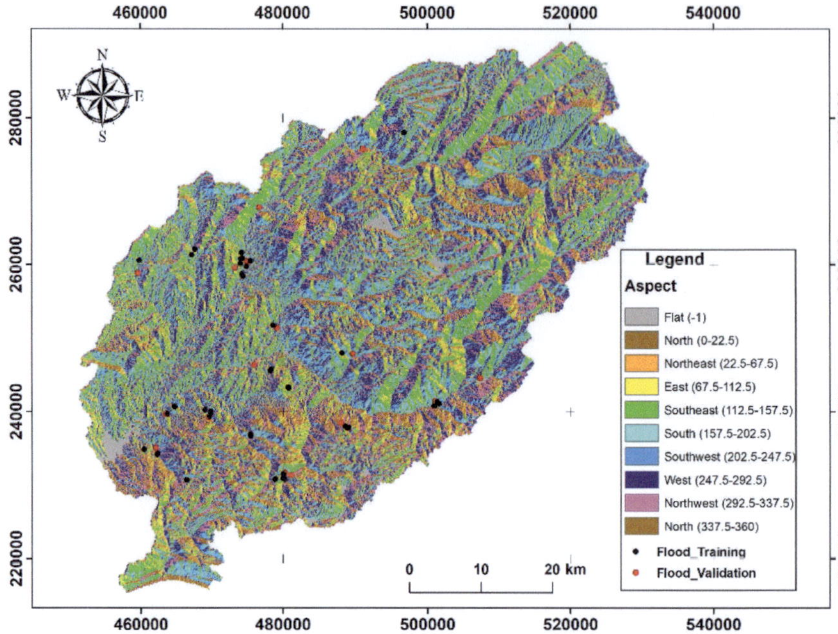

Fig. 5 Aspect map of the Oum Er Rbia watershed (Morocco)

6.3 Slope Aspect

The slope aspect map of a region presents the trajectory of the surface slope. The direction in which a slope faces the sun (aspect) has a significant impact on microclimate. The slope aspect is a significant factor impacting the occurrence of floods [22]. It affects hydrological processes [25] and then, impacts vegetation cover and soil moisture. In this research study, the aspect map was generated from the DEM with ten classes: flat, northeast, south, southeast, southwest, west, northwest, and north (Fig. 5).

6.4 Curvature

Curvature serves as a valuable geomorphological indicator in the context of flooding. Notably, flat terrain areas are particularly susceptible to a heightened risk of flooding. To depict plan curvature within a GIS (Geographic Information System) framework, the convention adopted was to classify negative values as incurved, zero as representing flat terrain, and positive values as indicative of convex area, as detailed in the study by [15]. It is worth mentioning that the resulting plan curvature map was categorized into nine distinct classes, as illustrated in Fig. 6.

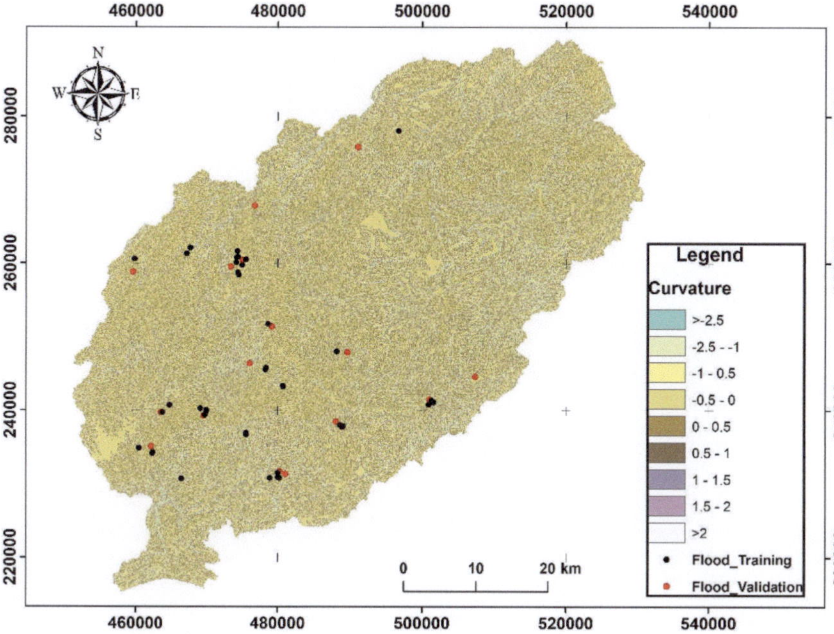

Fig. 6 Curvature map of the Oum Er Rbia watershed (Morocco)

6.5 *Topographic Wetness Index (TWI)*

The TWI is a topographic attribute that describes the geographical distribution of wetness conditions, soil moisture and erosion [26]. High TWI values have a high correlation with floodplain development. TWI can be calculated using the following equation:

$$TWI = \ln\left(\frac{\alpha}{\tan\beta}\right) \tag{2}$$

In this equation, α represents the total upslope area draining through a point (per unit contour length and tanβ represents the slope angle at that point. In the current study, the TWI map was created in SAGA-GIS and classified into nine classes: <2.2, 2.2–3.7, 3.7–5.2, 5.2–6.7, 6.7–8.2, 8.2–9.7, 9.7–11.2, 11.2–12.7, and >12.7 (Fig. 7).

6.6 *Stream Power Index (SPI)*

SPI presents the erosive power of water flow. The estimation of SPI is performed with (Eq. 3):

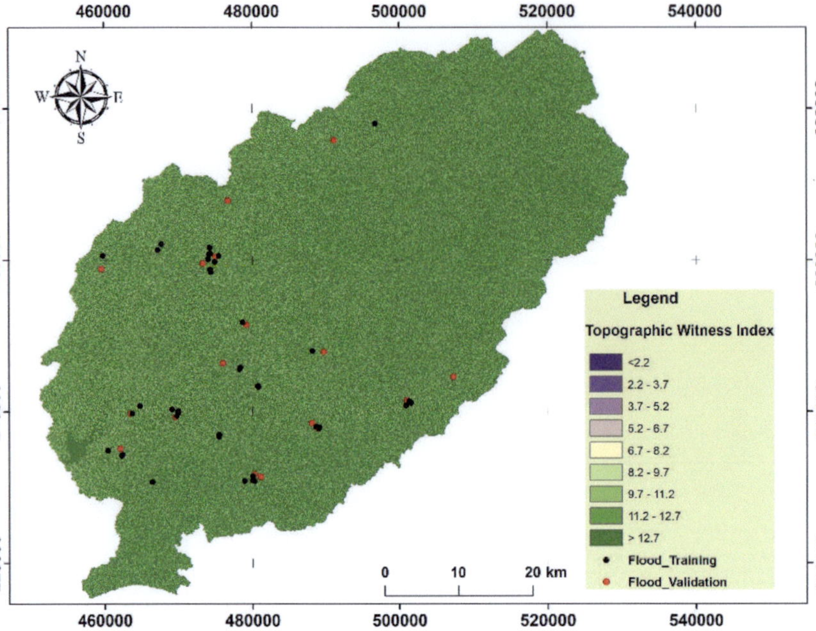

Fig. 7 Topographic witness index (TWI) map (Morocco)

$$\mathbf{SPI} = \mathbf{As} \times \mathbf{tan\beta} \tag{3}$$

where A_s is the catchment area and β is the local slope measured in degrees

The stream power index is defined as the significant performance of fluvial sediment transport and river channel erosion. In the context of this current study, the SPI map underwent a reclassification process, resulting in its categorization into nine distinct classes: <−6, −6 to −2.5, −2.5 to −0.4, −0.4 to 1.7, 1.7–3.8, 3.8–5.9, 5.9–8, 8–10, and >10, as visually represented in Fig. 8.

6.7 Topographic Position Index (TPI)

We computed the topographic position index (TPI) using SAGA GIS 2.3.2, employing a Digital Elevation Model (DEM) as the source data. TPI quantifies the difference in elevation between a focal cell and its neighbouring cells within the DEM.

In this context, positive TPI values designate locations that sit at a higher elevation than the average elevation of their immediate surroundings, often signifying ridges. Conversely, negative TPI values indicate locations that are situated at lower elevations compared to their surroundings, typically representing valleys. TPI values close to zero correspond to either flat areas, where the slope is nearly zero, or areas with a

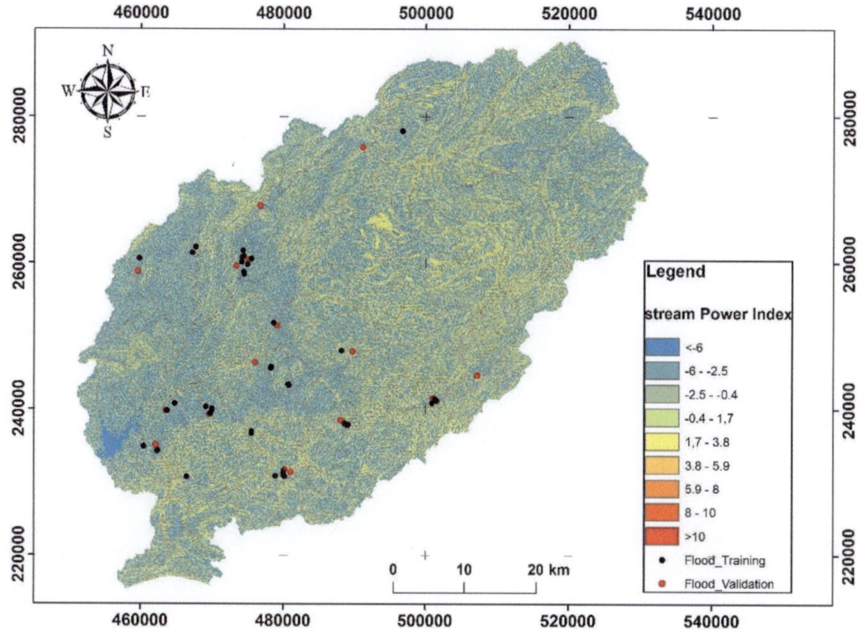

Fig. 8 Stream power index (SPI) map (Morocco)

consistent slope, where the slope of the point remains significantly different from zero. Figure 9 visualizes the distribution of TPI values.

6.8 Topographic Ruggedness Index

The TRI was created to express the level of elevation variation between the adjacent cells in DEM. It determines the difference in elevation values between a centre cell and the eight cells that surround it. Following that it squares each of the eight elevation difference values, making them all positive, and averages the squares. The topographic ruggedness index is then calculated using the square. TRI map is shown in (Fig. 10).

6.9 Distance from the River

In general, areas near rivers are more prone to flooding in the river basin as water flows from higher elevations and accumulates in lower elevation. Areas other terrestrial water bodies, such as ponds, and dams are also likely to flood during heavy rains because the terrain in their vicinity is nearly flat. Based on the meteorological and

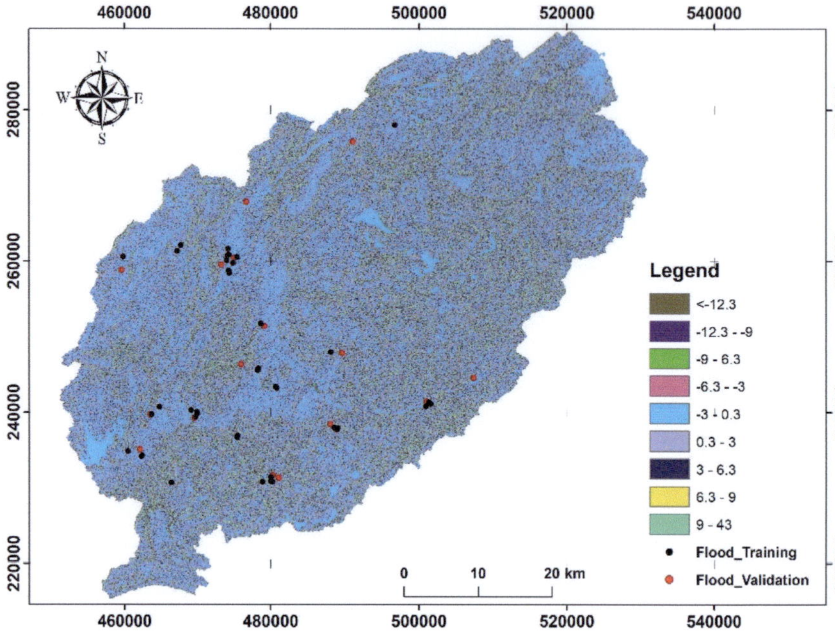

Fig. 9 Topographic position index (TPI) map (Morocco)

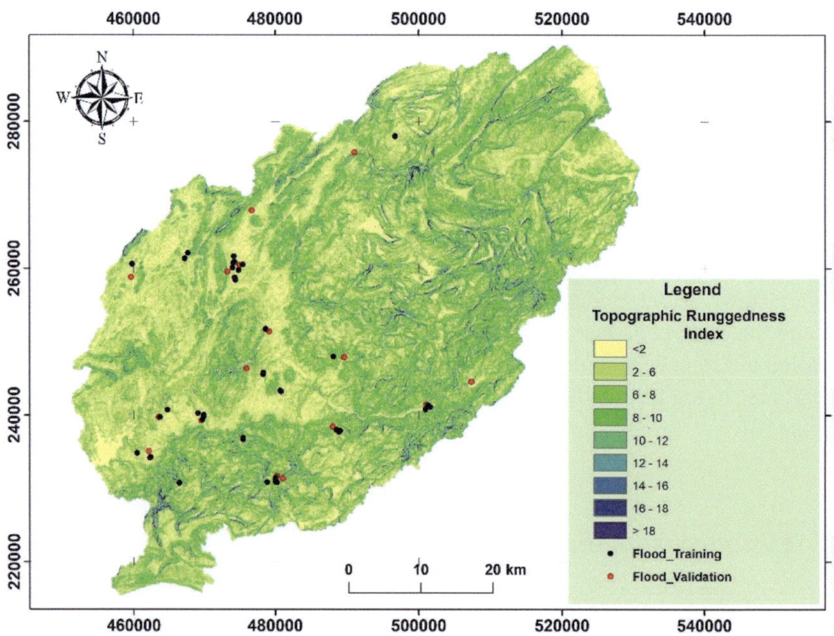

Fig. 10 Topographic ruggedness index (TRI) map (Morocco)

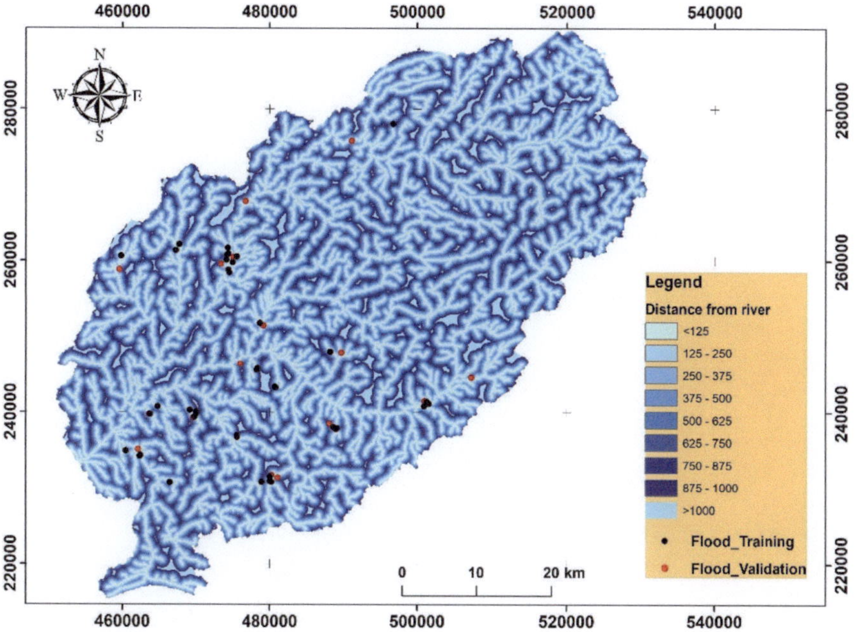

Fig. 11 Distance (in m) from rivers map (Morocco)

topographical conditions, pluvial flash floods can occur far water bodies. In the current study, nine classes of buffer have been developed at a buffer distance of 100 m from the river (Fig. 11).

6.10 *Distance to Roads*

Distance to roads is also an essential consideration in vulnerability to flooding mapping. Impermeable roads and surrounding urban surfaces have a significant impact on flood levels. They reduce the terrain's infiltration capacity and cause runoff. For instance, nine classes of buffers have been developed at a buffer distance of 100 m from the road (Fig. 12).

6.11 *Stream Density*

Stream density is one of the important factors influencing flood occurrences. The region characterized by greater stream density is more likely to respond to rainstorms and more prone to flash flooding. The stream density map was reclassified into

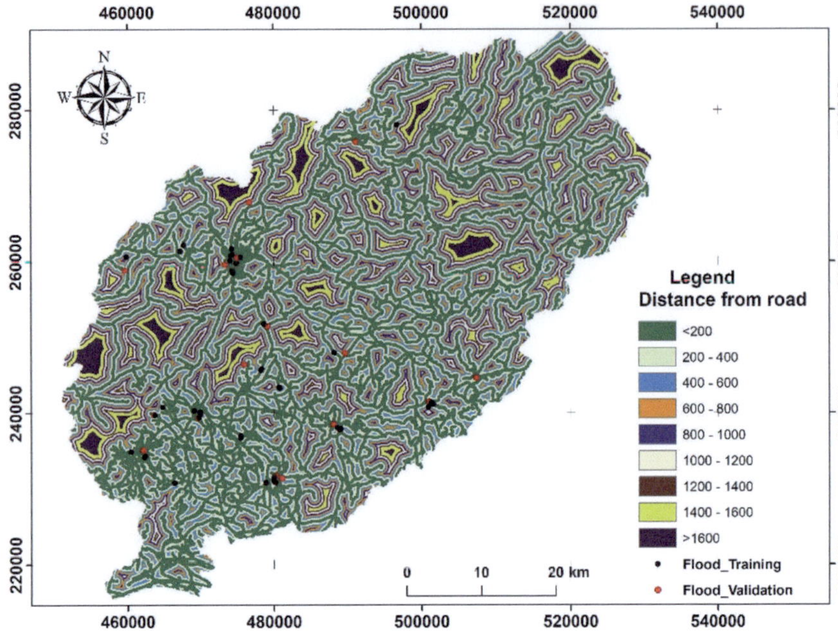

Fig. 12 Distance (in m) from road map (Morocco)

nine classes: <0.4, 0.4–0.8, 0.8–1, 1–1.2, 1.2–1.4, 1.4–1.8, 1.8–2, 2–2.2, and >2.2 (Fig. 13).

6.12 *Rainfall*

Rainfall is a major influencing factor modelling and the primary cause of flood generation. Several previous studies [10] established a relationship within rainfall and flooding. The precipitation map was created from 18- year records of rainfall data of the TRMM (Tropical Rainfall Measuring Mission). The rainfall map was classified into nine classes from <225 to >645 mm (Fig. 14).

6.13 *Landuse/Landcover*

Land use plays has an important effect on flooding, which influencing directly or indirectly some hydrological process such as evapotranspiration, infiltration, and runoff. The land use of the investigation area was identified by the supervised classification method in GIS from the Landsat OLI image. The land use types identified in the Oum Er Rbia watershed are water, vegetation, construction, and barre soil (Fig. 15).

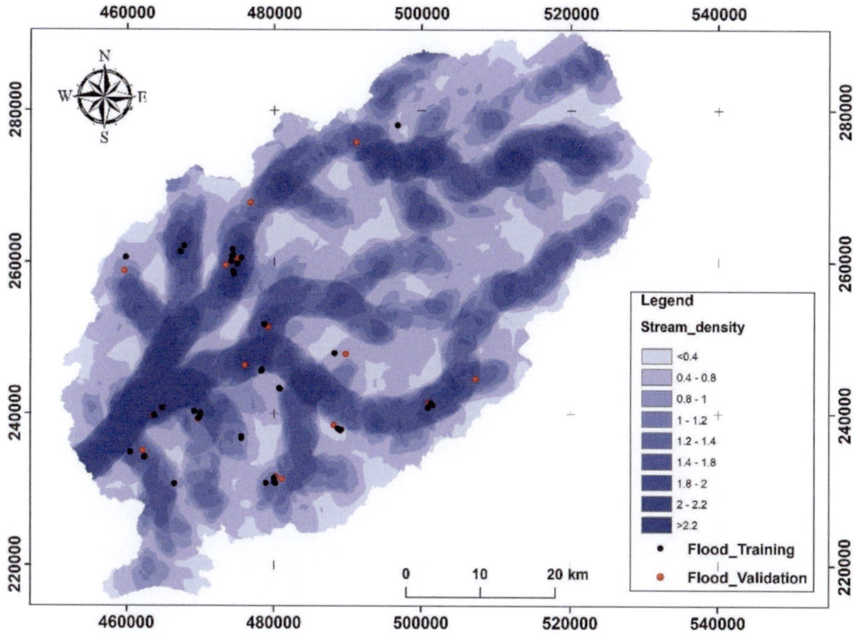

Fig. 13 Stream density map (Morocco)

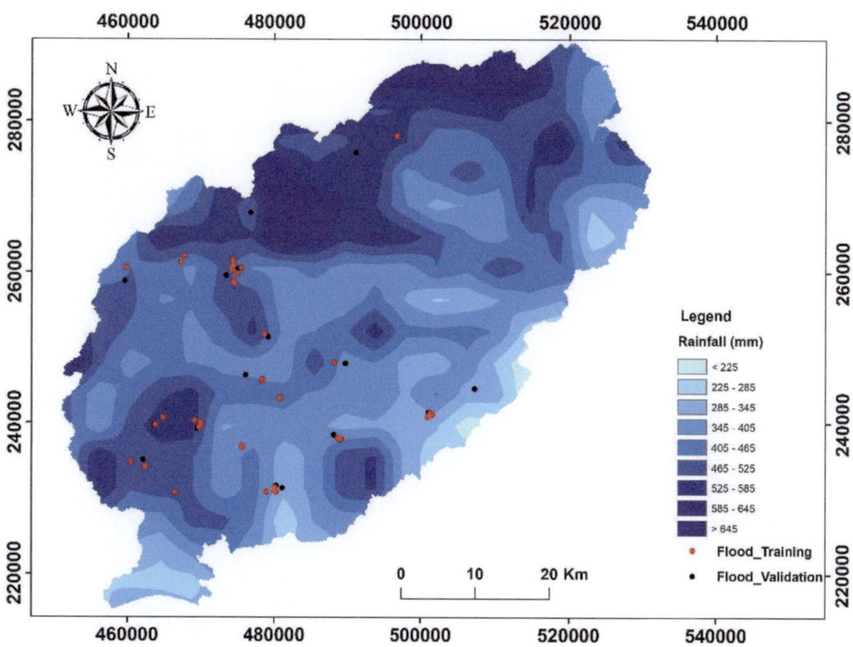

Fig. 14 Rainfall map of the study area (Morocco)

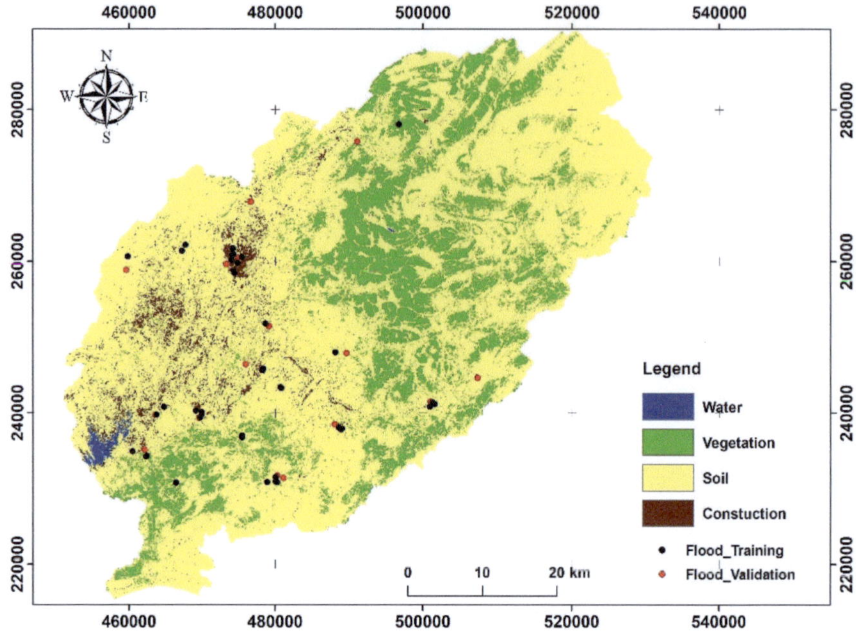

Fig. 15 Land use map of the study area (July 2019) (Morocco)

6.14 *Normalized Difference Vegetation Index*

The normalized difference vegetation index is another causative factor. According to [14], negative values indicate water and positive values indicates vegetation, implying that NDVI is negatively related to flooding: Higher NDVI values suggest a lower likelihood of flooding, whereas lower NDVI values indicate a higher flood risk. In this study NDVI values range from −0.33 to 0.91 (Fig. 16).

NDVI value is calculated using the common formula (Eq. 4) from the Landsat 8:

$$NDVI = \frac{NIR - Red}{NIR + Red} \tag{4}$$

6.15 *Lithology*

Lithology or geological characterization is an important factor for flood susceptibility mapping. The area with highly resistant rock and highly permeable subsoil has lower drainage density [6]. Lithological map was created by digitizing the geological map of Morocco (1/1000000-scale). The lithology of the study area was distinguished into 11 categories or formations: 41% of the area of the basin is covered by limestone and

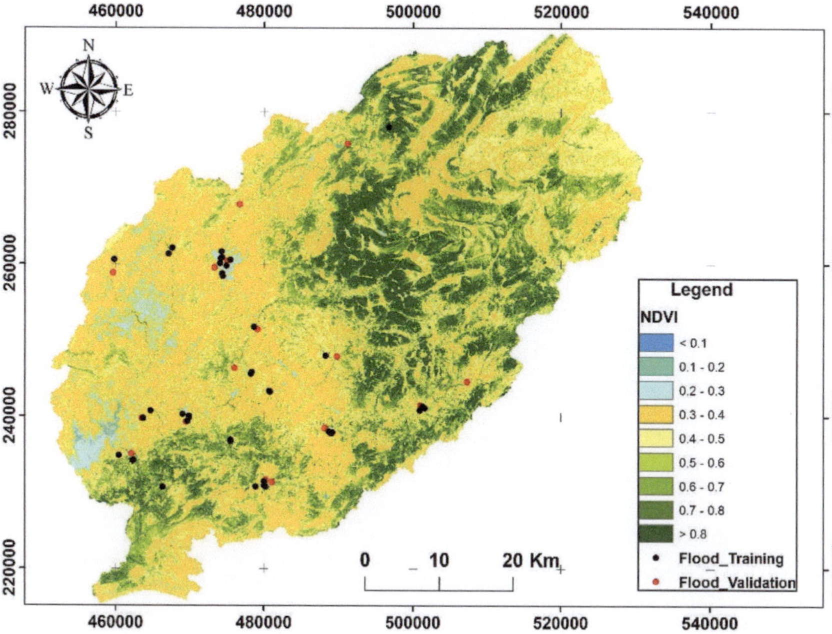

Fig. 16 NDVI map of the study area (Morocco)

dolomite of Jurassic in terms of the Middle Atlas precisely in the cause area, 21% of lithology was included by the magmatic rock mostly is the basalt. The Triassic formation constituted an impermeable level. The schist of carboniferous presents 12% of the area, in general, this formation has covered the west of the basin around of Khenifra region (Fig. 17).

7 Flood Modelling Susceptibility

7.1 Frequency Ratio

FR model identifies the quantitative relationship among the spatial distribution of the floods and flood causative factors [28]. The frequency ratio for each class of conditioning factor (i.e., elevation, slope, etc.........) is determined by the division of the flood occurrence ratio by the area ratio. Each factor frequency ratio was calculated using Eq. 5:

$$FR = \frac{Pourcentage\ of\ flood\ (PL)}{Pourcentage\ of\ the\ class\ of\ each\ conditionnig\ factor\ (PD)} \tag{5}$$

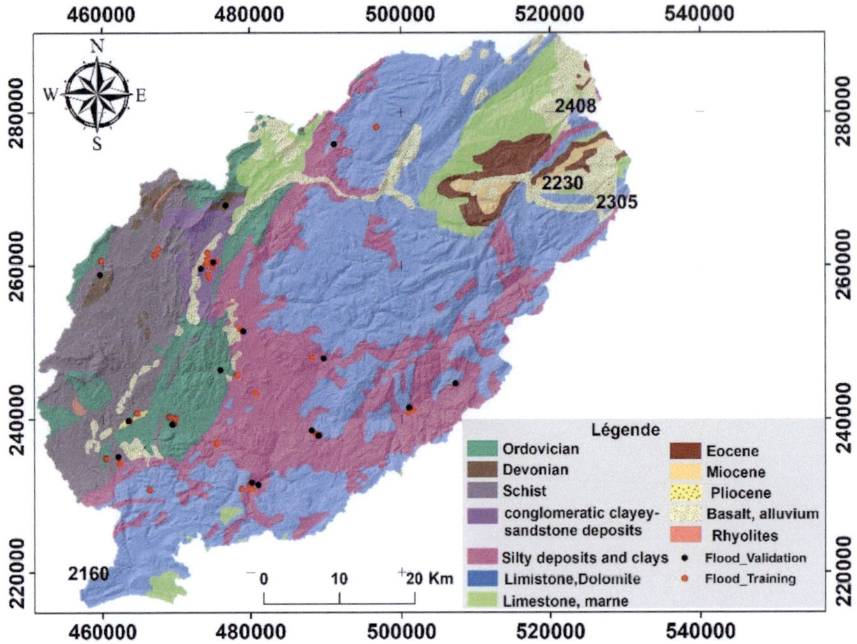

Fig. 17 Lithological map of the study area (Morocco)

Higher bivariate probabilities (greater than 1) indicate a stronger correlation within flood incidence and flood causative factors, while lower probabilities (less than 1) indicate a weaker correlation.

7.2 *Weighting Factor*

The weighting factor model assigns weights to each class factor based on the statistical index value. Çevik and Topal [7] describe the Wf method as a modified form of the statistical index. The flood inventory map is initially rasterized before being compared to various rasterized factors. SI values are obtained for each flood pixel in each layer using the following equation:

The corresponding values for each pixel in every layer are subsequently added up. The weighting factor values for each layer are calculated in the range of 1 to 10, using the following equations:

where:

Si represents the weight of each class I of the causative factors;

NF_i: is the number of flood pixels in a thematic class I;

NF_i: is the total number of pixels in a thematic class I;

NF_t: is the number of pixels of all floods;

NF_t: is the total number of all pixels.

$$T(SI) = \sum_{i=1}^{n} (SI) \times S.pix \tag{6}$$

$$Wf = \frac{TSI\ (value\ ) - \ Min(TSI\ value\ )}{Max(TSI\ value\ ) + Min(TSI\ value)} \times 10 \tag{7}$$

where:

T(SI) represents the total value of flooding pixels for a given class of factors, with n indicating the number of parameter classes. Wf is the weight factor for each layer of maps (final weight).

The TSI indicates the minimum and maximum values of total weights among layers.

Min (TSI) the minimum value of the overall weight within layers.

Max (TSI) the maximum value of the overall weight within layers.

The Wf values for each layer are multiplied by Si values of the factor classes. Finally, all factors are combined, and the flood susceptibility map is generated using the Wf method.

7.3 Weight of Evidence

Weight of evidence is a quantitative approach that uses the Bayes rule to predict the probability of events. This method is prevalent due to its use of the Bayesian probability model to determine the relative importance of causative factors through statistical means [11, 15].

This method calculates the relative weights of evidence coefficients according to the association between each type of factor and flood absence or flood presence. The computation requires both positive weight (W+) and negative weight (W−). The weights (W−). The method estimates the weight for each flood causative factor (A) based on the presence or absence of floods (B) in the given region area [11] as follows:

$$Wi+ = Ln \frac{P\ \{B|A\}}{P\{B|A\}} \tag{8}$$

$$Wi- = Ln \frac{P\{\overline{B}|A\}}{P\{B|\overline{A}\}} \tag{9}$$

P is the probability, and lm is the natural logarithm. B and $\overline{B}$ indicate the presence and the absence of the conditioning factors, respectively. A represents the presence of a flood, while (A) indicates the absence of a flood.

A positive weight (W+) indicates that the condition factor is present at flood locations; and its values indicate a positive correlation between the presence of causative factor and floods [19]. A negative weight (W−) demonstrates the absence of causative factor and the level of negative correlation [19].

The weight contrast © is used in flood susceptibility mapping to measure and reflect the spatial relationship between conditioning factors and flood occurrences. Pourghasemi et al. [17] found that C is positive for positive relationships and negative for negative spatial relationships. The standard deviation of W is calculated as follows (Eq. 10).

$$S\,(C) = \sqrt{S^2(W+) + S^2(W-)} \tag{10}$$

where, S2 (W+) and S2 (W−) are the variance of the positive and negative weights, respectively. The variance of the weights can be calculated using the following Eqs. 8 and 9:

$$S^2(W+) = \frac{1}{N\{B \cap A\}} + \frac{1}{B \cap \overline{A}} \tag{11}$$

$$S^2(W-) = \frac{1}{\{\overline{B} \cap \overline{A}\}} + \frac{1}{\{\overline{B} \cap \overline{A}\}} \tag{12}$$

8 Result and Discussion

8.1 Frequency Ratio Model

This relationship between floods locations and each conditioning is depended on the ratio of the area where floods occurred to the total area. A value of 1 indicates an average correlation. Values higher than 1 indicate a stronger correlation. Figure 18 shows that approximately 10% of the total flood of area in a high and very high susceptibility class, and located in two areas, the first is nearly to Ait Ishak locality, and the second area is about 4.5 km to Lehri locality. 25% of the basin area is moderate localized essentially in the Khenifra region, and 64% of the area is sited in high altitudes such as Oum Er Rbia spring and Aguelmam Azezgza.

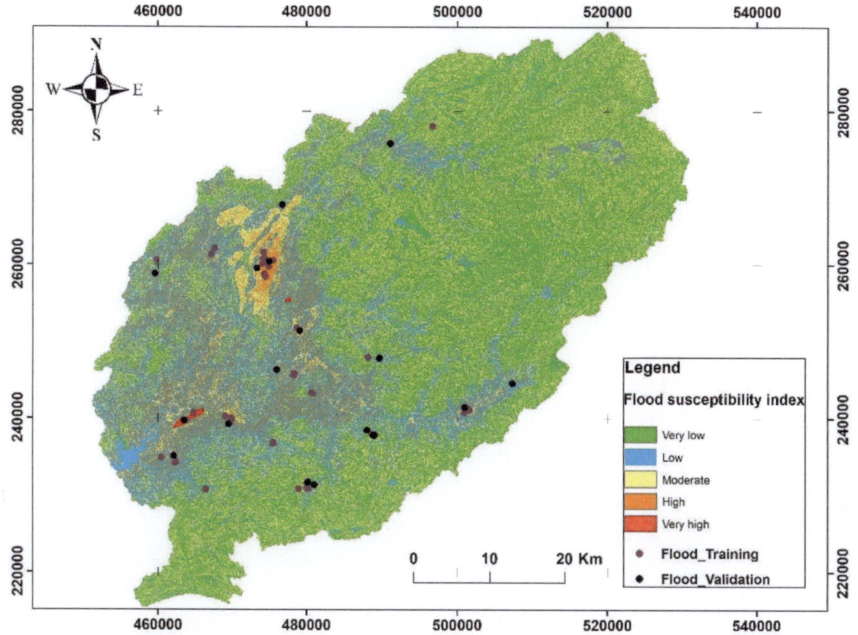

Fig. 18 Flood susceptibility map using the FR model (Morocco)

8.2 Weighting Factor Model

The present susceptibility map using Wf was classified into five classes: very low, low, moderate, high, and very high susceptibility classes (Fig. 19). Generally, the very high and high susceptibility represents 13% of the high Oum Er Rbia basin area and is located essentially in the Khenifra region and zones near the main rivers, 30% in moderate, 34% in low and 21% in very low flood hazard.

8.3 Weighting of Evidence

For the WoE model, the process to generate the maps susceptibility was similar to the FR, Wf model. The flood susceptibility maps (Fig. 20) were obtained after a combination of the weight of evidence for each factor. The analysis of maps illustrates approximately 31% of total area is classified as very high and high flood risk located in Khenifra, Ait Ishak, Lehri, Tighassaline, and Ouaoumana, 25% in moderate, 24% in low, and 19% in very low flood hazard.

The analysis of map susceptibility using frequency ratio illustrates that approximately 10% of entire flood area in a high and extremely high susceptibility class, and located in two areas, the first is near to Ait Ishak locality, and the second area is

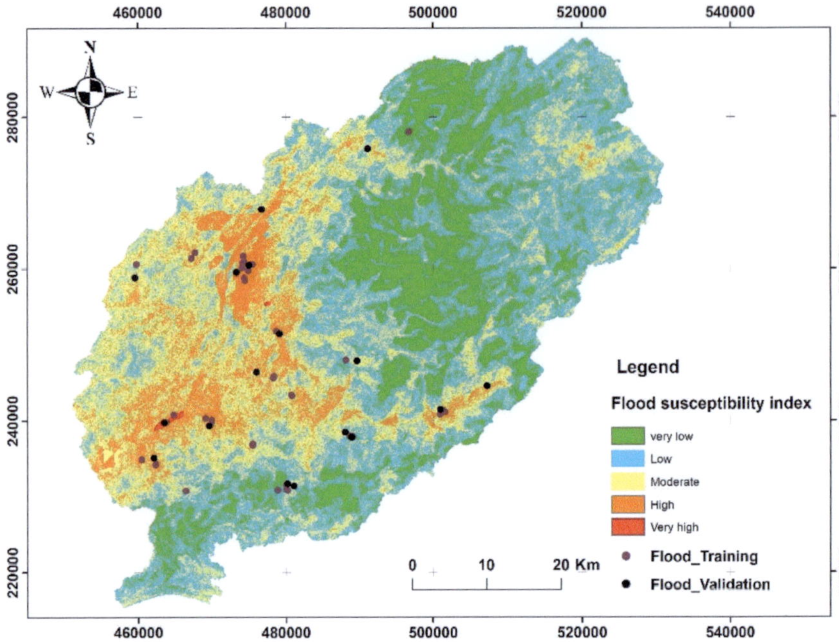

Fig. 19 Flood susceptibility using the WF model (Morocco)

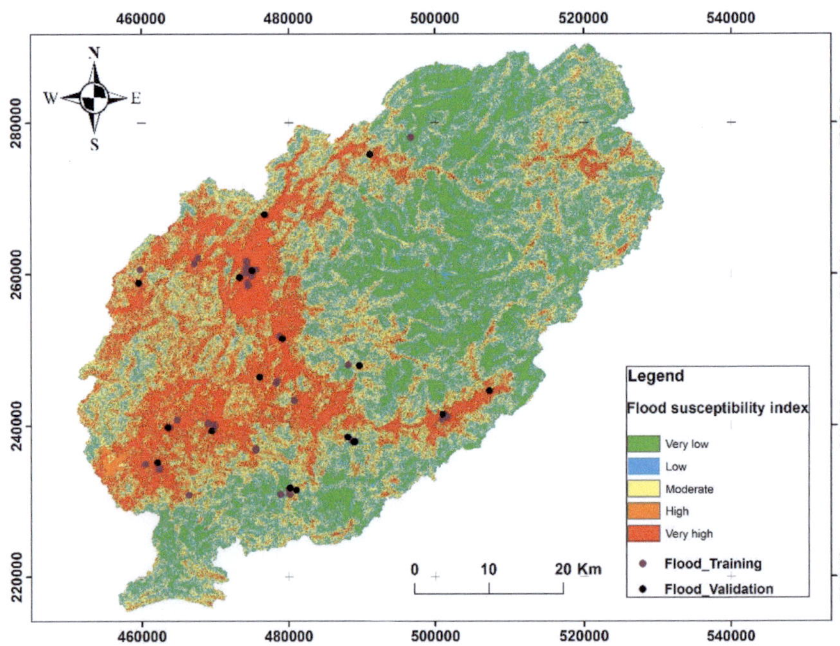

Fig. 20 Flood susceptibility map using WoE model (Morocco)

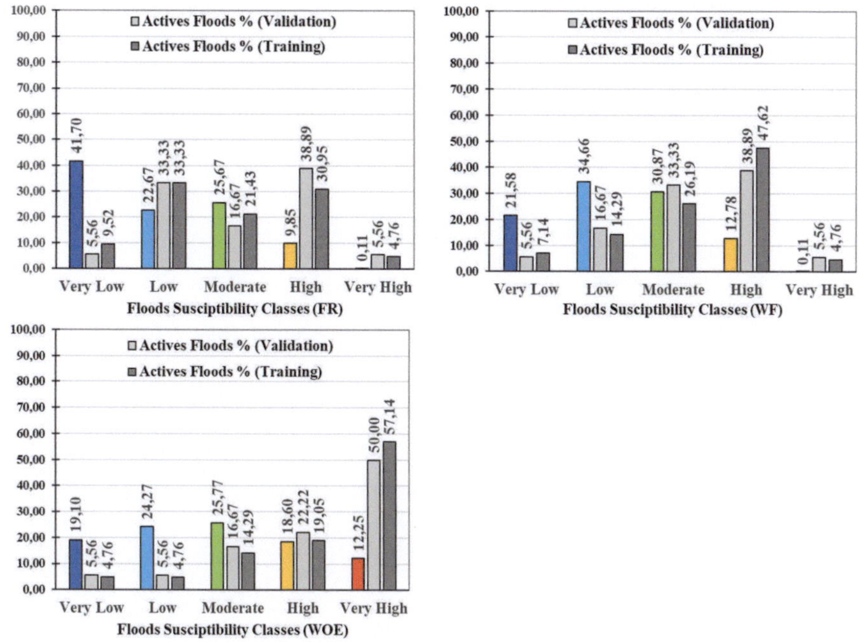

Fig. 21 Flood susceptibility classes of the models (Morocco)

about 4.5 km to Lehri locality. 25% of the basin area is moderately localized essentially in the Khenifra region, and 64% of the area is sited in high altitudes such as Oum Er Rbia spring and Aguelmam Azezgza. The present susceptibility map using the weighting factor was classified into 13% of the basin to the high and very high susceptibility of the total area of the high Oum Er Rabia basin, and located essentially in the Khenifra region and zones near the main rivers, 30% in a moderate, 34% in low and 21% in very low flood hazard (Fig. 21).

The analysis of maps illustrates approximately 31% of the total area in a high and very high flood risk located in Khenifra, Ait Ishak, Lehri, Tighassaline, and Ouaoumana, 25% in moderate, 24% in low, and 19% in very low flood hazard.

Consequently, the area under the curve (AUC) is simply used for the validation of flood susceptibility map. In the study area, the AUC values for frequency ratio, the weight of evidence, and the weighting factors model were 0.752, 0.768, and 0.733% respectively (Fig. 22), and the prediction accuracy of these models is 75.27%, 76.8%, and 73.33%.

According to the evaluation of the AUC results, the WoE is the most accurate prediction, and the best for identifying flood susceptibility map, followed by FR and the latest prediction id the WF. However, generally, the prediction accuracies of the three models satisfactory and reasonable. For the prediction rate, the weight of evidence was also the first with AUC = 0.751%, followed by the weighting factor with AUC = 0.732%, and frequency ratio with AUC = 0.713% (Fig. 23), these

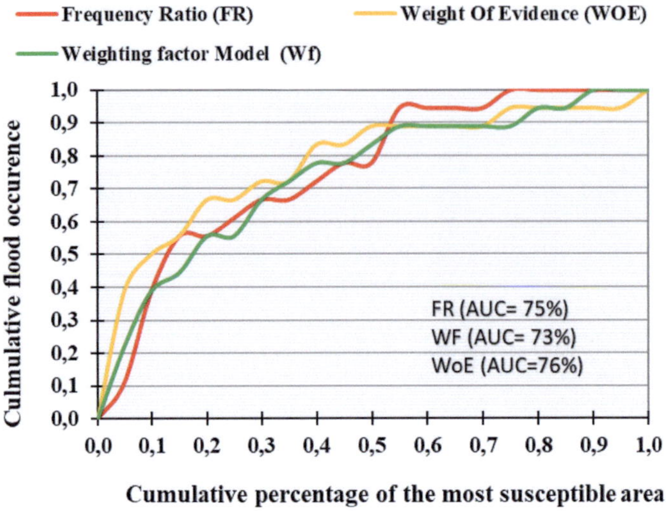

Fig. 22 The ROC curve value of the success rate

results showed that the weight of evidence is a preferment model. Therefore, we can state that the AUC values are statistically significant.

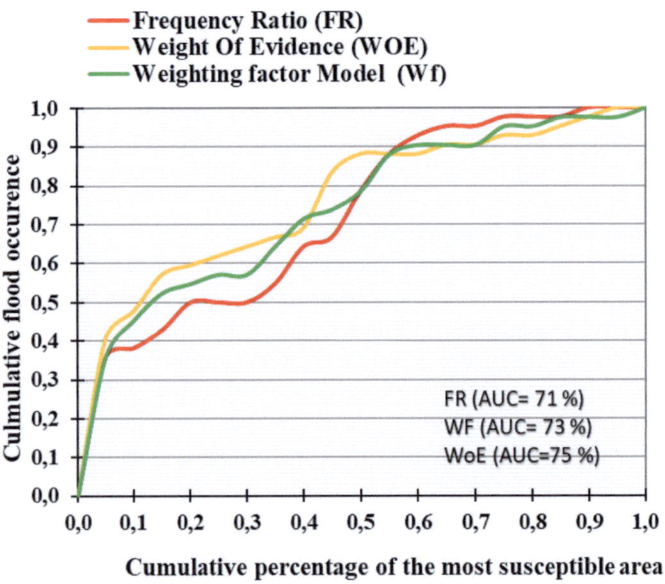

Fig. 23 The ROC curve value of the prediction rate

9 Conclusion

The current study aimed at assessing flood hazards and mapping flood-prone zones in the upper Oum Er Rbia watershed. To achieve this, the frequency Ratio (FR), Weight of Evidence (WoE) methods were integrated with remote sensing and GIS for flood risk mapping. Eight conditioning factors were considered to generate flood susceptibility maps: elevation, slope, curvature, aspect, Topographic Wetness Index (TWI), Stream Power Index (SPI), Terrain Roughnness Index (TRI), distance to rivers, distance to roads, lithology, rainfall, Normalized Difference Vegetation Index (NDVI), land use/land cover (LULC), and drainage density.

A total of 60 flood-affected points were identified through field surveys and analysis of Google Earth imagery. These points were randomly divided into two sets: 42 points were used to train the model, while the remaining 18 were reserved for validation. The flood risk areas were categorized into five susceptibility classes: very high, medium, low, and very low. The model's performance was evaluated using the ROC curve, with the FR model achieving an AUC of 75%, WF at 73%, and WoE at 76%, indicating good predictive accuracy for flood susceptibility.

The study also determined the most influential factors contributing to flood occurrence in the region. Based on the WF results, lithology, SPI, TWI, drainage density, land use, and slope were identified as the most significant factors, while TPI, aspect, elevation, and NDVI had moderate influence. Factors, while TPI, aspect, elevation, and NDVI had moderate influence. Factors such as distance to rivers, TRI, distance to roads, rainfall, and curvature had minimal impact on flood hazards.

This research highlights the effectiveness of the FR, WF, and WoE methods for flood susceptibility mapping within a GIS framework.

10 Recommendations

Flood susceptibility mapping is essential for assessing and managing flood risks. Accurate determination of flood prone areas is crucial to executing effective mitigation measures and informing the process of making decisions. For flood susceptibility mapping there are recommendations to follow for good accuracy. Among these recommendations:

- Explore the various factors contributing to flood susceptibility, including topography, land use/land cover, soil characteristics, hydrological conditions, climate, patterns, and human activities.
- An inventory containing a considerable number of flood locations.
- Choosing the high performance models
- Analyzing the most contributing factors in flood susceptibility and the redundancy factors.

- Develop Adaptive Flood management plan. Government and planners should develop adaptive flood management plan strategies that involve in response to new data and technologies. This includes continuously updating flood risk maps, incorporating real-time data, and revising zoning laws to reflect the most recent susceptibility findings.
- Investments in early warning systems, linked with susceptibility maps, should be a top.
- Flood resilient infrastructure designs be implemented, particularly in rapidly urbanizing areas. Green infrastructure such as permeable pavements, green roofs, and wetland restoration, should be encouraged in order to be absorb excess rainfall, reduce flood damage, and preserve natural drainage systems.

References

1. Al-Hinai H, Abdalla R (2021) Mapping coastal flood susceptible areas using Shannon's entropy model: the case of Muscat governorate, Oman. ISPRS Int J Geo-Inform 10:252. https://doi.org/10.3390/ijgi10040252
2. Althuwaynee OF, Pradhan B, Lee S (2012) Application of an evidential belief function model in landslide susceptibility mapping. Comput Geosci 44:120–135. https://doi.org/10.1016/j.cageo.2012.03.003
3. Aydin MC, Sevgi Birincioğlu E (2022) Flood risk analysis using gis-based analytical hierarchy process: a case study of Bitlis Province. Appl Water Sci 12:122. https://doi.org/10.1007/s13201-022-01655-x
4. Bibi K, Siddique F, Gul S, Atta-ur Rahman, Mustafa FB (2022) Methods and approaches of flood susceptibility assessment and mapping: a review in geographical perspective. In: Mustafa FB (ed) Methodological approaches in physical geography. Springer International Publishing, Cham, pp 53–69. https://doi.org/10.1007/978-3-031-07113-3_4
5. Burby RJ, Deyle RE, Godschalk DR, Olshansky RB (2000) Creating hazard resilient communities through land-use planning. Nat Hazard Rev 1:99–106. https://doi.org/10.1061/(ASCE)1527-6988(2000)1:2(99)
6. Çelik HE, Coskun G, Cigizoglu HK, Ağıralioğlu N, Aydın A, Esin Aİ (2012) The analysis of 2004 flood on Kozdere Stream in Istanbul. Nat Hazards 63:461–477
7. Çevik E, Topal T (2003) GIS-based landslide susceptibility mapping for a problematic segment of the natural gas pipeline, Hendek (Turkey). Environ Geol 44:949–962. https://doi.org/10.1007/s00254-003-0838-6
8. Charlton R, Fealy R, Moore S, Sweeney J, Murphy C (2006) Assessing the impact of climate change on water supply and flood hazard in Ireland using statistical downscaling and hydrological modelling techniques. Clim Change 74:475–491
9. Chowdhuri I, Pal SC, Chakrabortty R (2020) Flood susceptibility mapping by ensemble evidential belief function and binomial logistic regression model on river basin of eastern India. Adv Space Res 65:1466–1489. https://doi.org/10.1016/j.asr.2019.12.003
10. Elkhrachy I (2015) Flash flood hazard mapping using satellite images and GIS Tools: a case study of Najran City, Kingdom of Saudi Arabia (KSA). Egypt J Remote Sens Space Sci 18:261–278. https://doi.org/10.1016/j.ejrs.2015.06.007
11. Felicísimo ÁM (2003) Bonham-Carter, G. F. (1996): Geographic information systems for geoscientists. Modelling with GIS [WWW Document]
12. Guo E, Zhang J, Ren X, Zhang Q, Sun Z (2014) Integrated risk assessment of flood disaster based on improved set pair analysis and the variable fuzzy set theory in central Liaoning Province, China. Nat Hazards. https://doi.org/10.1007/s11069-014-1238-9

13. Khalki YE, Taïbi AN, Benyoucef A, Hannani ME, Hafid A, Mayoussi M, Zmou A, Ragala R, Geroyannis H (2007) Processus d'urbanisation et accroissement des risques à Beni Mellal (Tadla-Azilal, Maroc) : apports des SIG et de la télédétection. Mosella : revue du Centre d'études géographiques de Metz XXX, 147
14. Khosravi K, Shahabi H, Pham BT, Adamowski J, Shirzadi A, Pradhan B, Dou J, Ly H-B, Gróf G, Ho HL, Hong H, Chapi K, Prakash I (2019) A comparative assessment of flood susceptibility modeling using multi-criteria decision-making analysis and machine learning methods. J Hydrol 573:311–323. https://doi.org/10.1016/j.jhydrol.2019.03.073
15. Lee M-J, Kang J, Jeon S (2012) Application of frequency ratio model and validation for predictive flooded area susceptibility mapping using GIS. In: 2012 IEEE international geoscience and remote sensing symposium. https://doi.org/10.1109/IGARSS.2012.6351414
16. Mahmouhi E, Mahmouhi N, Wartiti M, Wissem A, Kemmou S, Bahi S (2016) The use of geographic information system for the extraction of physical characteristics of Assaka watershed: sub-basins of sayed and oum laachar wadis (southern Morocco). Int J Innov Appl Stud 16 N
17. Pourghasemi H, Moradi H, Aghda SF (2013) Landslide susceptibility mapping by binary logistic regression, analytical hierarchy process, and statistical index models and assessment of their performances. Nat Hazards: J Int Soc Prevent Mitigation Nat Hazards 69:749–779
18. Pourghasemi H, Pradhan B, Gokceoglu C (2012) Application of fuzzy logic and analytical hierarchy process (AHP) to landslide susceptibility mapping at Haraz watershed, Iran. Nat Hazards. https://doi.org/10.1007/s11069-012-0217-2
19. Pradhan B (2009) Flood susceptible mapping and risk area delineation using logistic regression, GIS and remote sensing. J Spat Hydrol 9:1–18
20. Rahmati O, Pourghasemi HR, Zeinivand H (2015) Flood susceptibility mapping using frequency ratio and weights-of-evidence models in the Golastan Province, Iran. Geocarto Int
21. Rudra RR, Sarkar SK (2023) Artificial neural network for flood susceptibility mapping in Bangladesh. Heliyon 9:e16459. https://doi.org/10.1016/j.heliyon.2023.e16459
22. Sadr MP, Maghsoudi A, Saljoughi BS (2014) Landslide susceptibility mapping of Komroud Subbas in using fuzzy logic approach. undefined
23. Saidi ME, Daoudi L, Aresmouk M, Fatima F, Boukrim S (2010) Les crues de l'oued Ourika (Haut Atlas, Maroc): Événements extrêmes en contexte montagnard semi-aride. Comunicações Geológicas t 97:113–128
24. Shafapour Tehrany M, Kumar L, Neamah Jebur M, Shabani F (2019) Evaluating the application of the statistical index method in flood susceptibility mapping and its comparison with frequency ratio and logistic regression methods. Geomat Nat Haz Risk 10:79–101. https://doi.org/10.1080/19475705.2018.1506509
25. Sidle R, Ochiai H (2013) Landslides: processes, prediction, and land use. https://doi.org/10.1029/WM018
26. Sun D, Yu Y, Goldberg M (2011) Deriving water fraction and flood maps from MODIS images using a decision tree approach. IEEE J Sel Top Appl Earth Observ Remote Sens 4:814–825. https://doi.org/10.1109/JSTARS.2011.2125778
27. Tehrany MS, Pradhan B, Jebur MN (2013) Spatial prediction of flood susceptible areas using rule based decision tree (DT) and a novel ensemble bivariate and multivariate statistical models in GIS. J Hydrol 504:69–79. https://doi.org/10.1016/j.jhydrol.2013.09.034
28. Tehrany MS, Pradhan B, Mansor S, Ahmad N (2015) Flood susceptibility assessment using GIS-based support vector machine model with different kernel types. CATENA 125:91–101. https://doi.org/10.1016/j.catena.2014.10.017
29. Yilmaz I (2010) Comparison of landslide susceptibility mapping methodologies for Koyulhisar, Turkey: conditional probability, logistic regression, artificial neural networks, and support vector machine. Environ Earth Sci 61:821–836. https://doi.org/10.1007/s12665-009-0394-9
30. Youssef A, Pradhan B, Jebur MN, El-Harbi HM (2014) Landslide susceptibility mapping using ensemble bivariate and multivariate statistical models in Fayfa area, Saudi Arabia. Environ Earth Sci. https://doi.org/10.1007/s12665-014-3661-3

Urban Heat Island in Algiers: Case of the Belcourt District

Sahraoui Bensaid, Halima Slimani, and Djihane Charif

Abstract Large cities are increasingly affected by the phenomenon of "Urban Heat Islands" (UHI), particularly during the dry summer months. This phenomenon especially affects the MENA region, which is the driest in the world and therefore the most vulnerable to climate change. The aim of this work is to identify UHI that affect the quality of urban life. The study was conducted during 2022 in a densely populated area of the Belouizdad (Belcourt) municipality (Algiers, Algeria). The study area is particularly dominated by the residential urbanization that represent about 75%. The plant element account for only 14 per cent of the study area.. To identify the urban heat islands (UHI), we used a simple, easy and reproducible empirical method developed by NEPSEN-E6. This approach consists of attributing a UHI score to each element of the ground surface: plant, buildings, parks and roads. The predominance of UHIs, underscore the importance of the mineral element (over 70%), creating a large zone of UHIs whows the imbalance between mineral and plant elements. The projection map shows these UHIs that can be reduced if mitigation measures are taken into account in the municipality's strategy (such as increasing green space) to address the effects of climate change.

Keywords Algiers · Belouizdad · Climate Changes · Geographic Information System · Urban Cool Island · Urban Heat Island

1 Introduction

In a context of the warming climate, heat waves and droughts will be higher, longer and more intense [8, 29]. Surveys show that heat island phenomenon severely affects large cities in the Maghreb and Middle East, favored by the arid climatic conditions of these regions [3, 10, 18, 23, 25].

S. Bensaid · H. Slimani (✉) · D. Charif
Laboratory of Plant Ecology and Environment, University of Sciences and Technology Houari Boumediene, Algiers, Algeria
e-mail: hslimani@usthb.dz

F. Khebour Allouche (ed.), *Environmental Remote Sensing and GIS in MENA Region*, Earth and Environmental Sciences Library,
https://doi.org/10.1007/978-3-031-86574-9_8

It is in urban areas where these episodes will be felt in the form of heat islands, leading to serious health consequences for the city-dwellers [2, 35].

The urban heat island (UHI) means a rise in localized temperature compared to the surrounding countryside and by a reduction of the difference between day and night temperatures. The urban overheating is the capacity of the city to keep heat stored during the day, due to anthropogenic, morphological and areal parameters [28].

In urban environment, changes in land use, the urban morphology, the properties of building materials, and the human activities will increase average global temperatures [19, 21, 28]. Furthermore, the urbanization to the detriment of green spaces contributes to increase the phenomenon of heat islands [5, 13, 16]. This urban heat island induces an exacerbation of day and summer nighttime temperatures and a lower daily thermal amplitude in the urban core compared to the countryside. This represents an artificial microclimate affecting the quality of the living environment of city dwellers [10]. Observations showed microclimatic temperature increases in the cities of 1–3 °C during the day and up to 12 °C at night [33].

Anthropogenic parameters come from human activities that depend on the used energy. This energy produces a lot of heat, adding to the ambient temperature. The morphological parameters such as height, size, location of buildings, have an effect on wind strength and circulation as well as exposure to solar radiation [7, 17, 34]. This solar radiance is thus trapped with a low release of heat. The areal parameters take into account the surfaces impermeability and the thermal capacity of the materials. The soil sealing and the reduction of green spaces, lead to loss of the surface water in city. Consequently, it can play a big role in air freshener by evaporation or evapotranspiration [9].

The thermal behavior of materials, in such operating conditions, depends on two parameters: its albedo and its thermal inertia. The Albedo is the fraction of solar radiation that is reflected from material. The thermal inertia deals with the capacity of a material to reach the environment temperature [26, 33]. The lower is the material's albedo, the higher is its thermal inertia.

Therefore, many cities implement heat island mitigation strategies based on planting vegetation [4, 30].

The aim of this study is to identify and map UHI to show the imbalance between mineral and plant elements. This work, done in the municipality of Belouizdad also known as Belouizdad (Algiers, Algeria), used an empirical approach using the basis of the UHI score of material.

2 Methodology

2.1 *Study Area*

2.1.1 Location and Population

The area was located in Algiers (the capital city of Algeria). It extends on approximately 16 km along the bay of Algiers. It is bounded on the north by the Mediterranean Sea, on the South by the Wilaya (administrative divisions of the country) of Blida, on the west by the Wilaya of Tipaza, and on the East by the Wilaya of Boumerdes (Fig. 1). The capital city is composed of 57 municipalities that has an area of 1,190 km^2 and has a total population of over 3 million people and an average population density of 4,090 inhabitants per square kilometer [22].

Algiers represents the largest city of the Mediterranean coast, the most populated and the most urbanized in Algeria. It is built on the foothills of the hilly land and the city foundation dates back to the IVth century before B.C. (Roman period).

The study was conducted during 2022, in the municipality of Belouizdad (also known Belcourt) which is part of the daïra (sub-prefecture level) of Hussein-Dey. The site is located approximately 3 km southwest of the Algiers city center. It is composed of four main districts: Belouizdad, Laâqiba, Hamma and Ruisseau.

This last district separated from the three others by the largest experimental garden (Hamma garden) in Algeria covering 32 ha. Belouizdad, the part closest to downtown area of Algiers, is made up of Haussmann buildings and is dominated by residential urbanization [24], with more than 70% of the area of the study site. By the end of 2020, the total population of the municipality of Belouizdad was 48,809 [1].

2.1.2 Vegetation

The highly urbanized municipality of Belouizdad includes the Hamma botanical garden (Fig. 2), which is the city's main plant feature and lungs. Created in 1832 by General Avisard as a garden of acclimatization, it covers a closed area of 32 ha.

The Hamma Garden is one of the most beautiful botanical gardens in the Mediterranean with nearly 3,000 identified species, represented by both local and introduced species from tropical and subtropical regions.

2.1.3 Climate Conditions

The city is characterized by mild Mediterranean climate with mild and rainy winter and warm and dry summer.

A climatic analysis has been made using the METEONORM 8.0.4 global database on city climatology. This database also contains algorithms for creating—from the measured values—weather files from any location on the globe.

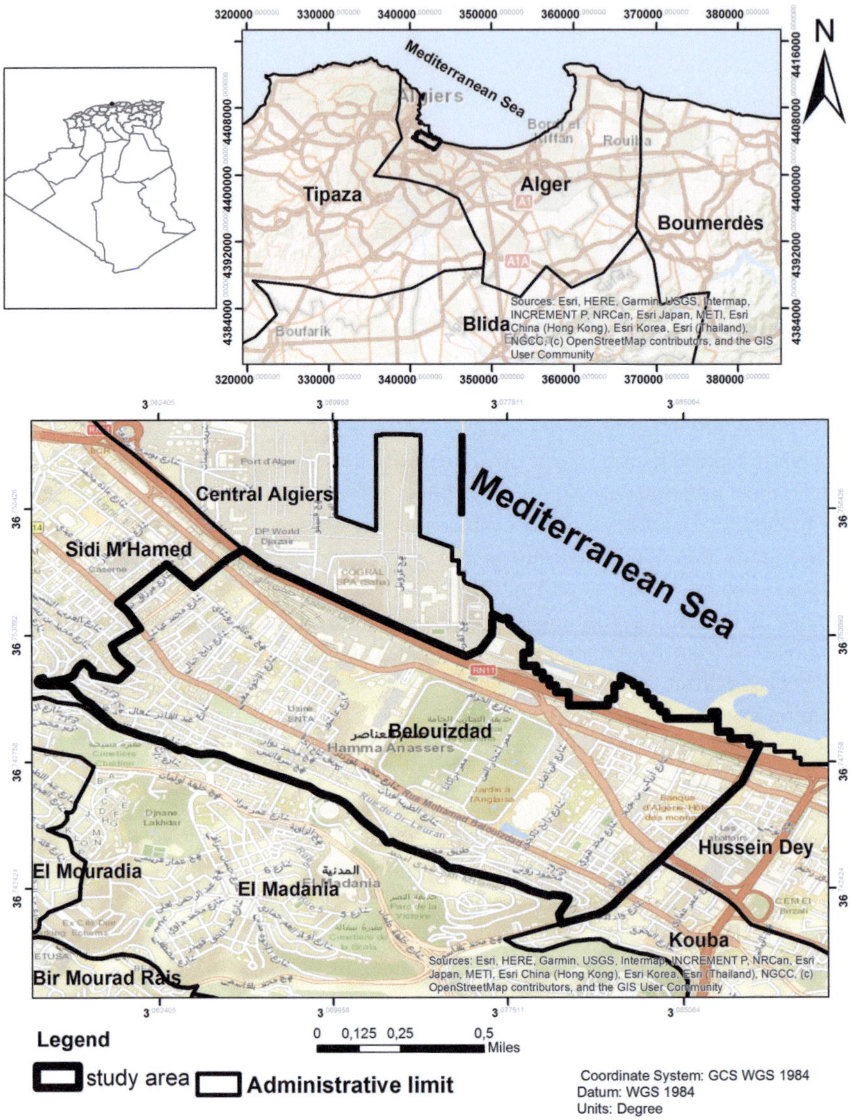

Fig. 1 Location map of study area (Slimani et al. 2022)

In winter, the mean annual temperature is around 11 °C and a relative humidity can be up to 79%. August is the hottest month with an average temperature of 32.6 °C (Fig. 3). In addition, the long periods of sunshine (Fig. 3) increase the air temperature and radiative balance.

The daily period of sunshine is important in the summer and can exceed fourteen hours in summer (Fig. 4). The diffuse and global solar radiation are also important

Fig. 2 Botanical Garden Hamma, Algiers (*source* Slimani 2023)

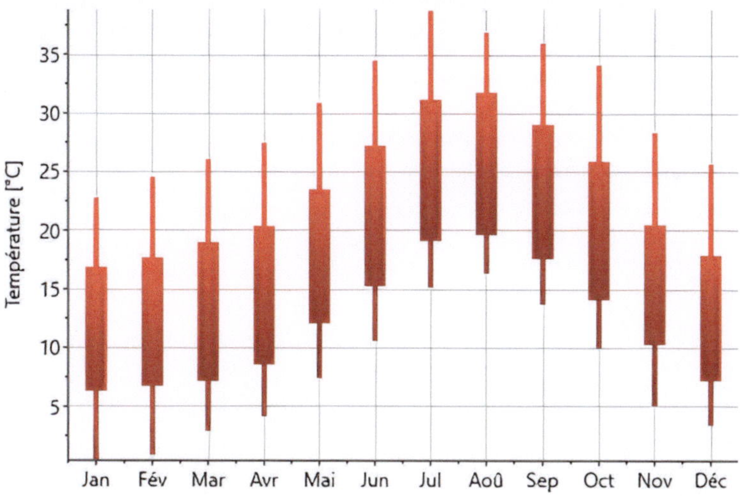

Fig. 3 The average monthly temperature in Algiers in 2022 (Meteonorme V8.2.0.) (Slimani et al. 2022)

in summer (Fig. 5) that reached maximum values in July (83 kWh/m^2 and 219 kWh/m^2 respectively).

The variability of monthly rainfall in Algiers (Fig. 6) shows three periods: a period of heavy rain (in November December and January), period with moderate amounts rain (in February March and April) and low rain period (May, June, July, August). Thus, summer is the driest season.

These temperature and humidity characteristics have a major impact on summer comfort.

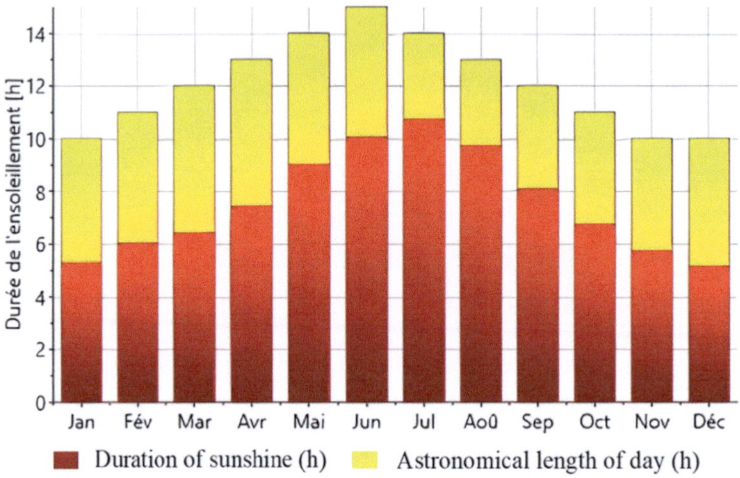

Fig. 4 The duration of sunshine in Algiers in 2022 (Meteonorme V8.2.0.) (Slimani et al. 2022)

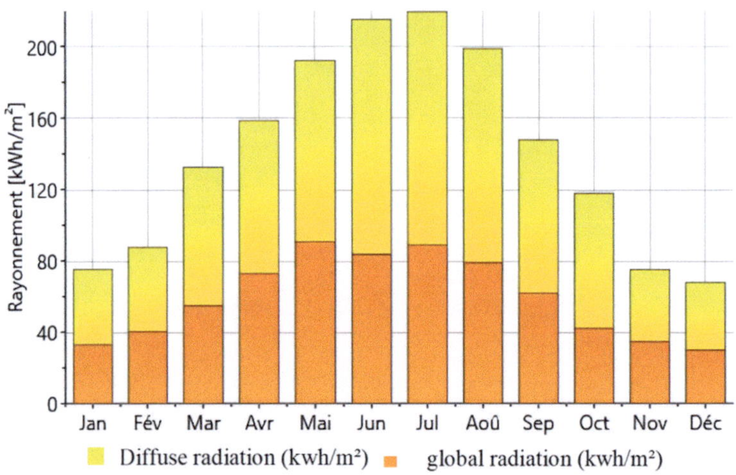

Fig. 5 The radiation in Algiers in 2022 (Meteonorme V8.2.0.) (Slimani et al. 2022)

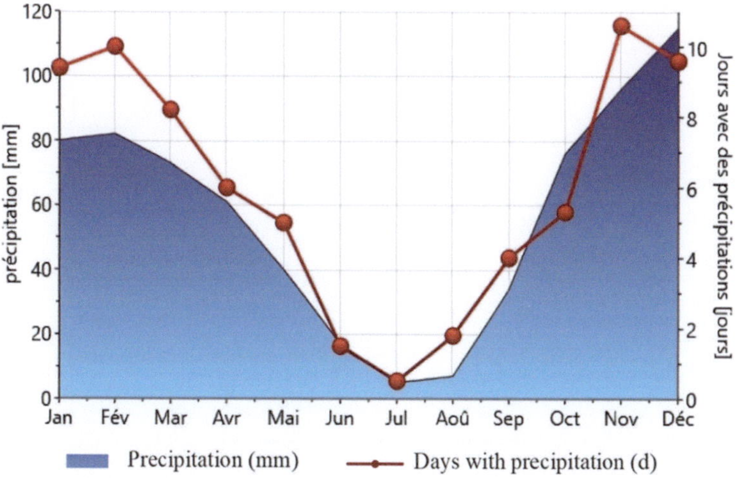

Fig. 6 The monthly rainfall in Algiers in 2022 (Meteonorme V8.2.0.) (Slimani et al. 2022)

2.2 Method

To identify the urban heat islands (UHI), we used a simple, easy and reproducible empirical method to identify and map these heat islands. The empirical models are increasingly being used because they meet operational requirements and correlate urban parameters with local climatic characteristics.

This study, carried out in 2022, used an empirical and simple approach developed by NEPSEN-E6 (https://nepsen.fr/). The latter is an independent consulting and engineering firm which is intended to ensure the transition and sustainable improvement of energy and environmental footprints, and the preservation of natural resources for 40 years and supports public and private stakeholders in implementing energy-efficient and environment friendly projects, to ensure sustainable and profitable results through its organization (https://nepsen.fr/).

That approach provides UHI scores to each element on the ground surface, in order to define empirically which zones will be the hottest and coolest. The UHI Score is a dialogue tool to easily address the issue of urban heat and cool islands.

The methodological approach taken is summarized in Fig. 7. We used Arcgis10.4 software and Google Earth satellite imagery (dated June 2022) in this work.

Once the layers have been created and the map digitized, the five parameters selected to identify heat islands (buildings, plants, parks and roads) are assigned by a UHI (http://www.e6-consulting.fr/ilots-de-chaleur-urbains-outil-score-icu/) score.

The UHI score tool proposes to classify different parameters in nine temperature bands (from 0 for coolest island to 1 for heat island) (Table 1). It takes into account the urban surfacing materials, their position in the shade or the sun, presence of vegetation, water, and various factors.

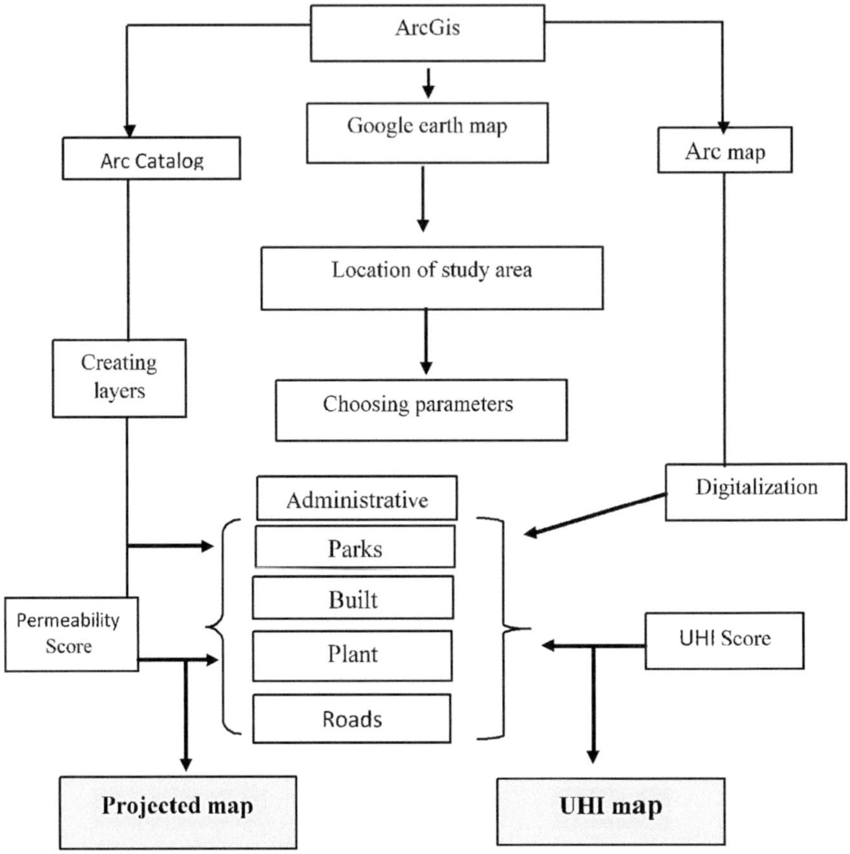

Fig. 7 The methodological approach

The classification by temperature bands is then evaluated to calculate the part of each band in the total area and thus propose a score. This classification can be carried out before and after work to improve the score and reduce the heat island of a region. The objective is to help us to achieve the UHI map.

This UHI score tool is completed by the water permeability score of a space (http://www.e6-consulting.fr/score-permeabilite/). It is considered that the permeability of certain urban elements that play a role as heat or cool islands [12] like albedo [27] and the color of materials [6], the place of shade, vegetation and water [31].

For a better characterization of heat islands, we also considered permeability. This approach is to give a permeability score to all parameters covering the ground, such as buildings, roads, etc. The permeability score is based on the same principles as the UHI score (an overall score varying between 0 and 1). A score of 0 means that a plot is entirely permeable, while a score of 1 means that a plot is 100% impermeable (e.g. roads).

Table 1 The UHI scores assigned to different parameters considered

Score	0	0.125	0.25	0.375	0.5	0.625	0.75	0.875	1
Parameters	Dense vegetation	Clear Vegetation	Less dense vegetation	Urban wasteland	Vegetated road	Built near vegetated road	Built without vegetation	Vegetated Parks	Bare land, Parks and road without vegetation

Due to a lack of data on the permeability of materials, the estimation of permeability was based on the presence or absence of vegetation that humidifies the air through its evapotranspiration capacity [32] and on the presence or absence of bare land and wasteland.

The combination of these two factors (UHI score and permeability) allows to identify the heat islands (current state), create the projected map and compare current and projected profiles of UHI scores.

3 Results and Discussion

The two data collection surveys (UHI score and permeability score for buildings, roads, car parks and vegetation) were used to achieve heat islands and cool islands maps. The urban reading of these score maps leads to achieve the summary map of the urban islands of the study area (Fig. 8).

The heat islands represented in particular by roads, builds and parks, occupy a significant surface area in relation to the whole municipality. On the other hand, the urban cool islands (UCI) which are associated with the presence of vegetation, occupy a smaller surface area than the UHI.

Roads (23%) and parks (2.73%) that present two possibility, without or with vegetation, have the highest scores because their high thermal capacity. Builds are also divided into two scores: with a surrounding vegetation and without vegetation. Almost the whole building is represented by a large heat island (almost 70%). Four scores were also identified based on vegetation type and density, represented by dense vegetation, clear vegetation, urban wasteland and bare land. These correspond to the experimental garden as well as to the vestiges of the old forest, which underwent urban sprawl. Results show a predominance of heat islands that emphasizes the state of the city and confirms a dominance of the urban mineralized character, creating a large urban heat island area. However, cold islands are poorly represented, with just 21%. Vegetated surfaces outside the administrative boundaries of Belouizdad (Fig. 9), play also an important role in heat mitigation and should be taken into account in the estimation of urban heat islands. These cool islands must be preserved and maintained, because in contrast with impermeable surface, vegetated area represent a cooler surface temperature and could thus contribute to mitigating the phenomenon of the urban heat island.

The heat islands have to be mitigated using a good urban structure by refreshment and active vegetation: maintaining and reinforcing open spaces (wastelands), tree planting, permeabilization of soils and materials, creation of water bodies, lightening of surfaces (road surfaces and car parks) and reduction of anthropogenic heat production (solar energy, etc.) [20].

In this case, we could obtain in the future a city with predominance of the green belt network and permeable and innovative surface materials (Fig. 10) to ensure the sustainability of the city's development.

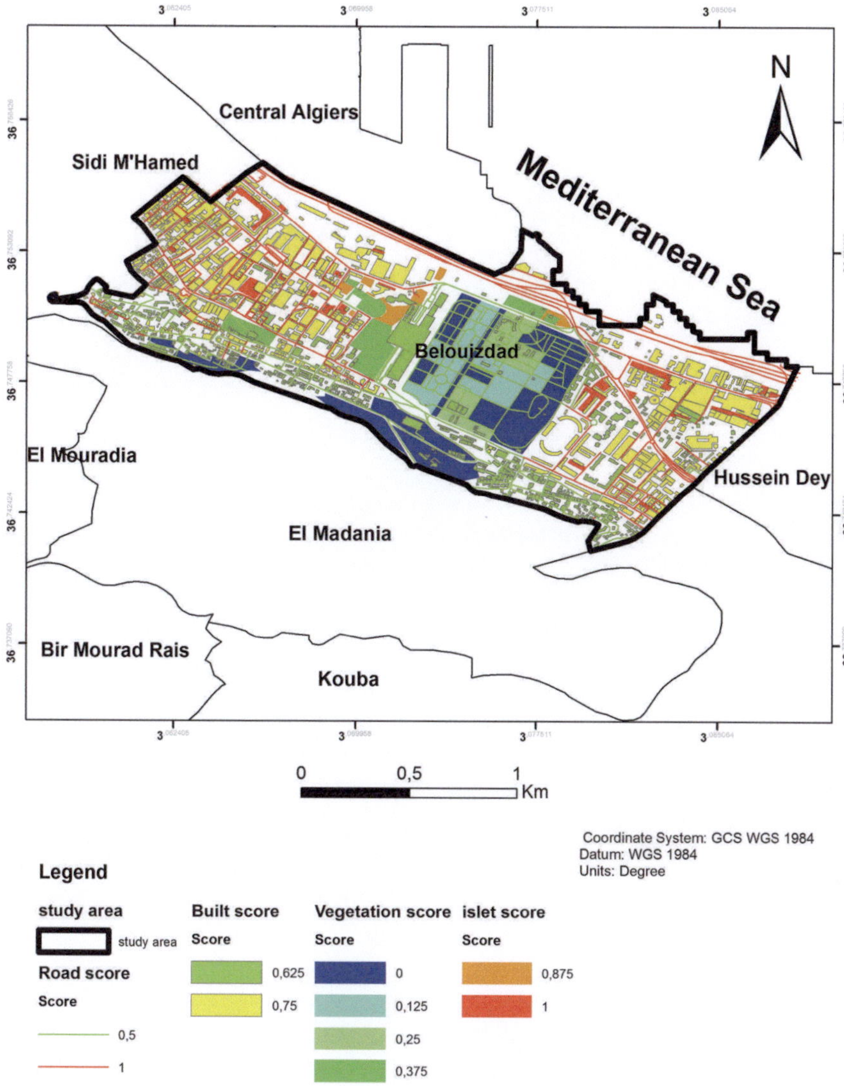

Fig. 8 Heat island map of Belouizdad municipality in 2022, Algiers (Slimani et al. 2022)

When the score tends towards zero it shows a profile dominated by vegetation as well as permeable materials. While it tends towards one, it means that it is a profile marked by a maximum heat island. A comparison of these two profiles shows that the greening of streets, roads car parks and increase permeability generate urban cool islands (Fig. 11). The urban city has many places that are suitable for planting. The greening cities can mitigate the effect of global warming [15]. The UHI have an important environmental and health effects.

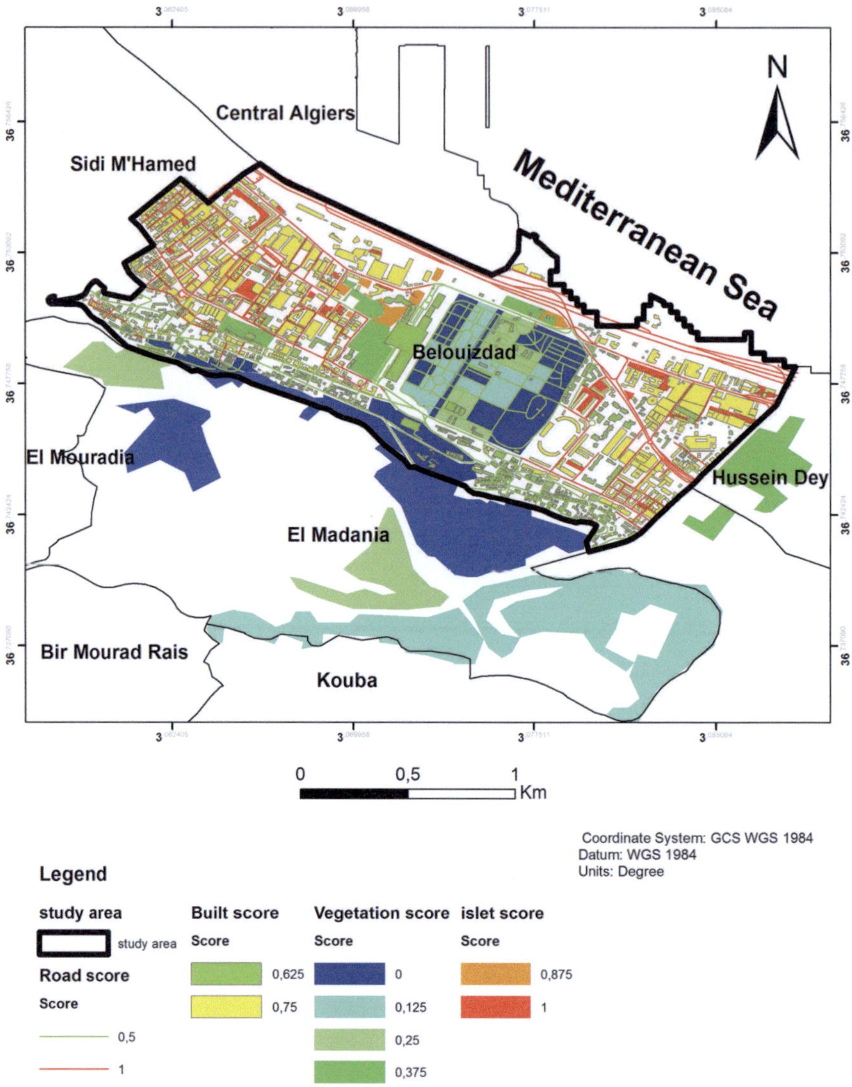

Fig. 9 Characterization of the urban heat island and cool islands in Belouizdad (Algiers) and its proximity in 2022 (Slimani et al. 2022)

Indeed, numerous studies worldwide show that in urban areas, the high density of the built fabric have an impact on the UHI and may lead to loss of biodiversity (deterioration in the outdoor and indoor air quality and increased demand for energy and drinking water [2, 8, 11, 16, 20]. It can also impact human health [14] by causing discomforts, weaknesses, even exacerbating preexisting chronic diseases and increased mortality during heat waves.

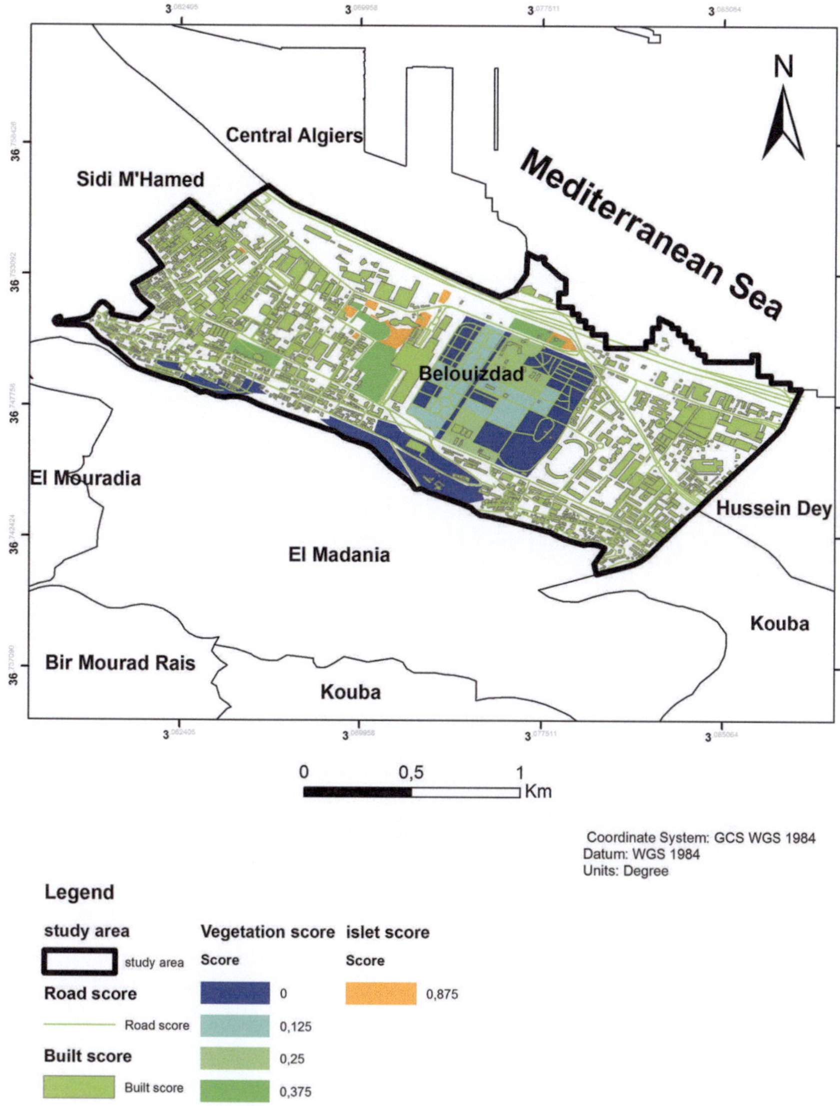

Fig. 10 Projected map of Belouizdad municipality (Slimani et al. 2022)

The municipality of Belouizdad is currently saturated by urbanization, with a high urban population. In addition, the lack of availability land has encouraged construction in prohibited zones [24] leading to an increase in UHI.

To reduce UHIs and increase UCIs, we need to apply laws that promote sustainable development and regulate inappropriate construction. The urban fabric needs to be gradually reorganized by demolishing old buildings and replacing them with effective greener alternatives. The idea is to create green spaces and to plant trees along main

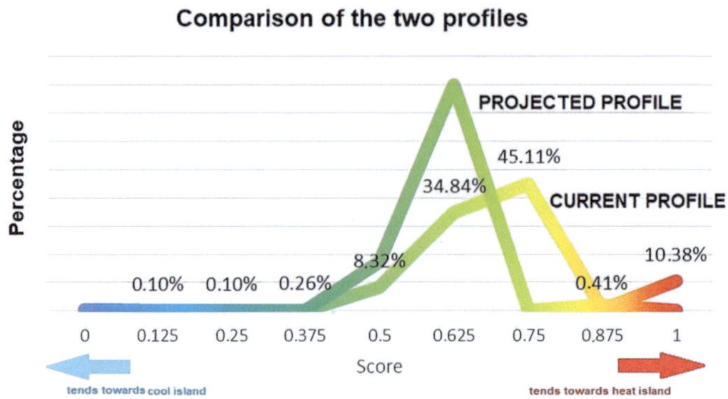

Fig. 11 Comparison of current and projected profiles UHI score of Belouizdad (Slimani et al. 2022)

and subsidiary roads to reduce heat islands. Improving surface permeability [11], green roofs, cool pavements and the use of high-albedo materials can also mitigate UHIs [32].

4 Conclusion

In the context of climate change, heat waves that have projected will increase in frequency and intensity. The UHI are mainly urban phenomenon which cities expect much higher air temperature. The study has allowed us to apply an empirical method to characterize the heat islands through the example of the municipality of Belouizdad.

Three dates were chosen: 2022; 2011 and 2001 for a diachronic approach.

Data analysis shows an expansion of the mineral element to the detriment of natural areas over a 21-year period. However, various coastal development projects have been realized or are still underway.

In addition, the cartographic results show that coastal development has not been in line with the green and blue corridor. The dense, ever-expanding urban fabric leaves no room for ecological communication as conceived through the Blue-Green Frame.

The projects Developments were designed without taking into account ecological notions of connectivity.

5 Recommendation

Faced with this alarming situation, municipal leaders have to take actions in order to mitigate the intensity of UHI including the creation of cool islands to adapt to a future climate and frequent and severe heat waves. The influence of vegetation on UHI appears to be a major tool for adaptation to climate change. The world health organization suggests creating green spaces to help reduce social inequalities and improve air quality.

Heat island studies should be further improved, by taking into account all relevant parameters, and expanded to neighboring municipalities, in order to provide a decision-making tool for the municipality and ensure effective action to reduce the impact of UHIs.

The main recommendations to combat urban heat islands are as follows:

- increase the presence of nature and water in new development projects;
- Strengthen the favourable conditions to mitigate the effects of heat waves
- Optimize the organization of space

The Specific actions include the creation of parks in the City, promoting permeable soil surfaces, ensuring an appropriate management of water, and all solutions based on the nature can be a major asset to fight against UHI. The refreshing power of plants is encouraged by the shading and evaporation. Water also helps to reduce the heat storage by evaporation.

It is also recommended to take into account the urban form in the numerous restoration operations of old buildings and the urban fabric. This could lead to reduce the heat island intensity of urban space.

In view of climate change, it is imperative that the southern shore of the Mediterranean includes this UHI issue in their urban development strategy. The implementation of a new urban paradigm in line with the sustainable city, with a positive impact on climate change and well-being of the citizens.

References

1. ANIREF (2022) Budget statistical yearbook. 1st edition: 142. https://dgb.mf.gov.dz/wp-content/uploads/2022/06/Yearbook-2020.pdf
2. Arnfield AJ (2003) Two decades of urban climate research: a review of turbulence, exchanges of energy and water, and the urban heat island. Int J Climatol J Roy Meteorol Soc 23(1):1–26. https://www.coolrooftoolkit.org/wp-content/uploads/2014/05/2-Decades-of-Urban-Climate-Research...-2003.pdf
3. Bouketta S (2023) Urban Cool Island as a sustainable passive cooling strategy of urban spaces under summer conditions in Mediterranean climate. Sustain Cities Soc 99:104956
4. Bowler DE, Buyung-Ali L, Knight TM, Pullin AS (2010) Urban greening to cool towns and cities: a systematic review of the empirical evidence. Landscape Urban Plan 97(3):147–155. https://www.sciencedirect.com/science/article/abs/pii/S0169204610001234

5. Brou TY, Rwanyiziri G, Uwayezu E (2024) Urban dynamics and urban heat islands development in Kigali, Rwanda. East Afr Rev 59(59). https://journals.openedition.org/eastafrica/4500
6. Calis G, Yildizel SA, Keskin US (2022) Investigation of color pigment incorporated roller compacted high performance concrete as a mitigation tool against urban heat island. Case Stud Construct Mater 17:e01479. https://www.sciencedirect.com/science/article/pii/S2214509522006118
7. Colombert M (2008) Contribution to the analysis of various means to take into account urban climate in urban planning. 541. https://www.theses.fr/2008PEST0233
8. De Munck C, Pigeon G, Masson V, Meunier F, Bousquet P, Tréméac B, Merchat M, Poeuf P, Marchadier C (2012) How much can air conditioning increase air temperatures for a city like Paris, France? Int J Climatol 33(1):210–227. https://www.cnrm-game-meteo.fr/IMG/pdf/demunck_2013_ijc.pdf
9. Deng F, Yang Y, Zhao E, Xu N, Li Z, Zheng P, Han Y, Gong J (2022) Urban heat island intensity changes in Guangdong-Hong Kong-Macao greater bay area of China revealed by downscaling MODIS LST with deep learning. Int J Environ Res Public Health 19(24). http://www.ncbi.nlm.nih.gov/pubmed/36554882
10. Fahed J, Kinab E, Ginestet S, Adolphe L (2020) Impact of urban heat island mitigation measures on microclimate and pedestrian comfort in a dense urban district of Lebanon. Sustain Cities Soc 61:102375. https://www.sciencedirect.com/science/article/abs/pii/S2210670720305965
11. Falolou LF, Orekan V, Houssou CS, Agbossou EK (2020) Characterization of heat islands in the municipality of porto-novo and its surroundings. Int J Sci High Technol 20(No. 2):442–456. https://doi.org/10.52155/ijpsat.v20.2.1852
12. Ferrari A, Kubilay A, Derome D, Carmeliet J (2020) The use of permeable and reflective pavements as a potential strategy for urban heat island mitigation. Urban Clim 31:100534. https://www.sciencedirect.com/science/article/abs/pii/S2212095519300926
13. Foissard X (2015) The urban heat island and climate change: application to the Rennes conurbation. Géographie 233. https://theses.fr/2015REN20027
14. Heaviside C, Macintyre H, Vardoulakis S (2017) The urban heat island: implications for health in a changing environment. Curr Environ Health Rep 4(3):296–305. http://www.ncbi.nlm.nih.gov/pubmed/28695487
15. Husni E, Prayoga GA, Tamba JD, Retnowati Y, Fauzandi FI, Yusuf R, Yahya BN (2022) Microclimate investigation of vehicular traffic on the urban heat island through IoT-Based device. Heliyon 8(11):e11739. http://www.ncbi.nlm.nih.gov/pubmed/36425428
16. Li L, Zhan W, Hu L, Chakraborty TC, Wang Z, Fu P, Wang D, Liao W, Huang F, Fu H, Li J, Liu Z, Du H, Wang S (2023) Divergent urbanization-induced impacts on global surface urban heat island trends since 1980s. Remote Sens Environ 295:113650. https://www.sciencedirect.com/science/article/abs/pii/S0034425723002018
17. Machard A, Abi Moussa T, Alessandrini J-M, De Oliveira F, Devys M (2024) Summer thermal comfort, vulnerability of the building stock to overheating and behaviour adaptation to extreme heat: state of the art. Sci Tech Centre Construct 91. https://cstb.hal.science/hal-04695333
18. Mazzeo D, Matera N, Peri G, Scaccianoce G (2023) Forecasting green roofs' potential in improving building thermal performance and mitigating urban heat island in the Mediterranean area: an artificial intelligence-based approach. Appl Therm Eng 222:119879. https://www.sciencedirect.com/science/article/abs/pii/S1359431122018099
19. Morice CP, Kennedy JJ, Rayner NA, Jones PD (2012) Quantifying uncertainties in global and regional temperature change using an ensemble of observational estimates: the HadCRUT4 data set. J Geophys Res Atmos 117(D8). https://doi.org/10.1029/2011JD017187
20. Nuruzzaman M (2015) Urban heat island: causes, effects and mitigation measures—a review. Int J Environ Monitor Anal 3(2):67. http://www.sciencepublishinggroup.com/j/ijema
21. Oke TR (1995) The heat island of the urban boundary layer: characteristics, causes and effects. Wind climate in cities, pp 81–107. https://doi.org/10.1007/978-94-017-3686-2_5
22. ONS (2020) National statistics office of the wilaya of Algiers, p 91

23. Pyrgou A, Hadjinicolaou P, Santamouris M (2020) Urban-rural moisture contrast: regulator of the urban heat island and heatwaves' synergy over a mediterranean city. Environ Res 182:109102. http://www.ncbi.nlm.nih.gov/pubmed/31927298
24. Rabehi W, Guerfi M, Mahi H, Rojas-Garcia E (2019) Spatiotemporal monitoring of coastal urbanization dynamics: case study of Algiers' Bay, Algeria. J Indian Soc Remote Sens 47(11): 1917–1936. https://www.researchgate.net/publication/335620851
25. Rodríguez LR, Ramos JS, Delgado MCG, Domínguez ÁS (2022) Implications of the Urban Heat Island on the selection of optimal retrofitting strategies: a case study in a Mediterranean climate. Urban Clim 44:101234. https://www.sciencedirect.com/science/article/abs/pii/S2212095522001523
26. Rullier F (2012) The role of urban landscaping in a context of global warming. The issue of urban heat application in Montreal, p 68. https://dumas.ccsd.cnrs.fr/dumas-00749024
27. Santamouris M, Fiorito F (2021) On the impact of modified urban albedo on ambient temperature and heat related mortality. Solar Energy 216:493–507. https://www.sciencedirect.com/science/article/abs/pii/S0038092X21000475
28. Solecki WD, Rosenzweig C, Parshall L, Pope G, Clark M, Cox J, Wiencke M (2011) Mitigation of the heat island effect in urban New Jersey. Environ Hazards 6(1):39–49. https://www.academia.edu/download/49960053/Mitigation_of_the_heat_island_effect_in_20161029-9547-ph0f0h.pdf
29. Soubeyroux J-M, Dubuisson B, Josse P, Régimbeau M, Ribes A, Sorel M (2023) Climate change and recent extreme events in the east of France. French For Rev 74(2):145–151. https://doi.org/10.20870/revforfr.2023.7588
30. Spronken-Smith RA, Oke TR (1998) The thermal regime of urban parks in two cities with different summer climates. Int J Remote Sens 19(11):2085–2104. https://doi.org/10.1080/014311698214884
31. Susca T, Zanghirella F, Del Fatto V (2023) Building integrated vegetation effect on microclimate conditions for urban heat island adaptation. Lesson learned from Turin and Rome case studies. Energy Build 295:113233. https://www.sciencedirect.com/science/article/pii/S0378778823004632
32. Vardhu VAK, Sharma DA (2023) Classification, mitigations and methods to detect UHI: a review. Int J Sci Res Eng Manage 7(2):26. https://www.researchgate.net/publication/368410910
33. Verbeke S, Audenaert A (2018) Thermal inertia in buildings: a review of impacts across climate and building use. Renew Sustain Energy Rev 82:2300–2318. https://doi.org/10.1016/j.rser.2017.08.083
34. Yang J, Shi B, Xia G, Xue Q, Cao S-J (2020) Impacts of urban form on thermal environment near the surface region at pedestrian height: a case study based on high-density built-up areas of Nanjing City in China. Sustainability 12(5):1737. https://doi.org/10.3390/su12051737
35. Zhao L, Oppenheimer M, Zhu Q, Baldwin JW, Ebi KL, Bou-Zeid E, Guan K, Liu X (2018) Interactions between urban heat islands and heat waves. Environ Res Lett 13(3):034003. https://doi.org/10.1088/1748-9326/aa9f73

Application of Time Series Models for Annual and Monthly Streamflow Forecasting in Northern Algeria

Khadidja Boukharouba

Abstract Precise prediction models are essential for water resources management and planning because they provide a valuable support tool for decision-makers. Unfortunately, the random variability of hydrological variables is a serious constraint that can affect the final performance of any water action plan and hence, limits the interventions of decision makers. Actually, this can entail a significant risk when it comes to developing a water resources strategy, because the achievement of the objectives depends on the random behavior of such hydrological variables. The consequences could be very serious from a water resources management point of view, and the socio-economic and environmental impact would be even more so. The present chapter fits in this context; it presents some of our recent works on stochastic modeling for prediction of water resources in Algeria. This is done by focusing on BOX JENKINS, a particularly interesting type of models that can deal with random variability of hydrological variables. The principle of this type of model lies on the analysis of the observed time series to predict the future behavior of the generating mechanism. The objective here, is to look for the possibility of substituting these generating mechanisms by some mathematical models that behave in a similar way and then, use them to predict the future behavior of the concerned hydrological variables. These models are mostly required in water resources design and management because they provide the opportunity to foresee the risk, in part and therefore, offer the possibility to reduce its effects. In this way, considerable help is available to decision-makers in water resources so that they can develop their strategies with more safety. The data used here, are the annual and monthly stream flow time series observed over 25 years at four hydrometric stations, all located on operational dams in the north of Algeria. The obtained results are some ARMA, ARIMA and SARIMA models which are parsimonious, easy to use and whose predictions are satisfactory, since the corresponding error statistics are acceptable.

Keywords Box-Jenkins · Modeling · Streamflow · Time series · Prediction · Algeria

---

K. Boukharouba (✉)
LCA Laboratory, Faculty of Nature and Life Sciences, University of Biskra, Biskra, Algeria
e-mail: khadidja.boukharouba@univ-biskra.dz

F. Khebour Allouche (ed.), *Environmental Remote Sensing and GIS in MENA Region*, Earth and Environmental Sciences Library,
https://doi.org/10.1007/978-3-031-86574-9_9

1 Introduction

Time series analysis in hydrology concerns many areas of everyday life and more precisely the water domain [31]. Among the many applications of such analyses, it is important in the hydrological context to mention at least two: simulation and forecasting [18]. These two applications are very important in water resources planning and management [1, 38], because they can provide some decision support tools [9].

In practice, the inputs and outputs of any hydrological system are considered as random variables measurable at fixed points in time, thus forming time series at daily, weekly, monthly or annual time scales [37, 48], this is the case for precipitation or stream flows in a watershed. The random variability in hydrology is a real constraint for water resources planners and managers because the risk and the uncertainty amounts are important. Hence, this randomness not only affects the final performance of any water action plan, but also limit decision-makers in their interventions [15]. For this reason, the idea of studying the underlying stochastic process is justified [55] and [47]. The idea is to analyze and understand the past behavior of this process, through its time series records, in order to be able to equate it and predict its future behavior [25] and [6].

In this approach, the types of models that can be considered are numerous [8, 19, 35, 38, 58]. In this chapter, a particularly interesting type of stochastic models is investigated: the stochastic linear Box-Jenkins or ARMA (AutoRegressive and Moving Average) models [13, 14] because they are suitable to represent stream flows as occurring in nature. Generally, the autoregressive (AR) part may be the stream flows because of their internal dependence structure, whereas the moving average (MA) part may be the precipitations because of their randomness. The usual modeling process assumes that observed time series are stationary. However, most of hydrological series are non-stationary, i.e. they grow over time with some instability in the level of the mean or the variance [36]. Generally, it is always possible to find a mathematical transformation or a filter that can make them stationary. In this case, the ARMA models are called ARIMA (Auto-Regressive Integrated Moving Average) whereas for the seasonal case, they are called SARIMA (which is a seasonal ARIMA) [14, 26].

The predictive power of ARIMA models has greatly been improved by the multiple technological breakthroughs in fields like mathematics, computers and Geographical Information Systems, among others. Their application for ends of prediction has consequently been largely extended to many other fields. Grandón et al. [29], Zhong et al. [57] and Toshaliyeva [51] are only a few recent examples of using the aforementioned models for prediction in applied energy and industry, Othman et al. [45], Jafarian-Namin [34] and Deshmukh [23] in climatology. In operational hydrology, ARIMA models have been largely applied, especially in stream flow forecasting. Tayyab et al. [50] applied a Box-Jenkins an AR model to the prediction of the Jinsha River stream flows in China and compared the obtained predictions to those of Artificial Neural Networks. Zhang (2003) used a hybrid ARIMA and neural network model to stream flow time series forecasting. Valipour et al. [52] used Box-Jenkins SARIMA and ARMA to modelling and forecasting the monthly inflow of Dez dam reservoir in Turkey and came to conclude that SARIMA model prediction performance is better than ARMA. Boukharouba et al. [12] used ARMA and SARIMA models to modelling and forecasting of stream flows in northern Algeria and compared their prediction performance to the linear discrete Kalman Filter.

Rana et al. (2017) applied two different Box-Jenkins models ARIMA and ARMA for monthly stream flow forecasting in Doyane station. They found that ARIMA model yield better predictions compared to ARMA, and concluded that ARIMA model could be used for one month ahead stream flow prediction at the concerned site. Parvaze et al. [46] tested the performance of the deseasonalised ARIMA modelling technique to evaluate its suitability for streamflow prediction in the flood-prone Kashmir Valley in India. Modelling and forecast of monthly peak flow was performed for three selected discharge stations of the river Jhelum. The deseasonalised ARIMA model was found to be effective in capturing the seasonal variations and providing accurate forecasts of monthly peak flows. More recent works on stream flow prediction using these models can be found in [54], [49] and [10].

Generally, a precise prediction of the stream flow is significantly important because it is directly related to any economic, environmental or social development [53]. In the present chapter, Box-Jenkins approach is investigated for modeling and prediction of annual and monthly stream flows in northern Algeria. Four interesting sites are selected because they are situated on some operational dams. The interest of observing the history of these four-time series consists of detecting certain regularities that can be expressed by some adequate models. The aim is to formulate a mathematical model that can substitute the underlying generating stochastic process (unknown). Formulating a model for the time series, is the mathematical operation by which we want to acquire information about the theoretical generating process which nature is stochastic and dynamic in the same time. The problem then, is to find the practical model that will approach the theoretical process, as closely as possible, and then to estimate it. Once this step is completed, the simulation or forecasting of the studied variable becomes an automatic operation [24, 30].

The aim of this chapter is firstly, to know how much random are the stream flows in Northern Algeria; the specific purpose here is to study whether the concerned time series contain or not some recognizable structured parts that can be extracted from the total variability, and secondary, to equate the time series (models) in order to make their prediction.

2 Box-Jenkins Theoretical Background

Box and Jenkins [13] proposed a unified procedure for the treatment of stationary and ergodic stochastic processes, which is still used as a reference [52]. The proposed method consists in modeling univariate time series using the so-called Box-Jenkins processes or, equivalently the Auto Regressive and Moving Average (ARMA) models.

The procedure aims at formulating a model that represents a time series as faithfully as possible. It relies on the least squares principle for model estimation and on the concept of parsimony for model selection [5]. In this approach, one hypothesis and four steps are considered. The following theoretical developments are essentially based on [13, 55] and [37].

Hypothesis: the time series X_t is supposed to be generated from another series ε_t that follows a Gaussian distribution of zero mean and standard deviation σ_ε, in the form of:

$$X_t = \mu + \varepsilon_t + \psi_1 \varepsilon_{t-1} + \ldots \tag{1}$$

ε_t must verify the following conditions:

$$\begin{cases} \varepsilon_t \rightsquigarrow N(0, \sigma_\varepsilon^2), \\ E(\varepsilon_t) = 0, \forall t, \\ \gamma_h = Cov\big(\varepsilon_t, \varepsilon_{t+h}\big) = \begin{cases} \sigma_\varepsilon^2, si\, h = 0 \\ 0\ si\, h \neq 0 \end{cases} \end{cases} \tag{2}$$

where ε_t is a white noise, which is the perfect example of a stationary series, since it has a zero mean, stable variance, and which elements are totally independent, μ is the mean of the series X_t.

The analysis of time series by Box-Jenkins method means to search for the series ε_t and the coefficients ψ_i that allows to switch from the series of ε_t to the X_t one.

Steps: Step 1 seeks to obtain the stationarity of the data by transformations and differentiations. Step 2 seeks to identify the order of the AR and MA parts of the model by interpreting the graphs of the autocorrelation and partial autocorrelation functions. Step 3 consists in estimating the parameters of the selected model based on the maximum likelihood concept. Step 4 seeks to validate the model by analyzing its residuals using some statistical tests. If the model does not perform well, the procedure must be repeated starting from step 2.

2.1 Stationarization and Deseasonalization

2.1.1 Transformation and Differentiation

ARMA models are quite restrictive because they assume that the observations are stationary, that the model parameters are constant, and that the errors variance is constant too. Assuming that the observations are stationary means that the corresponding mean and variance are constant.

If the observation series are non-stationary in terms of mean, the best method to return to stationarity is the differentiation operation. Hence, the original series is replaced by the series of adjacent differences of one or two lags. The obtained differentiated time series is considered as an i*ntegrated* version of a stationary series. In this case, the ARMA model becomes an *Integrated* ARMA "Autoregressive *Integrated* Moving Average" model or simply ARIMA.

On the other hand, if the observation series are non-stationary in terms of variance, some mathematical transformations (such as the logarithm or the square root) are required to meet the stationary condition. The retained transformation needs to be processed before the series is differentiated.

In case of logarithmic transformation, for example $Y_t = \log X_t$, by applying a filter of type (1 − B), where B is the back shift operator, the transformed series Y_t will represent the logarithmic differenciations of the series.

$$(1 - B)\log X_t = \log X_t - \log X_{t-1} \tag{3}$$

A one (1) order differentiation supposes that the difference between two successive values of Y_t is constant.

$$Y_t - Y_{t-1} = \mu + \varepsilon_t \tag{4}$$

where μ is the model constant, and represents the mean difference in Y_t. Such model is an ARIMA(0,1,0). It can be considered as a linear increasing in time. If μ equals zero (0), the series is stationary.

A difference of order 2 means that we are not working on the raw differences, but on the differences of differences. Thus, the second difference of Y_t at time t is equal to $(Y_t - Y_{t-1}) - (Y_{t-1} - Y_{t-2})$, or simply $Y_t - 2Y_{t-1} + Y_{t-2}$.

Such model is an ARIMA(0,2,0) and the corresponding prediction equation is as follows:

$$Y_t - 2Y_{t-1} + Y_{t-2} = \mu + \varepsilon_t \tag{5}$$

or:

$$Y_t = \mu + 2Y_{t-1} - Y_{t-2} + \varepsilon_t \tag{6}$$

Similarly, if a series has a strong seasonal component, this can be eliminated by differentiating the series not at order one, but at the order of seasonality. Thus, for monthly series, we can use $(1 - B^{12})$ to eliminate the annual effect.

For example, for a monthly data series with a seemingly linear trend, we combine the two filters Δ and Δ_{12} to remove the trend and a seasonality period of 12. This is equivalent to working on the series:

$$Y_t = \Delta\Delta_{12} X_t = \Delta(X_t - X_{t-12}) = X_t - X_{t-1} - X_{t-12} + X_{t-13}$$

$$Y_t = X_t - X_{t-1} - X_{t-12} + X_{t-13} \quad (7)$$

In this case, if the series X_t contains N values, the transformed series Y_t contains only N-13, i.e. the analysis of the ARMA(p, q) model is done on a series shortened by more than one year.

2.1.2 Differentiation Order

For a stationary time series, all values fluctuate around the mean value and the corresponding auto-correlation function declines quickly to zero. When a time series is differentiated, very often the optimal order of differentiation is that for which the standard deviation is minimal. An increase in the standard deviation should be considered as a sign of an over-differentiation.

2.2 *Determination of the Model Type and Order*

After the series has been stationarized (seasonality and trend have been removed) we are interested in identifying a stationary ARMA(p,q) process and attempt to recognize the type of process we are dealing with and identify the AR and MA terms needed to correct the residual autocorrelations. This analysis is based on the examination of the auto-correlation and partial auto-correlation functions.

At this stage, it is essential to establish a hypothesis of maximum number of coefficients to be used, i.e. the values p_{max} and q_{max} of the ARMA(p,q) model.

To do this, we use the ACF (ρ_h) and PACF (ϕ_h) functions and observe them visually, knowing that their properties are as follows:

MA(q): $\rho_h = 0 \; if \; h > q$
AR(p): ρ_h gradually tends to 0 and
AR(p): $\phi_h = 0 \; if \; h > p$
MA(q): ϕ_h tend gradually tends to 0

2.2.1 Autoregressive (AR) Parameters

Thus for a pure AR(p), we observe a PACF threshold for $h \leq p$, ϕ_h is nonzero and it becomes zero for h > p.

An abrupt extinction of partial autocorrelation coupled with a more gradual decline in autocorrelation is thus a characteristic signature of an autoregressive process. In particular, the partial auto-correlation of lag p is equal to the coefficient AR(p) estimated in a model containing p AR terms.

The lag at which the partial autocorrelation disappears indicates the number of autoregressive terms to include.

2.2.2 Moving Average (MA) Parameters

The ACF auto-correlation function plays the same role for moving average processes as the partial auto-correlation function for autoregressive processes. If the auto-correlation is significant at lag q but no longer at lag $q + 1$, this indicates that q moving average terms should be added to the model.

We can see that for a pure MA(q), the behavior is the same as a pure AR(p), but this time using ACF and q as cutoff value.

2.2.3 ARMA Parameters

For ARMA model, a diagnosis must be made by observing the ACF and PACF parts separately.

2.2.4 $SARIMA(P,d,q)(P,D,Q)_S$ Parameters

In addition to the parameters (p), (d) and (q) of a simple ARIMA(p,d,q), a Multiplicative Seasonal ARIMA ($SARIMA(p,d,q)(P,D,Q)_S$) introduces a seasonal autoregressive parameter (P), seasonal differentiation (D), and seasonal moving average (Q) parameters [17]. For example, the model SARIMA(0,1,3)$(0,1,2)_s$ describes a model with no autoregressive parameters, 3 normal moving average parameters and 2 seasonal moving average parameter. The calculation of these parameters occurs after differentiation once by 1 normal lag, and once by 1 seasonal lag. The general criteria concerning the selection of parameters to be estimated (based on ACF and PACF) also apply to seasonal models.

To sum up, the main tools used in the identification step are the plot of the time series, together with the autocorrelation (ACF) and partial autocorrelation (PACF) plots. The decision is not simple and the most atypical cases require numerous experiments with different models, in addition to the experience of the researcher.

2.3 *Model Parameters Estimation*

The estimation of the p and q parameters of an ARMA(p, q) model, under the assumption that they are finite but not known, can be done by different methods in the time domain such as the ordinary Least Squares method or the Maximum Likelihood method [13].

2.4 Validation

This is the last step in the modeling procedure, it consists of verifying the hypotheses of the fitted models by some tests: the first one concerns the estimated parameters, while the second concerns the models residuals. The most parsimonious model is then selected using some criteria based upon the residual variance [5].

2.4.1 Tests of Parameters Signification

Under the assumption that the white noise is Gaussian, we can show that the estimators $\widehat{\phi}_i$, and $\widehat{\theta}_j$ are approximately Gaussian. Therefore, we can perform some tests on the parameters using Student's law (T):

$$\frac{\widehat{\phi}_i}{\sqrt{Var(\widehat{\phi}_i)}} \to T \tag{8}$$

For an ARMA(p, q) model, one can notice that the test on a parameter ϕ_p:

$$H_0 : \phi_p = 0 \quad \text{against} \quad H_1 : \phi_p \neq 0$$

Is equivalent to the test on models:

$$H_0 : ARMA(p, q) \quad \text{against} \quad H_1 : ARMA(p-1, q)$$

since p is the parameter of the last term of the autoregressive part of the model. We show that under the assumption H_0, T follows a Student's law with (n–1) degrees of freedom. At the α confidence level (typically equal to 5% and $T_c = 1.96$):
if $T < T_c$, accept $H_0 \Rightarrow \phi_p = 0$
if $T > T_c$, reject $H_0 \Rightarrow \phi_p \neq 0$

2.4.2 Gaussian White Noise Tests

White noise is a process without any internal dependence structure and, therefore, completely unpredictable. The following tests are used on model residuals in order to verify their independence and normality. In addition, residuals must have zero mean. Most of the following test details can be found in [7, 37] and [28].

Test for nullity of the mean

Let T be the number of available data (after removing the delays corresponding to the AR and MA terms): If the process ofε_t, $t \in Z$, is an independent identically distributed i.i.d (0,σ_ε^2), we must have its average $\overline{\varepsilon}_t$ than converges to zero as follow:

$$\overline{\varepsilon}_t = \frac{1}{T}\sum_{t=1}^{T}\widehat{\varepsilon}_t \underset{T\to\infty}{\to} 0 \tag{9}$$

By application of the limit central theorem, we can have:

$$\frac{\overline{\varepsilon}_t}{\widehat{\sigma}_\varepsilon}\sqrt{T} \underset{T\to\infty}{\to} \mathcal{N}(0,1) \tag{10}$$

From then on, we can test the nullity of the mean of the residuals by constructing the confidence interval for $\underset{-t}{\varepsilon}$ at the standard of 95%.

$$P\left\{\overline{\varepsilon}_t \in \left[\frac{-1{,}96\widehat{\sigma}_\varepsilon}{\sqrt{T}}, \frac{+1{,}96\widehat{\sigma}_\varepsilon}{\sqrt{T}},\right]\right\} = 0.95 \tag{11}$$

Auto-correlation test

If the residuals ε_t, $t \in Z$ are white noise, there should be no auto-correlation in the series. We can then examine the plots of the Autocorrelation function (ACF) and Partial Auto-correlation Function (PACF): we must verify that there is no auto-correlation or partial auto-correlation that is significantly non-zero for the studied process. This task can be checked by the "Portemanteau" test [28].

Homoscedasticity test

Omoscedasticity is the hypothesis of equality of variances in different subgroups of model residuals [28]. For m subgroups this hypothesis H_0 is as follows:

$$H_0 = \sigma_1^2 = \sigma_2^2 = \cdots = \sigma_m^2 \tag{12}$$

hence:

- for each subgroup i of the residual series the variance σ_i^2 is:

$$\sigma_i^2 = \sum_{i=1}^{m}(y_{ij} - y_i)^2/\nu_i \tag{13}$$

$i = 1,2,\ldots,m; j = 1,2,\ldots,n_i;n_i$ are the number of values in subgroup$i$; $\nu_i = n_i - 1$ and$y_i = \sum_{i=1}^{m} y_{ij}/n_i$.

- Variance calculation

$$\sigma^2 = \sum_{i=1}^{m}\nu_i\sigma_i^2/\nu \tag{14}$$

where $\nu = \sum_i \nu_i$ and Q' is formulated as follows:

$$\acute{Q} = \nu \ln \sigma^2 - \sum_i \nu_i \ln \sigma_i^2 \tag{15}$$

Under the assumption H_0, $\acute{Q}$ is approximately distributed according to the Chi-square distribution with (m–1) degrees of freedom. Anyhow, the approximation is improved if we divide $\acute{Q}$ by the constant C given below:

$$C = 1 + \left(\frac{1}{3(m-1)}\right)\left(\sum_i \frac{1}{\nu_i} - \frac{1}{\nu}\right) \tag{16}$$

so as to abtain Q:

$$Q = \frac{\acute{Q}}{C} \tag{17}$$

If $Q < \chi^2_{0.95}(m-1)$, the null hypothesis H_0 of homoscedasticity is accepted with 95% confidence.

Chi-square test for Normality check

To check if the residuals process ε_t is Gaussian, several tests such as Anderson's test and Jarque and Bera's test can be used, but the simplest way is to fit a Normal distribution and then apply the goodness of fit test of Chi-square. The steps of the test are the following:

The residual series is first divided into equal intervals, so that each interval contains on average the same number of residuals. The number of intervals K is calculated according to the equation:

$$K = 1 + 3.3 \log \log N \tag{18}$$

where N is the total number of residuals. In this case the number of residuals in each one of the K intervals will be γ:

$$\gamma = N/K \tag{19}$$

This is not the case in general, because the number of residuals is different for each interval. If n_i is the number of residuals in each interval $i(i : 1 \ldots k)$ then:

$$\sum n_i = \sum \gamma = N \tag{20}$$

The Chi-square (χ^2) random variable is defined as:

$$\chi^2 = \sum_{i=1}^{k} \frac{(n_i - \gamma)^2}{\gamma} (calculated\ Chi\ square) \tag{21}$$

This random variable is tabulated and the corresponding theoretical values are given, for a q probability of being exceeded and a number μ of degrees of freedom, by the following:

$$\chi^2_{(q,\mu)} \rightarrow Table(Theoretical\ Chi - squared) \tag{22}$$

$$\mu = K - p - 1 \tag{23}$$

where p is the number of parameters defining the probability density function, which is equal to 2 in this case, because it concerns the Normal distribution.

If for a probability $g(= 1 - q)$ of not being exceeded the following condition is satisfied:

$$\chi^2_{(g)}\ calculated < \chi^2_{(g)}\ theoretical \tag{24}$$

Then, the fitted Normal distribution is accepted with 95% confidence level (probability $g = 0.95$).

2.4.3 Models Comparison and Selection Test

During the estimation step, based on the autocorrelation and partial autocorrelation functions, several model structures are tentatively chosen. Based on the variance of the model residuals in addition to the criterion of parsimony, many statistics are provided by the literature to help select the best model relative to the other possible ones [22, 33]. After calculation, the lower value of these statistics indicates the best model to be selected.

Among others, those statistics are the Akaike Information Criterion (AIC), Akaike Information Criterion corrected ($AICC$), and Shwartz criterion (BIC); and the corresponding expressions are given by Eq. (25), Eq. (26) and Eq. (27) respectively [12].

$$AIC = N\log(\sigma_\varepsilon^2) + 2k \tag{25}$$

$$AICC = N\log(\sigma_\varepsilon^2) + \frac{2N(k+1)}{N-k-2} \tag{26}$$

$$BIC = N\log\left(\sigma_\varepsilon^2\right) + k\log(N) \tag{27}$$

where σ_ε^2 is the residual variance, k the number of model parameters and N the total number of data.

2.5 Prediction

2.5.1 Transformation of the Series

It is necessary during the prediction phase to consider the transformation undergone by the data at the beginning, we then speak about "the prediction recoloring". For example, if the transformation of the original series was a linear filter (such as first differences), predictions of the filtered stationary series are calculated first and then, they are reconstructed by inversion of the filter, to obtain the predictions of the initial series [28].

2.5.2 ARMA Process Predictor

We consider an ARMA(p,q) process as detailed in [13, 37] and [52].

$$X_t = \phi_1 X_{t-1} + \cdots + \phi_p X_{t-p} + \varepsilon_t - \theta_1 \varepsilon_{t-1} - \cdots - \theta_q \varepsilon_{t-q} \tag{28}$$

where parameters $\phi_i, \theta_j, i = 1 \dots p; j = 1 \dots q$ are reals and ε_t is a White noise which variance is σ_ε^2. If the Wold theorem is applied to the process X_t, $t \in Z$ and considerate the corresponding moving average form MA(∞):

$$X_t = \sum_{i=0}^{\infty} \psi_i \varepsilon_{t-i}, \ \psi_0 = 1 \tag{29}$$

It follows that the best prediction that can be made of X_{t+1}, given all available information up to date t, mentioned as $\widehat{X}_t(1)$ is given by:

$$\begin{aligned} \widehat{X}_t(1) &= E(X_t, X_{t-1}, X_{t-2}, \dots, X_0) \\ &= E(\varepsilon_t, \varepsilon_{t-1}, \varepsilon_{t-2}, \dots, \varepsilon_0) \end{aligned}$$

$$\widehat{X}_t(1) = \sum_{i=1}^{\infty} \psi_i \varepsilon_{t+1-i} \tag{30}$$

Then the prediction error is given by the realization in $t + 1$ of the innovation which is not known at time t.

$$X_{t+1} - \widehat{X}_t(1) = \varepsilon_{t+1} \tag{31}$$

More generally for a prediction at the horizon k we have:

$$\widehat{X}_t(k) = \sum_{i=k}^{\infty} \psi_i \varepsilon_{t+k-i} \tag{32}$$

Hence,

$$X_{t+k} - \widehat{X}_t(k) = \sum_{i=0}^{k-1} \psi_i \varepsilon_{t+k-i} \tag{33}$$

Let us determine a confidence interval for the prediction under the assumption of the residuals ε_t normality. Hence, it is shown that:

$$\frac{X_{t+k} - \widehat{X}_t(k)}{Var\left[X_{t+k} - \widehat{X}_t(k)\right]^{1/2}} \underset{T\to\infty}{\rightarrow} \mathcal{N}(0, 1)$$

Considering the following:

$$E\left\{\left[X_{t+k} - \widehat{X}_t(k)\right]^2\right\} = E\left[\left(\sum_{i=0}^{k-1} \psi_i \varepsilon_{t+k-i}\right)^2\right] = \sum_{i=0}^{k-1} {\psi_i}^2 \sigma_\varepsilon^2$$

It follows that:

$$\frac{X_{t+k} - \widehat{X}_t(k)}{\sigma_\varepsilon\left[\sum_{i=0}^{k-1} {\psi_i}^2\right]^{1/2}} \underset{T\to\infty}{\rightarrow} \mathcal{N}(0, 1) \tag{34}$$

We can therefore, construct a 95% confidence interval IC in the form of the following:

$$IC = \left[\widehat{X}_t(k) \mp t^{\alpha/2}\left(\sum_{i=0}^{k-1} {\psi_i}^2\right)^2 \sigma_\varepsilon\right] \tag{35}$$

where $\alpha = 0.05$.

2.5.3 Evaluation of Model Performance

In order to evaluate the goodness of fit and hence the prediction performance of a given model, the most commonly used statistics are: the root mean square error (RMSE) as given in Eq. (36) equivalently the (MSE) Eq. (37), the mean error (ME) as given in Eq. (38). These statistics are calculated between predictions (S_p) of the model and observations (S_0) during the validation period; they give a measure of the model prediction performance [28, 43].

$$RMSE = \sqrt{\frac{1}{N} \sum (S_0 - S_p)^2} \tag{36}$$

$$MSE = \frac{1}{N} \sum (S_0 - S_p)^2 \tag{37}$$

$$ME = \frac{1}{N} \sum (S_0 - S_p) \tag{38}$$

where N is the total data, S_0 is the observed stream flow, and S_p is the predicted stream flows.

Another criterion to evaluate the model performance is the Nash efficiency (NE) (Mc Cuen et al. 2006). It indicates how much the predicted values match the observed data: the more NE value is near to 100%, the better is the performance of the model. Hence, a Nash efficiency value ranging from 80 to 90% indicate a good performance, while a value of 90% or over indicates a satisfactory performance.

$$NE = 1 - \frac{\sum (S_0 - S_p)^2}{\sum (\overline{S_0} - \overline{S_p})^2} \tag{39}$$

where $\overline{S_o}$ and $\overline{S_p}$ are respectively, the average of the observed and predicted stream flows.

3 Application of Box-Jenkins Modeling Method

In this section, Box-Jenkins method has been applied to the annual and monthly stream flows observed at four hydrometric stations in northern Algeria, all located on operating dams. The coordinates and destination of these dams are given in Table 1 and Table 2, respectively. The observed data were provided by the National Agency for Hydraulic Resources (ANRH) of Algiers, Algeria. The total observation period lies from 1925 to 1992 constituting 69 years, about 70% of them were used for model training (estimation) and 30% of them for testing (validation).

The Box-Jenkins method requires the condition of normality and stationarity. This condition means that the mean of the series and its variance must be constant, otherwise the series must undergo a transformation. In addition, the method is based on the internal dependence structure of the time series, so the use of the method is completely unjustified for independent series.

Table 1 Coordinates of the four considered hydrometric stations in Northern Algeria

N°	Code of the station	Station	Latitude (N)	Longitude (E)	Elevation (m)
01	160,403	Beni-Bahdel	38 56 60	04 26 00	645
02	111,504	Bouhanifia	39 24 80	02 65 10	295
03	050,902	Ksob	39 80 30	02 48 12	570
04	160,701	Mefrouche	38 72 80	04 03 25	1110

Source National Agency for Hydraulic Resources (ANRH) of Algiers

Table 2 Destination of the four corresponding dams in Northern Algeria

No	Dam	Wilaya	Capacity (Hm^3)	Regularized volume (Hm^3/year)	Destination
01	Béni-Bahdel	Tlemcen	55.5	48	AEP-Oran IRR. Maghnia perimeter
02	Bouhanifia	Mascara	48.4	Transfert	Transfert Fergoug and AEP-Hacine
04	Ksob	M'sila	15.6	20	IRR. Ksob perimeter
03	Mefrouche	Tlemcen	14.5	17	AEP- Tlemcen

Notation AEP: Drinking Water Supply, IRR: irrigation
Source National Agency for Hydraulic Resources (ANRH) of Algiers

The main steps of the modeling process can be summarized as follows:

(1) The data must be transformed to satisfy the stationarity and normality assumption, usually working with centered and reduced data.
(2) The identification of the model order, which is mainly based on the visual inspection of the autocorrelation and partial autocorrelation function (ACF and PACF respectively) of the time series, assuming that this order is finite but not known.
(3) Estimation of the model parameters, and
(4) Validation of the model by testing the produced errors.

Finally, it should be remembered that the series of data is divided into two parts, while the first one is used for model estimation, and the second is used for model validation.

The annual and monthly stream flows time series modeling is performed to obtain respectively Autoregressive and Moving Integrated Average (ARIMA) models for the annual step and Seasonal ARIMA called SARIMA models for the monthly step. These models will be used for prediction.

Whether the time scale is annual or monthly, the main steps of the modeling remain the same. First, for reasons of redundancy, only one typical modeling example is given below. It shows all the required details of the modeling procedure of the annual stream flows observed at Beni Bahdel station. Then, all the results obtained for annual and monthly scales are given, but in a summarized way. Finally, a synthesis is given to highlight the main findings of the present work.

3.1 A Typical Modeling Example: Case of BENI BAHDEL Annual Data

The statistics of the annual stream flows time series observed at BENI BAHDEL station between 1924 and 1992 are summarized in Table 3. It presents measures of central tendency, variability and the overall view of the time series of concern. Standard skewness and kurtosis are of particular interest because they are used to determine whether the distribution is Gaussian or not. A value of these parameters outside the interval [−2, +2] indicates a significant departure from normality [32]. In our case, the standard kurtosis is within the interval while the skewness is not.

3.1.1 Stationarization

The plot of the original time series as shown in Fig. 1 is indicative of a trend presence in the end of the data, which indicates the existence of non-stationarity [21]. The calculation of the first and the second regular differences of the series allows to get the stationarity condition and the calculation of the standard deviation allows to choose the optimal order of differentiation [20].

The calculation details for the first 5 years differentiated series are given in Table 4, while the differentiated series statistics for the entire data set are given in Table 5.

Table 3 Statistics of the annual stream flows observed at Béni bahdel station from 1924 to 1992

Parameter	Value
Observations number	69
Mean	65.65
Variance	1502.69
Standard deviation	38.76
Minimum	8.26
Maximum	177.69
Standard skewness	2.77
Standard kurtosis	0.55

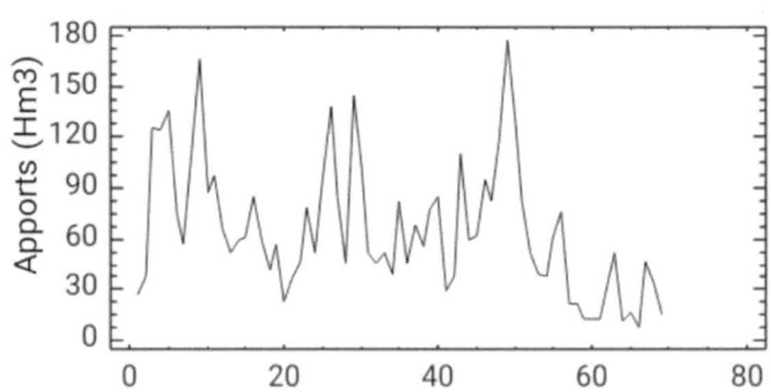

Fig. 1 Annual stream flows observed at Beni bahdel station from 1924 to 1992

Table 4 First and second order regular differentiations of Beni bahdel annual stream flows (only the first 5 data)

N°	SF	DIFF (SF,1)	DIFF (SF,2)
1	26,878		
2	38,631	11,753	
3	125,258	86,627	74,874
4	123,925	−1,333	−87.96
5	135,425	11.5	12,833

Notation SF: stream flow; DIFF(SF,i): ith regular differentiation of SF

Table 5 Descriptive statistics of differentiated data at Beni bahdel

Stream Flow (SF) and differentiation	N	Mean	Variance	St. dev
SF	69	65.66	1502.69	38.76
DIFF(SF,1)	68	–0.17	1304.74	36.76
DIFF(SF,2)	67	–0.46	3099	55.66

Notation DIFF(SF,i): ith regular differentiation of SF, St. dev: standard deviation

The minimum standard deviation is obtained by the first regular difference (d = 1). The obtained series is more stable with a variance of 1304.74 instead of 1502.69 for the initial data, and the trend is removed from the data. The result is a stationary series with a minimum standard deviation value of 36.76, which verifies the hypothesis of stationarity of the model.

To test the normality assumption, we fitted a normal distribution to the differentiated data and measured the goodness of fit which is confirmed by the Chi-square test (test statistic = 18.4706, P = 0.42508).

3.1.2 Identification

The visual examination of the autocorrelation function (ACF) and the partial autocorrelation function (PACF) plots for the differentiated stream flows (Figs. 2 and 3) indicates one significant peak at lag 2 for the ACF, as well as one significant peak at lag 2 for the PACF.

The results of the Box-Pierce test based on the first 22 autocorrelations show that at 90% confidence level, the series is not completely random and the calculated test statistic is 32.259 with a probability P = 0.07 less than 0.1.

The visual examination of the ACF and PACF behavior suggests the consideration of the following possible models with constant:

(1) Arima(2,1,2)
(2) Arima(2,1,1)
(3) Arima(2,1,0)
(4) Arima(1,1,2)
(5) Arima(0,1,2)

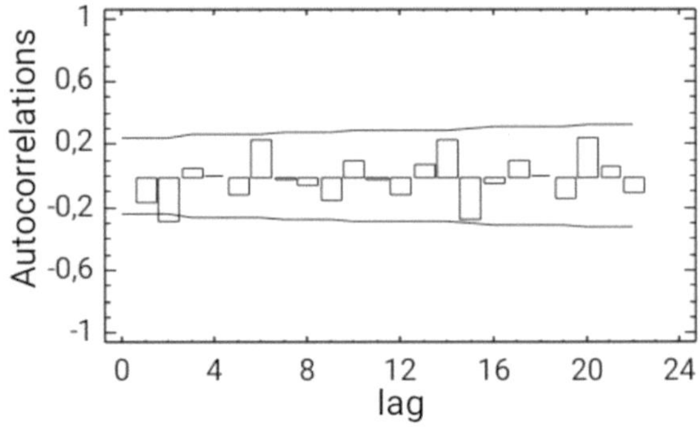

Fig. 2 Autocorrelation function of differentiated annual stream flows at Beni bahdel

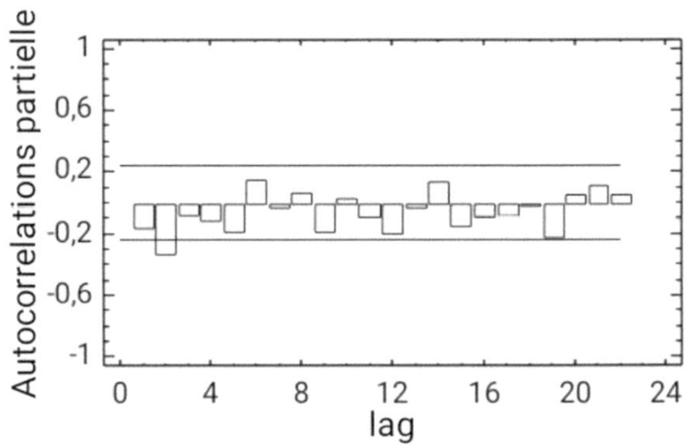

Fig. 3 Partial autocorrelation function of differentiated annual stream flows at Beni bahdel

3.1.3 Model Selection and Estimation

For models parameters estimation, it should be reminded that 69 data were used. The first 39 data were used for model estimation, while the last 20 ones were used for model validation.

The analysis and comparison of the fitted models, as presented in Table 6 shows that the five fitted models all produce completely independent residuals with a stable mean and variance between the estimation and validation periods. However, ARIMA(0,1,2) model produces the minimum residuals variance with a value of 809.399 for the validation period against a value of 1170.7 for the estimation period.

The Akaike Information Criterion (AIC) together with AICC and BIC are calculated for the five fitted models in order to select the most suitable and parsimonious one. As shown in Table 7, the minimum value of these criteria correspond to ARIMA(0,1,2).

The parameter estimates of the selected ARIMA(0,1,2) model with and without a constant and the statistical significance of its terms are summarized in Tables 8 and 9. The terms corresponding to a value of $P < 0.05$ are statistically significantly different from zero at 95% confidence level.

Table 6 Model residuals analysis for the annual scale

N°	Fitted model	Residual variance		RI	ME	HT
		Estimation period	Validation period			
(1)	ARIMA(2,1,2)	1287.4	798.7	OK	OK	OK
(2)	ARIMA(2,1,1)	1267.6	822.5	OK	OK	OK
(3)	ARIMA(2,1,0)	1284.2	852.2	OK	OK	OK
(4)	ARIMA(1,1,2)	1226.3	884.8	OK	OK	OK
(5)	ARIMA(0,1,2)	1170.7	809.4	OK	OK	OK

Notation RI: Residuals independence test; ME: test of mean equality; HT: homoscedasticity test; OK: positive test

Table 7 Performance criteria AIC, AICC and BIC values for different ARIMA model structures

N°	Model structure	Residual variance	AIC	AICC	BIC
1	ARIMA(2,1,2)	798.7	469.1	211.1	207.5
2	ARIMA(2,1,1)	822.5	469.1	208.7	206.6
3	ARIMA(2,1,0)	852.2	469.6	208.5	205.8
4	ARIMA(1,1,2)	884.8	474.2	211.4	208.8
5	ARIMA(0,1,2)	809.4	466.0	206.4	204.3

Therefore ARIMA(0,1,2) model is selected for the BENI BAHDEL annual time series

Table 8 Parameter estimation for ARIMA(0,1,2) with constant

Parameter	Estimation	St. dev	t	P
MA(1)	0.459,571	0.133,983	3.43,007	0.001,053
MA(2)	0.462,918	0.130,167	3.55,633	0.000,708
Mean	0.199,681	0.687,542	0.290,427	0.772,414
Constant	0.199,681			

Table 9 Parameter estimation of ARIMA(0,1,2) without constant

Paramètre	Estimation	Ecart- type	t	P
MA(1)	0.466,639	0.138,751	3.36,314	0.001,286
MA(2)	0.472,472	0.135,975	3.47,468	0.000,909

For this model, the estimated variance of the white noise is 1238.96 with 65 degrees of freedom which means an estimated standard deviation of 35.1989.

Here, the estimated variance of the white noise is 1220.21 with 66 degrees of freedom and the corresponding estimated standard deviation is 34.9315.

Table 10 Goodness of fit for ARIMA(0,1,2)

Statistic	Estimation period	Validation period
MSE	1170.67	809.39
MAE	27.48	23.77
ME	3.58	−21.53

Notation MSE: mean square error, MAE: Mean absolute error, ME: Mean error

The P value for the MA(1) and MA(2) terms is less than 0.05 so they are significantly different from zero, at 95% confidence level with values different from one. In Table 7, the P = 0.77 value for the mean (constant) is much greater than 0.05 so it is statistically not significantly different from zero and can therefore be removed. The ARIMA(0,1,2) model without constant is hence retained.

The performance of the ARIMA(0,1,2) model regarding the goodness of fitting the data is assessed by analyzing the corresponding residuals (errors). Table 10 summarizes the different statistics of these residuals between the estimation and the validation periods.

Each one of these statistics is based on the one-year-ahead prediction error, which is the difference between the data value at time t and its prediction at time $t-1$. The first two statistics (MSE and MAE) give a measure of the error magnitude, while the last one (ME) gives the bias. These error statistics are calculated for both of estimation and validation periods. For the considered model, these residuals statistics are lesser for the validation period than the estimation one. Hence, the fitted model can be considered as adequate and good for performing predictions.

3.1.4 Model Validation

For model validation, it is necessary to verify the hypothesis that the residuals of the selected model are a white Gaussian noise with zero mean. To do this, the autocorrelograms test (ACF and PACF) and the Portmanteau test (Box-Pierce) are used to test the independence of the residuals, the Student's t-test is used to test the nullity of their mean and the Chi-square test at 95% confidence level is used to test their normality [14, 37, 39].

ACF and PACF plots for model residuals, as given by Figs. 4 and 5, respectively, show that all coefficients are included in the confidence interval (CI) plotted at 95% confidence level. This means that they are statistically insignificant. Hence, the model residuals are independent.

Box-Pierce independence test is calculated here based on the first 22 residual autocorrelations. The results show a calculated value of the Q-statistic = 24.3376 with a value of P = 0.2279 which is greater than 0.10, so we cannot reject the null hypothesis that the residuals are independent with a confidence interval of 90% or more.

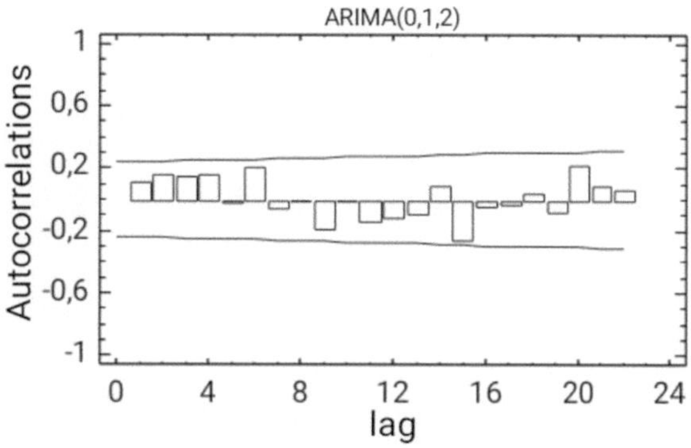

Fig. 4 Autocorrelation function of ARIMA(0,1,2) residuals

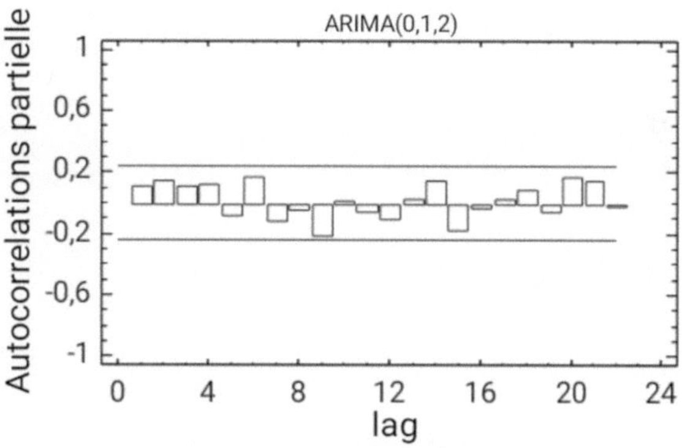

Fig. 5 Partial autocorrelation function of ARIMA(0,1,2) residuals

Student test is applied on model residuals to verify the nullity of their mean. The mean of the residuals is −3.79, their median is −11.59 and their variance is 1030.72 with a standard deviation of 32.10. The results of Student's t-test show a statistic t = −0.97 and a P = 0.33 value which is greater than 0.05. This implies that the null hypothesis of the residuals mean nullity cannot be rejected at 95% confidence level.

Chi-square test is applied to check if the residuals of the ARIMA(0,1,2) model follow a Normal distribution. A Gaussian distribution fit was, therefore, performed and the quality of its fit was measured by the Chi-square test at 95% confidence level. The results of the test show a value of t = 33.91 for the test statistic with a value for P = 0.0129 which is greater than 0.01. Therefore, the null hypothesis of residuals Normality cannot be rejected at 95% confidence level.

Definitively, the results of the previous tests applied to the residuals of the ARIMA(0,1,2) model allowed for the verification of the Box-Jenkins method hypotheses.

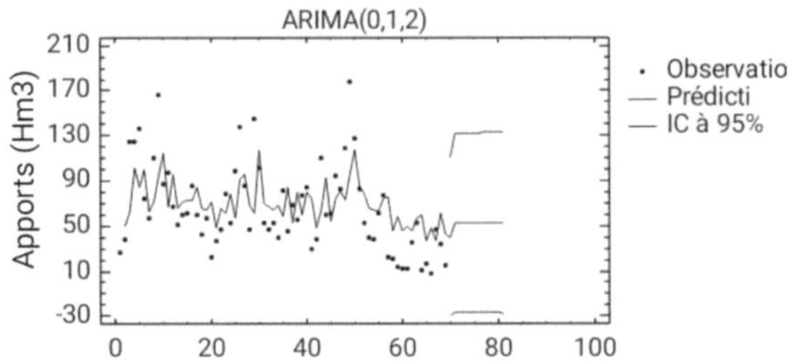

Fig. 6 Annual stream flow predictions at Beni bahdel

Table 11 Model performance for the annual scale

N°	Model structure	MSE	RMSE	NE
1	ARIMA(2,1,2)	798.7	28.3	0.59
2	ARIMA(2,1,1)	822.5	28.7	0.58
3	ARIMA(2,1,0)	852.2	29.2	0.59
4	ARIMA(1,1,2)	884.8	29.7	0.56
5	ARIMA(0,1,2)	809.4	28.3	0.61

3.1.5 Model Performance Evaluation

Graphically, Fig. 6 give the predictions obtained by the ARIMA(0,1,2) model at 95% confidence level. Predictions are shown in as a continuous line together with the observed data as square points. It is evident that the obtained predictions do not closely follow the general pattern of the observations and important differences between them are obvious.

This is confirmed by the Nash efficiency NE value in Table 11. Even if the performance of the selected model ARIMA(0,1,2) is the best among the fitted models, with the higher value of NE($= 0.61$) and lower value of RMSE (28.3), the obtained predictions for this model are not good because its efficiency is only 61%.

3.2 Summary of the Annual Scale Results

The main Box-Jenkins modeling results for the annual scale are summarized in the following tables. Table 12 gives the selected ARIMA (p,d,q) model for each one of the four time series, together with the percentage of the corresponding explained variance; Table 13 gives the adequation of those models to fit the data; whereas Table 14 gives the evaluation of models performance. Finally, Table 15 shows the check for stationarity hypothesis for model residuals.

According to Table 12, it is clear that the structured part for the annual data represents—on average—only 30% of the total variability. Consequently, the remaining

Table 12 Selected models and the corresponding explained variance in percent for the annual scale

Station	Model	σ_M^2 (%)
Béni bahdel	ARIMA(0,1,2)	21.00
Bouhnifia	ARIMA(1,0,3)	25.00
Ksob	ARIMA(2,0,0)	39.50
Mefrouche	ARIMA(1,0,3)	36.25
Average		30.43

Notation $\sigma^2{}_M$ (%): percentage of variance explained by the model

70% variability is random, so impossible to predict. These results show the magnitude of the random part for the annual data. This is very much expected because the annual data are generally independent and do not show a significant correlation structure within the data, which is also consistent with the literature [6, 15].

Nevertheless, the performance of the obtained models can be appreciated by analyzing the prediction errors between the estimation and the validation period. The adequacy of those models to fit the observed data is checked by some statistics such as the Mean Square error (MSE), Absolute Mean error (AME) and Mean Error (ME). These statistics are calculated and the results grouped in Table 13 indicate that, on average, these statistics are lower during the validation period, particularly for the MSE, which alone represents a reliable criterion.

Meanwhile, the prediction performance of the fitted models for the validation period is evaluated by the MSE, RMSE and the NE criterion as shown in Table 14. Even if every one of these models is selected on the basis of the minimum RSME and highest NE, among many other possible models, it is clear that their efficiency does not exceed 73%, in the best case of Ksob. This result confirms once more that for the concerned sites, the biggest part of variability for the annual stream flows is random and hence not predictable.

Finally, Table 15 gives the results of checking for residuals stationarity, which is the second hypothesis of Box-Jenkins models. The aim is to verify the Mean and variance constancy over different segmentations of the residuals. This test is done for both of the estimation and validation periods. It is realized via the Mean Equality test and the Homoscedasticity test. Definitively, the test is positive for both of the mean and the variance of the four model residuals which is indicative of their stationarity.

3.3 Summary of the Monthly Scale Results

As for the annual scale, the main results obtained for the monthly stream flows are summarized in Table 14. It shows the selected SARIMA (p,d,q) model for each one of the four time series, together with the percentage of the corresponding explained variance. Table 15 presents the adequation of those models to fit the data whereas

Table 13 Test of model adequation for the annual scale

N°	Station	Model	MSE		MAE		ME	
			Estimation	Validation	Estimation	Validation	Estimation	Validation
1	Beni Bahdel	ARIMA(0,1,2)	1170.67	809.39	27.48	23.77	3.58	−21.52
2	Bouhanifia	ARIMA(1,0,3)	3118.66	4164.25	42.05	56.52	61.56	113.57
3	Ksob	ARIMA(2,0,0)	477.72	191.00	13.87	12.40	−0.97	−3.46
4	Mefrouche	ARIMA(1,0,3)	5211.81	2379.47	53.66	42.35	6.02	37.92
	Average		2494.71	1886.03	34.26	33.76	17.54	31.62

Notation MSE: Mean Square Error; MAE: Mean Absolute Error; ME: Mean Error

Table 14 Evaluation of model performance for the annual scale

N°	Station	Model structure	MSE	RMSE	NE
1	Beni Bahdel	ARIMA(0,1,2)	809.4	28.3	0.67
2	Bouhanifia	ARIMA(1,0,3)	4164.2	64.5	0.60
3	Ksob	ARIMA(2,0,0)	191.0	13.8	0.73
4	Mefrouche	ARIMA(1,0,3)	2379.0	48.7	0.51

Table 15 Test of residuals stationarity for the annual scale

N°	Station	Model	Mean equality test		Homoscedasticity test	
			Estim	Valid	Estim	Valid
1	Beni Bahdel	ARIMA(0,1,2)	Ok	Ok	Ok	Ok
2	Bouhanifia	ARIMA(1,0,3)	Ok	Ok	Ok	Ok
3	Ksob	ARIMA(2,0,0)	Ok	Ok	Ok	Ok
4	Mefrouche	ARIMA(1,0,3)	Ok	Ok	Ok	Ok

Notation Estim: estimation period;Valid: validation period; Ok: positive test of stationarity

Table 16 gives the evaluation of the models performance. Finally, Table 17 shows the check for stationarity hypothesis for the models residuals.

Table 16 Selected Models and the corresponding explained variance in percent for the monthly scale

N°	Station	Model	σ_M^2 (%)
01	Béni bahdel	SARIMA(1,1,1)x(2,1,1)$_{12}$	62
02	Bouhanifia	SARIMA(1,1,1)x(0,1,1)$_{12}$	71
03	Ksob	SARIMA(0,0,3)x(0,1,1)$_{12}$	56.2
04	Mefrouche	SARIMA(1,0,0)x(0,1,1)$_{12}$	44.6
	Average		58.45

Notation σ_M^2 (%): percentage of variance explained by the model

Table 17 Test of model adequation for the monthly scale

N°	Station	Model	MSE		MAE		ME	
			Estim.	Val.	Estim.	Val.	Estim.	Val.
01	Béni bahdel	SARIMA(1,1,1)x(2,1,1)$_{12}$	47.34	15.85	3.3	2.46	0.13	−0.39
02	Bouhanifia	SARIMA(1,1,1)x(0,1,1)$_{12}$	155.9	64.03	5.57	4.13	1.20	−0.10
03	Ksob	SARIMA(0,0,3)x(0,1,1)$_{12}$	13.01	12.98	2.41	2.13	−0.31	0.83
04	Mefrouche	SARIMA(1,0,0)x(0,1,1)$_{12}$	3.59	1.18	0.99	0.66	0.05	−0.23
	Average		54.85	23.51	3.06	2.34	0.52	0.03

Notation MSE: Mean Square Error; MAE: Mean Absolute Error; ME: Mean Error; Estim: estimation period; Valid: validation period

According to Table 16, it is obvious that the structured part in the monthly data that could be extracted and modeled by SARIMA models represents—on average—almost 60% of the total variability. This result is almost double of the annual scale (30%). It shows how important the effect of seasonal variability is on the internal dependence structure of the monthly data. Actually, this is due to the existence of the seasonal component which creates a kind of tighter dependence within the data. Hence, only 40% remaining variability may be thought of random (not predictable). This result is expected and also consistent with the literature [15]. It can be concluded that the obtained SARIMA models represent more faithfully the variability of the considered monthly stream flows.

The performance of the obtained SARIMA models can be appreciated by analyzing the prediction errors between the estimation and the validation periods. The adequacy of those models to fit the observed data is checked by the statistics MSE, AME and ME. These statistics are calculated and the results grouped in Table 17. These results indicate that these statistics are all lower during the validation period compared to the estimation one which confirms the adequacy of these models in fitting the monthly data and shows their performance in doing so.

Furthermore, the prediction performance of the fitted models is evaluated using the MSE, RMSE and NE criterion for the validation period as shown in Table 18. It must be remembered here, that every one of these models is selected, among many other possible models, on the basis of the minimum RSME and the highest NE. These results show that the efficiency of the models is fairly good for the monthly scale. Actually, all of the four selected models have a NE value over than 80% with a maximum value of 89% for Mefrouche. This result means that the structured variability of the monthly stream flows is more important than the random one. This is due to the internal seasonal dependence of such data. Consequently, the obtained models are more representative of the monthly stream flows and may be used for prediction in the concerned sites in the limit of their performance.

Finally, Table 19 gives the results of checking for model residuals stationarity. Model residuals are stationary if there is a constancy of the mean and the variance over the estimation and validation periods. This is checked by the Mean equality test and the homoscedasticity test respectively. The results of stationarity test are positive for all of the four SARIMA models which is indicative of their residuals stationarity.

Table 18 Evaluation of model performance for the monthly scale

N°	Station	Model structure	MSE	RMSE	NE
1	Beni Bahdel	$SARIMA(1,1,1)x(2,1,1)_{12}$	15.85	3.9	0.87
2	Bouhanifia	$SARIMA(1,1,1)x(0,1,1)_{12}$	64.03	8.0	0.89
3	Ksob	$SARIMA(0,0,3)x(0,1,1)_{12}$	12.98	3.6	0.86
4	Mefrouche	$SARIMA(1,0,0)x(0,1,1)_{12}$	1.18	1.1	0.89

Table 19 Test of residual stationarity for the monthly scale

N°	Station	Model	Mean equality test		Homoscedasticity test	
			Estim.	Val.	Estim.	Val.
01	Béni bahdel	SARIMA(1,1,1)x(2,1,1)$_{12}$	Ok	Ok	Ok	Ok
02	Bouhanifia	SARIMA(1,1,1)x(0,1,1)$_{12}$	Ok	Ok	Ok	Ok
03	Ksob	SARIMA(0,0,3)x(0,1,1)$_{12}$	Ok	Ok	Ok	Ok
04	Mefrouche	SARIMA(1,0,0)x(0,1,1)$_{12}$	Ok	Ok	Ok	Ok

Notation Estim: estimation period; Valid: validation period; Ok: positive test of stationarity

3.4 Synthesis

The problem in the present work was to answer the question: are stream flows in the Northern Algeria completely random? The answer to this question is provided by Box-Jenkins models. Indeed, these models allow to dissociate the random variability from the structured one within a time series. This can be done taking into account the stochastic nature of the hydrological variable of concern. The recommended method is a procedure that is essentially based on the internal dependence structure of the data, and its theoretical foundations are based on the assumptions of normality and stationarity.

The obtained results led to the building of some mathematical models for both of annual and monthly time lags and the extent of random variability was clearly highlighted. For the annual scale, the average variability explained by the models is close to 30% against nearly 60% for the monthly scale. This means that the random variability within the data is about 70% for the annual scale, against only 40% for the monthly one. This result shows the magnitude of the risk incurred if the obtained ARIMA and SARIMA models are used for simulation or forecasting. This difference between the annual and monthly models is due to the internal dependence structure of the data. Actually, the internal dependence structure of the data is reinforced by the introduction of the seasonal effect for the monthly data, which decreases the total random variability.

Regarding stationarity, which is a condition for Box-Jenkins models, the results showed that annual data are generally stationary, normally distributed with a weak autocorrelation in their internal dependence structure, while monthly data are, on the contrary, characterized by a seasonal instability in variance, by a non-normal distribution and by a strong regular and seasonal autocorrelation in their internal dependence structure. These results are consistent with those of the literature [27, 52].

For both of the annual and monthly scales, the present work led to the construction of a Box-Jenkins model for each one of the four hydrometric stations. These models are representative of the generating process as closely as possible since the corresponding prediction error statistics are the minimum. They are simple and parsimonious and the predictions they provide are acceptable within the limits of their performance and the rate of their explained variance. The obtained results go hand in hand with those of the literature [2, 52, 56].

4 Conclusion

In the present work, the use of Box-Jenkins models, which are essentially based on the internal dependence structure of observed data, is fully justified because this type of model is particularly well suited to streamflow contributions in any hydrological system and not only in northern Algeria. Indeed, the contribution of groundwater flow can be considered as the auto-regressive (AR) part, while random precipitation can be considered as the moving average (MA) one.

The main findings of this modeling work are some mathematical Box-Jenkins models for both annual and monthly stream flows in northern Algeria. Those models show the magnitude of the random variability of such hydrological variable.

For the annual scale, where data are generally independent, the application of the Box-Jenkins method showed that the fitted models explain on average only 30% of the total variability. This shows the magnitude of the random variability that is about 70%.

Whereas for the monthly scale, the seasonal contribution decreases the total random variability because it reinforces the internal dependence structure of data. The calculations proved it and the fitted models explain 60% of the total variability against only 30% of the annual data. This decreases the total random variability to 40% for the monthly scale, versus 70% for the annual one.

Thus, the monthly models are more suitable to provide predictions than the annual models because they are more representative of the stream flows in northern Algeria. The obtained predictions are acceptable in the limit of the models performance, and hence, the obtained models might be used for prediction of stream flow at these sites.

5 Recommendations

Finally, the Box-Jenkins approach is recommended if one is looking for a quick, simple and parcimonious model that takes into account the stochastic nature of stream flows. Actually, the results of such models are quite satisfactory and might be used for either simulation or prediction [12].

One of the key advantages of this type of models is their methodology and the corresponding systematic searching steps (identification, estimation, validation) for any suitable model. Another advantage is their high potential of modeling either stationary or non-stationary time series, in addition to their forecasting performances [16].

However, this type of model presents some limitations and constraints such as: (a) the assumption of normality and stationarity, whereas time series with a smaller scale than one year are neither normal nor stationary. In addition, a mandatory step is to make transformations and differentiations to fulfill these conditions. (b) This type of model is based on the short-term dependence, whereas for hydrological processes a long-term dependence is to be considered [11, 41, 44]. (c) Except for simple models such as AR(1) or ARMA(1, 1) the most models do not have a physical meaning and are used in a black box concept [40]. (d) For a good result, this type of models require at least 50–100 observations [3]. Consequently, this condition for the yearly hydrologic time series will be problematic because the observation period is generally less than 50 years [27].

Acknowledgements Many thanks go to the National Agency for Hydraulic Resources (ANRH) of Algiers for all the provided data.

References

1. Adamowski JF (2008) Development of a short-term river flood forecasting method for snowmelt driven floods based on wavelet and cross-wavelet analysis. J Hydrol 353:247–266. https://doi.org/10.1016/j.jhydrol.2008.02.013
2. Adhikary SK, Rahman M, Gupta AD (2012) A stochastic modelling technique for predicting groundwater table fluctuations with time series analysis. Int J Appl Sci Eng Res 1:238–249. https://doi.org/10.6088/ijaser.0020101024
3. Adnan RM, Yuan X, Kisi O et al (2017) Application of time series models for streamflow forecasting. Civil Environ Res 9(3):56–63. https://core.ac.uk/download/pdf/234678525.pdf
4. Adnan RM, Yuan X, Kisi O et al (2017) Streamflow forecasting of Astore River with seasonal autoregressive integrated moving average model. Eur Sci J 13(12):145–156. https://doi.org/10.19044/esj.2017.v13n12p145
5. Akaike H (1974) A new look at the statistical model identification. IEEE Trans Automatic Control 19(6):716–723. https://ieeexplore.ieee.org/abstract/document/1100705
6. Anderson OD (1976) Time series analysis and forecasting: the BOX JENKINS approach. London: Butterworth. https://archive.org/details/timeseriesanalys0000ande
7. Anderson TW (1971) The statistical analysis of time series. John Wiley & Sons, NewYork. https://doi.org/10.1002/9781118186428
8. Ari Yair BA, Lukumon O, Muhammad B et al (2022) Rainfall prediction: a comparative analysis of modern machine learning algorithms for time-series forecasting. Mach Learn Appl. https://doi.org/10.1016/j.mlwa.2021.100204
9. Awwad HM, Valdés JB, Restrepo PJ (1994) Streamflow forecasting for Han River Basin, Korea. J Water Resour Plng Mgmt 120(5):651–673. https://doi.org/10.1061/(ASCE)0733-9496(1994)120:5(651)
10. Azizipour M (2024) Prediction of monthly inflow to Karkhe reservoir using ARIMA model. Iran J Ecohydrol 10(4):595–606. https://doi.org/10.22059/ije.2023.369183.1777
11. Beran J (1994) Statistics for long memory processes. Chapman and Hall, New York. https://doi.org/10.1201/9780203738481
12. Boukharouba K, Kettab A (2016) Comparison of prediction performances between Box-Jenkins and Kalman filter models—case of annual and monthly sreamflows in Algeria. Desalin Water Treatment J 57(36):17095–17101. https://doi.org/10.1080/19443994.2015.1109558

13. Box PG, Jenkins PM (1970) Time series analysis. Holden Day, San Francisco. https://doi.org/10.1057/9781137291264_6
14. Box GE, Jenkins GM, Reinsel GC (2011) Time series analysis: forecasting and control. John Wiley & Sons, NewYork. https://doi.org/10.1057/9781137291264_6
15. Bras RL, Rodriguez I (1993) Random functions and hydrology. Addison-Wesley, Massachusetts. ISBN:0486676269, 9780486676265
16. Brown LC, Mac Berthouex P (2002) Statistics for environmental engineers. CRC press. https://doi.org/10.1201/9780367802691
17. Chang X, Gao M, Wang Y, Hou X (2012) Seasonal autoregressive integrated moving average (SARIMA) model for precipitation time series. J Math Stat 8(4):500–505. https://doi.org/10.3844/jmssp.2012.500.505
18. Chatefield C (2001) Time series forecasting. Chapman and Hall, New York. https://doi.org/10.1201/9781420036206
19. Chen CS, Liu CH, Su HC (2008) A nonlinear time series analysis using two-stage genetic algorithms for streamflow forecasting. Hydrol Process 22:3697–3711. https://doi.org/10.1002/hyp.6973
20. Clement MP, Hendry DF (2000) Forecasting economic time series, vol 1. Cambridge University continentale, pp 57–70. https://doi.org/10.1017/CBO97805115992866
21. Cox DR (1981) Statistical analysis of time series: some recent developments. Scand J Stat 8:93–115. https://www.jstor.org/stable/4615819
22. Dabral PP, Murry MZ (2017) Modelling and forecasting of rainfall time series using SARIMA. Environ Process 4:399–419. https://doi.org/10.1007/s40710-017-0226-y
23. Deshmukh KH, Bamnote GR, Deshmukh PS (2024) Predicting climatic parameters using the ARIMA time series and LSTM deep learning models for Vidarbha Region. Int Res J Adv Eng Hub (IRJAEH) 2(05):1255–1264. https://doi.org/10.47392/IRJAEH.2024.0173
24. Droesbeke J, Fichet B, Tassi P (1989) Séries chronologiques: théorie et pratique des modèles ARIMA. Economica, Paris. https://www.persee.fr/doc/hism_0982-1783_1992_num_7_1_1684
25. Geetha A, Nasira GM (2016) Time-series modelling and forecasting: modelling of rainfall prediction using ARIMA model. Int J Soc Syst Sci 8(4):361–372. https://doi.org/10.1504/IJSSS.2016.10002401
26. Ghafoor A, Hanif S (2005) Analysis of the trade pattern of Pakistan: past trends and future prospects. J Agric Soc Sci 1:346–349. https://eurekamag.com/research/004/397/004397505.php
27. Gocheva-Ilieva SG, Ivanov AV, Voynikova DS et al (2014) Time series analysis and forecasting for air pollution in small urban area: an SARIMA and factor analysis approach. Stochastic Environ Res Risk Assess 28:1045–1060. https://doi.org/10.1007/s00477-013-0800-4
28. Gourieroux C, Monfort A (1990) Séries temporelles et modèles dynamiques. Economica, Paris. https://www.sudoc.fr/001589032
29. Grandon TG, Schwenzer J, Steens T et al (2024) Electricity demand forecasting with hybrid classical statistical and machine learning algorithms: case study of Ukraine. Appl Energy 355:122249. https://doi.org/10.1016/j.apenergy.2023.122249
30. Hamilton JD (1994) Time series analysis. Princeton University Press, Princeton. https://archive.org/details/timeseriesanalys0000hami
31. Hipel KW, McLeod AI (1994) Time series modelling of water resources and environmental systems. Elsevier, Amsterdam, Netherlands. eBook ISBN: 9780080870366
32. Hirsch RM, Helsel DR, Cohn TA et al (1993) Statistical analysis of hydrologic data. In Handbook of Hydrology. New York: McGraw-Hill. https://api.semanticscholar.org/. Corpus ID:128385891
33. Hurvich CM, Tsai CL (1989) Regression and time series model selection in small samples. Biometrika 76:297–307. https://doi.org/10.2307/2336663
34. Jafarian-Namin S, Shishebori D, Goli A (2024) Analyzing and predicting the monthly temperature of Tehran using ARIMA model, artificial neural network, and its improved variant. J Appl Res Indus Eng 11(1):76–92. https://doi.org/10.22105/jarie.2023.356297.1502

35. Karthikeyan L, Nagesh KD (2013) Predictability of nonstationary time series using wavelet and EMD based ARMA models. J Hydrol 502:103–119. https://doi.org/10.1016/j.jhydrol.2013.08.030
36. Kendall M, Ord JK (1990) Time series. John Wiley & Sons, New-York. https://archive.org/details/timeseries0003kend
37. Kottegoda NT (1980) Stochastic water resources technology. Macmillan Press, London. https://doi.org/10.1007/978-1-349-03467-3
38. Koutroumanidis T, Sylaios G, Zafeiriou E et al (2009) Genetic modeling for the optimal forecasting of hydrologic time-series: application in Nestos River. J Hydrol 368(1–4):156–164. https://doi.org/10.1016/j.jhydrol.2009.01.041
39. Koutsoyiannis D, Montanari A (2007) Statistical analysis of hydroclimatic time series: uncertainty and insights. Water Resour Res 43(5):1–9. https://doi.org/10.1029/2006WR005592
40. Koutsoyiannis D, Huaming Y, Georgakakos A (2008) Medium-range flow prediction for the Nile: a comparison of stochastic and deterministic methods. Hydrological Sci–J–des Sci Hydrol 53(1):125–132. https://doi.org/10.1623/hysj.53.1.142
41. Koutsoyiannis D (2002) The Hurst phenomenon and fractional Gaussian noise made easy. Hydrol Sci J 47(4):573–596. https://doi.org/10.1080/02626660209492961
42. Lettenmaier DP, Wood EF (1993) Séries temporelles et modèles dynamiques. Economica, Paris. https://worldcat.org/oclc/919625398
43. Lin JY, Cheng CT, Chau KW (2006) Using support vector machines for long-term discharge prediction. Hydrol Sci J 51:599–612. https://doi.org/10.1623/hysj.51.4.599
44. Otache MY, Zhijia L, Mohammad B (2008) Analysis of long-term dependence phenomenon in Benue River flow process and its hypothesis testing. Chin J Oceanol Limnol 26:313–322. https://doi.org/10.1007/s00343-008-0313-z
45. Othman SA, Jameel HH, Abdulazeez ST (2024) Comparative analysis of univariate SARIMA and multivariate VAR models for time series forecasting: a case study of climate variables in Ninahvah City, Iraq. Math Prob Comput Sci 61:27–49. https://doi.org/10.51408/1963-0113
46. Parvaze S, Kumar R, Khan JN, Allaie SP (2024) Forecasting monthly peak flows for the flood-prone River Jhelum in the Kashmir Valley, India. Water Pract Technol. https://doi.org/10.2166/wpt.2024.225
47. Pena D, Tiao GC, Tsay RS (2001) A course in time series analysis. John Wiley & Sons, NewYork. https://doi.org/10.1002/9781118032978
48. Reinsel G, Box G, Jenkins G (1994) Time series analysis: forecasting and control. Prentice Hall, Englewood Cliffs, NJ. https://doi.org/10.1057/9781137291264_6
49. Sharma RK, Kumar S, Padmalal D et al (2023) Streamflow prediction using machine learning models in selected rivers of Southern India. Int J River Basin Manage. https://doi.org/10.1080/15715124.2023.2196635
50. Tayyab M, Zhou J, Zeng X et al (2016) discharge forecasting by applying artificial neural networks at the Jinsha River Basin, China. Eur Sci J 12:1857–7881. https://api.semanticscholar.org/corpus
51. Toshaliyeva S (2024) Using the Arima model to forecast the share of railways in the industry. In: E3S web of conferences, vol 531. EDP Sciences, p 02024. https://doi.org/10.1051/e3sconf/202453102024
52. Valipour M, Banihabib ME, Behbahani SMR (2013) Comparison of the ARMA, ARIMA, and the autoregressive artificial neural network models in forecasting the monthly inflow of Dez dam reservoir. J Hydrol 476:433–441. https://doi.org/10.1016/j.jhydrol.2012.11.017
53. Yang CG, Lin ZH, Yu ZB et al (2010) Analysis and simulation of human activity impact on streamflow in the Huaihe river basin with a Large-Scale hydrologic model. J Hydrometeorol 11:810–821. https://doi.org/10.1175/2009JHM1145.1
54. Yap ZN, Musa S (2023) Stream flow forcasting on pahang river by time series models, ARMA, ARIMA and SARIMA. Recent Trends Civil Eng Built Environ 4(1):331–341. https://doi.org/10.30880/rtcebe.2023.04.01.035
55. Yevjevich V (1971) Stochastic processes in hydrology. Fort Collins, Colorado. ISBN: 0918334012, 9780918334015

56. Yurekli K, Kurunc A, Ozturk F (2005) Application of linear stochastic models to monthly flow data of Kelkit Stream. Ecol Model 183:67–75. https://doi.org/10.1016/j.ecolmodel.2004.08.001
57. Zhong W, Zhai D, Xu W et al (2024) Accurate and efficient daily carbon emission forecasting based on improved ARIMA. Appl Energy 376:124232. https://doi.org/10.1016/j.apenergy.2024.124232
58. Zou H, Yang Y (2004) Combining time series models for forecasting. Int J Forecast 20(1):69–84. https://doi.org/10.1016/S0169-2070(03)00004-9
59. Zounemat KM, Teshnehlab M (2008) Using adaptive neuro-fuzzy inference system for hydrological time series prediction. Appl Soft Comput 8(2):928–936. https://doi.org/10.1016/j.asoc.2007.07.011

Spatio-Temporal Variation in Urban Vegetation in Maghrebian Cities Since the XXst Century Using Remote Sensing

Aude Nuscia Taïbi, Abdelali Gourfi, Salima Salhi, Mustapha El Hannani, Saïd Boujrouf, Faiza Allouche-Khebour, Sahraoui Bensaid, Halima Slimani, Rania Ajmi, and Hawa Bara

Abstract In Maghreb, vegetation cover, a crucial factor for the sustainability of urban ecosystems, is undergoing significant change in cities. After the French colonial period giving key consideration to public vegetation in the new towns built such as in Sousse, Algiers or Marrakesh, the importance attached to greenery in cities is not clearly defined in independent Tunisia, Algeria and Morocco, and today consideration of urban vegetation in Maghrebian spatial planning policy is still nascent. By taking a spatialized approach involving satellite remote sensing we show an overall decline in vegetation cover in the cities of Algiers, Sousse and Marrakech between the 1970s and the 2020s with two distinct periods; a swift and fairly steady decline in greenery until 2000, slowing significantly and even reverses slightly after this date. The spatial distribution of vegetation across the cities has also changed, much more uniform in 1973 than during the later periods where greenery became increasingly concentrated within clearly defined areas. A neighborhood-based analysis of the vegetation cover makes it possible to connect the contrasting spatial distribution and variation observed with the public spatial and urban planning policies since

A. N. Taïbi · A. Gourfi (✉) · S. Salhi · M. El Hannani · H. Bara
ESO, UMR 6590, UNIV Angers, CNRS, SFR Confluences, UFR LLSH, Angers, France
e-mail: gourfigrn@gmail.com

A. N. Taïbi
e-mail: nucia.taibi@univ-angers.fr

M. El Hannani
e-mail: mustapha.elhannani@univ-angers.fr

S. Boujrouf
Faculté des lettres et des Sciences Humaines, Laboratoire des Études sur les Ressources, Mobilités et Attractivités (LERMA), Université Cadi Ayyad, Marrakech, Morocco

F. Allouche-Khebour · R. Ajmi
Green Team, Institut Supérieur Agronomique Chott Meriem, Université de Sousse, Sousse, Tunisie

S. Bensaid · H. Slimani
Laboratoire Écologie et Environnement, Université des Sciences et Techniques Houari Boumedienne d'Alger, Algiers, Algérie

F. Khebour Allouche (ed.), *Environmental Remote Sensing and GIS in MENA Region*, Earth and Environmental Sciences Library,
https://doi.org/10.1007/978-3-031-86574-9_10

the 1970s, as well as since the colonial period and their implications in terms of socio-environmental justice and socio-spatial segregation generated by the spatial organization of vegetation in Algiers, Sousse and Marrakesh. Today, the proportion of green space per capita shows that these cities have an overall average (of 3.1 m^2 per capita for Marrakesh), falling short of the minimum 10 m^2 per capita recommended by UN-Habitat.

Keywords Urban Vegetation · Spatio-Temporal Variation · Urban Planning Policies · Remote Sensing · Algiers · Sousse · Marrakesh · Maghreb

1 Introduction

More than half of the world's population lives in urban areas today [38] often beset by environmental problems with water and air pollution and deteriorating ecosystems [23], which are undermining the quality of life for inhabitants.

Compounding the stress of urban dwellers, the COVID pandemic has confirmed the need for "nature" in cities and the importance of a change in urban model for the sake of inhabitants' well-being [2, 8]. Greenery in cities is of the utmost importance as it provides goods and services and complete ecosystem functions that benefit city dwellers and the environment alike. In addition, it provides a whole host of advantages in terms of human well-being [1, 11], not least by mitigating the urban heat island (UHI) effect [22]—a threat to human health [31]—by reducing stress, improving healing time and cognitive abilities [29]. Such greenery also provides a range of ecosystem services including protection from some natural disasters—flooding for example [9, 27]—strengthening biological processes like pollination [40] and reducing water erosion [10, 36].

Vegetation cover is therefore a crucial factor for the sustainability of urban ecosystems [35]. And yet it is undergoing significant change in cities, not only as regards spatial extent but also density, structure and flower composition.

In Maghreb, the forms and functions of urban greenery as well as the importance attached to it have changed over time. The French colonial period marked a turning point in these changes in particular. Indeed, the North African colonies would turn out to be ideal urban testing grounds and the French protectorate in Morocco (1912–1956) and Tunisia (1881–1956) or colonizing state in Algeria (1830–1962) would set special store by vegetation in the new towns built, while still giving key consideration to the pre-existing scenery, the Medinas (old walled quarters) in particular, such as in Marrakesh [37], Algiers and Sousse.

That said, the cities of Marrakesh, Algiers, and Sousse prior to the creation of the new European neighbourhoods from the colonization outside the ramparts of the Medinas, were surrounded by countless gardens organized around water resources. In Marrakesh some centuries old orchards (Almohad Agdal and Menara gardens) are neighbouring a palm grove and its farmland irrigated from *khettaras*, underground channels carrying the water from the groundwater sheet of the Haouz plain [7, 14,

17]. In Algiers the medina (called Casbah) was surrounded by the ottoman *Fahs.* In Sousse the medina was surrounded by olive groves. Inside the walled ramparts, meanwhile, were *Arsas* ornamental gardens and kitchen gardens for the community, mainly in Marrakesh. Lastly, there were many *Riads,* traditional town houses built around a patio or enclosed inner garden surrounded by a gallery. All of these gardens primarily served a sustenance purpose, but they also provided "green space" in the cities and a setting for the the weekly picnic (*N'Ezha* in Marrakesh), or some religious festivals [17].

With the French protectorate and colonisation, the ideal city, the model city to which French urban planners, able to give free rein to their creativity, aspired at the turn of the twentieth century [34], was a modern, airy, green and open city where open green spaces and public gardens were given pride of place.

Maghreb, free from the regulatory red tape of mainland France, represented choice sites for developing new European cities providing a healthy living environment appealing to the settlers [19] who found the old walled part of the Medinas frightening on account of its "labyrinthine", "foul-smelling" and "swarming" quarters [37]. This French model of the colonial city was thus "projected" as a counterpoint to the local Arabo-Berber model of the Medinas so as to "reveal" Western civilization. The aim was to reflect what a "modern" colonial city should be in the Western sense, with perfect geometry combining functionality, viability and efficiency [17], by giving precedence to sports amenities, physical education in the open air and green spaces. This was so as to guarantee the physical and mental development of the French [7].

When independences came in Morocco, Algeria and Tunisia, new neighbourhoods were added to the outskirts of the three cities, amid fairly swift demographic growth. Listed as UNESCO World Heritages of Humanity today, the three medinas supports touristic activities, mainly in Marrakesh and Sousse, thanks to the world-class hotel facilities and sunny climate of the cities, as well as their gardens and golf courses—even though these are enormously water-reliant. In Morocco and Tunisia, the city's hotel facilities have certainly expanded rapidly in recent years, and the development of tourism represents for example a priority in terms of promotion for the Marrakesh Urban Community (CUM), with 72% of the overall investment amount (monograph paper for Marrakesh, non-dated). However, the three cities has warm, Mediterranean or arid bioclimates with average annual rainfall of just 281 mm for Marrakesh for instance [15], and a hot and dry summer. This all places them under regular water stress and significant pressure where its vegetation is concerned. Be that as it may, against the current climate backdrop, this vegetation is one way to reduce urban heat islands by helping to keep temperatures under control.

The importance attached to greenery in cities is not clearly defined in Morocco, Algeria and Tunisia today, despite incentives from the governments, for instance the Moroccan ministry for Housing and Urban Planning to encourage local authorities to draw up green plans, the stated intention being to improve residents' living environments by giving significant precedence to local green space. In Marrakesh, although leaders claim to have "crafted a proactive policy in terms of green space", with a positive impact on the city's national and international image [5], in 2015 its parks & gardens department said that it did not have a clear idea of the vegetation composition in green public spaces [37]. In Sousse, as well as in Algiers, the green spaces inventory is very recent and still incomplete. This shows that, in reality, consideration of greenery in Maghrebian spatial planning policies is still nascent. Whatever the official rhetoric might be, in most cases it is not until buildings, roads and car parks have been erected that a landscape architect is then called in to fill in the gaps [8]. Furthermore, the colonial legacy in terms of greenery, in the same way as urban design or architecture, has considerably shaped the city's landscapes and functioning.

The areas studied here are, therefore, Marrakesh in Morocco, Algiers in Algeria and Sousse in Tunisia.

The Marrakesh Urban Community (CUM), one of the municipalities in the region of Marrakesh-Safi (Fig. 1), extends over 230 km^2, with five *arrondissements* (districts) (Marrakech-Medina, Gueliz, Menara, Ennakhil and Sidi Youssef Ben Ali/SYBA) and 25 neighborhoods. It is home to 928,850 inhabitants with a density of 350 inhabitants per sq. km [25].

Sousse, a very old Phoenician city in the centre-east of Tunisia, includes 15 districts 5 of which are specifically studied in its centre: Sousse Medina, Sousse North, Sousse South, Sousse Jawhara and Sousse Riadh with a total area of 78.9 km^2 (Fig. 1).

Algiers old Punic city where the Romans, Arabs, Ottomans and French succeeded one another, is included in a wilaya comprising 13 administrative districts (Dairas) and 57 communes and covering an area of 808 km^2 for a total estimated population of more than 3 154 792 inhabitants.

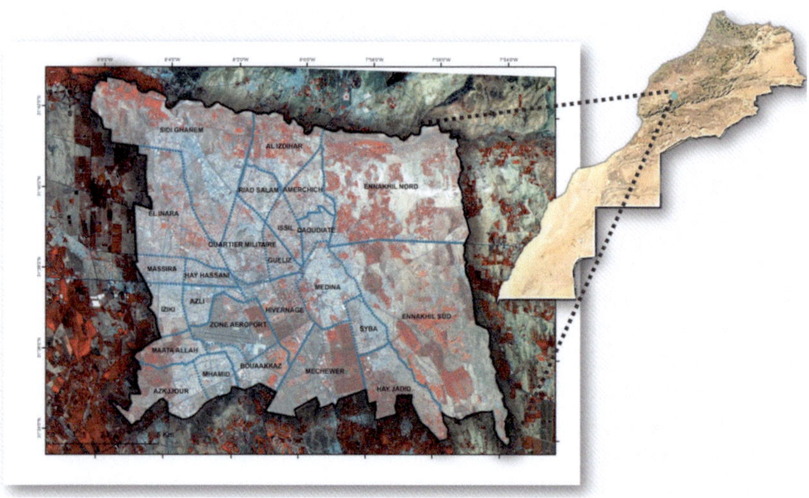

Fig. 1 Location of the Marrakesh Urban Community (CUM) and its different neighborhoods (the background image is a color composition of Sentinel-2A images) (Gourfi 2023)

In Marrakesh the Medina and Mechwer make up the city's old center, next to which are the neighborhoods of Hivernage and Gueliz as well as the military district during colonial times. The other neighborhoods were built following independence. SYBA, a 1960s working-class informal settlement, was built around ancient foundations and subsequently regularized. To the south, Hay Jadid went through a similar process after the 1990s. Daoudiate is a planned middle-class neighborhood from the 1970s, as are Issil and Amerchich from the 1980s. Mhamid, Massira, Iziki, Azli, Hay Hassani and Riad Salam are high-rise working-class estates planned in the 1990s and 2000s and which spill over into predominantly rural neighborhoods such as Sidi Ghanem, El Inara, Maata Allah, Azkjjour, Bouaakaz and Al Izdihar. The neighborhoods of Ennakhil Nord and Sud, which developed in the palm grove mainly after 2000, are home to villas and hotels and are relatively unurbanized.

We are therefore more interested in the vegetation in these cities and its variations from the 1970s.

Urban vegetation refers to all parks and gardens, whether private or public (in the meaning of open to the public or of ownership of the land), managed, abandoned or spontaneous, located within or near an "urban area". It also concerns isolated plants as well as large green formations (forested areas, hedges, fallow land and lawns for example) [37].

By taking a spatialized approach involving satellite remote sensing we are going to document the spatio-temporal variations in urban vegetation in the three cities and determine the factors behind these variations. The aim is particularly to understand the social, economic and political forces driving these trends and its effects on the urban space of the cities. We are also seeking to grasp the implications in terms of socio-environmental justice and socio-spatial segregation generated by this spatial organization of vegetation at the scale of the CUM.

2 Material and Methods

The method adopted in this research has three main steps. The first step is to determine the extent of the green spaces across the three cities and its variation over time from a set of Landsat satellite images dated 1980's and 2020's. During the second step, built-up areas (artificial surfaces) have been determined for a spatial comparison for the year 2020 using Landsat satellite images. For Marrakesh, as an example of maghrebian cities, the method have been developed in order to have details about the evolution of the city and its vegetation. The method has been applied to Landsat satellite images from 1973 to 2020 and a last step involves the use of NDVI data at 8-day intervals from 2000 to 2020, to confirm the results and to compare the variation in vegetation with rainfall rates for Marrakesh.

2.1 Data Source and Preprocessing

The Landsat satellite product (spatial resolution: 30–80 m) were used in this study, dated June or July 1984 or 1985 and 2020 or 2021 for the three cities (Table 1).

We added for Marrakech a collection of Landsat satellite images from 1973 to 2020 (1973, 1985, 1990, 1995, 2000, 2005, 2010 and 2015), mainly used to map vegetation areas changes, and Sentinel-2A satellite product (spatial resolution: 10–20 m) used mainly to map built-up areas in 2020 (Table 2). All the images were acquired from USGS Global Visualization Viewer (GloVis) [21].

All the images were taken during the months of June, July and August (dry period), in order to characterize only the permanent vegetation cover, which is irrigated during this arid spell. Only images with no or less cloud cover were selected to improve accuracy of vegetation (Landsat) and built-up (Sentinel-2A) classification detection.

An atmospheric correction was performed for each image using Line-of-sight Atmospheric Analysis of Spectral Hypercubes (FLAASH) in ENVI software. In addition, a radiometric calibration was used to calibrate image data to radiance, reflectance, and brightness temperatures using ENVI Calibrate Raster routine.

Table 1 Landsat satellite images used for the diachronic study

City	Present (Landsat 8)	Past (Landsat 1)
Algiers	07-29-2021	07-30-1984
Sousse	07-01-2021	06-09-1984
Marrakech	08-12-2020	07-27-1985

Table 2 Added satellite images used for Marrakech

Satellite	Sensor	No. of MS bands (nominal resolution)	Period
Landsat 1	MSS/RBV	4 (80 m)	27-03-1973
Landsat 5	MSS/TM	4 (30 m)	07-27-1985
			07-09-1990
			08-08-1995
			08-19-2005
			07-16-2010
Landsat 7	ETM +	6 (30 m)	08-13-2000
Landsat 8	OLI/TIRS	8 (30 m)	07-14-2015
			08-12-2020
Sentinel-2A		13 (10–20 m)	08-14-2020

2.2 Determination of Vegetation and Built-Up Areas from Satellite Images

We applied a Principal Component Analysis (PCA) on each Landsat corrected and preprocessed data set, followed by an unsupervised classification on all the principal components produced, and then supervised classifications with a maximum likelihood algorithm on the results, in order to spatialize the vegetation and built-up areas of the cities in the 1980's and 2020's.

For Marrakesh specifically, we also applied the supervised classification on all the principal components produced for the 10 dates from 1973 to 2020 to show the evolution of the vegetation areas of the CUM during that period.

A Principal component analysis (PCA) is a method based on the covariance and variance of the data set [33], transforming the data into a new coordinate system to reduce multidimensional data sets to lower dimensions [12]. The PCA method produces uncorrelated variables called principal components, and only the first principal components can represent significant variance compared with the original data [26].

The supervised classification method with the maximum likelihood algorithm (MLC) was applied in ArcGIS 10.7.1 to extract vegetation and built-up cover information. The MLC algorithm is one of the most commonly used supervised classification methods in remote sensing image processing. This technique is based on the possibility that a pixel belongs to a particular class. In this model, we assume that these possibilities are equal for all classes and that the input bands have normal distributions. The spectral distance method calculates the mean vector of each signature and the spectral distance between the measurement vector of the candidate pixel (samples). For vegetation detection, the samples were selected by viewing on screen [24], and also based on the multi-spectral false-color composite image consisting in making the vegetation appear in red (Fig. 2). For built-up detection, samples were selected through Google satellite imagery [39] (Fig. 3).

For Marrakesh, we used a hexagonal grid applied on the different classification results in order to extract spatial information on vegetation areas. In each polygon of 100 ha we quantify the extent of vegetation from 1973 to 2020, but also built-up and bare soil areas used occasionally in the study for 2020. Built-up areas are urbanized landscapes shaped by stone. Bare soils correspond to stony mineral areas that have not been built on.

In addition, we also used the annual time series of rainfall data from the Marrakesh weather station to analyze the variation in rainfall in the city in relation to the vegetation data results for the period between 1973 and 2020 (Fig. 4).

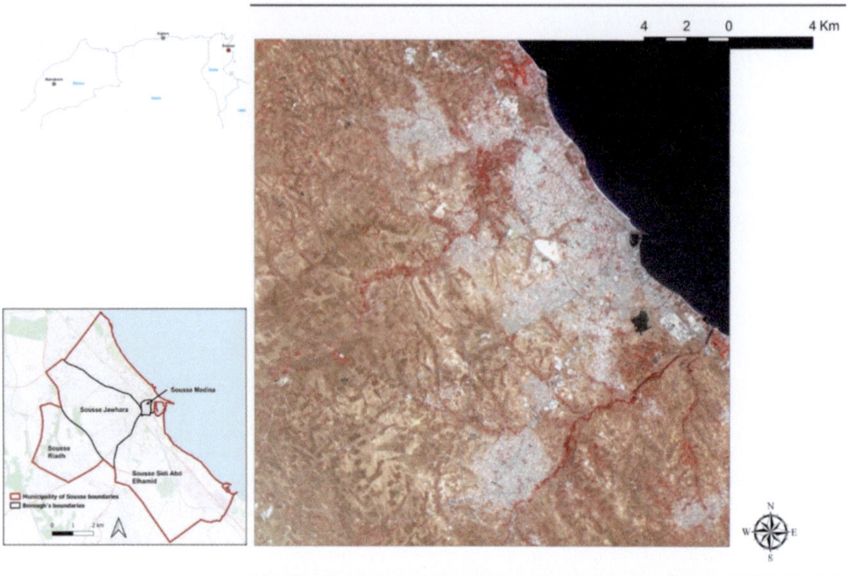

Fig. 2 Location of the Sousse city (from 2020 Landsat images) and the studied five of Sousse's 15 boroughs: Sousse-Khzema, Sousse Medina, Sousse Sidi Abd Elhamid, Sousse Jawhara, and Sousse Erriadh (Gourfi 2023)

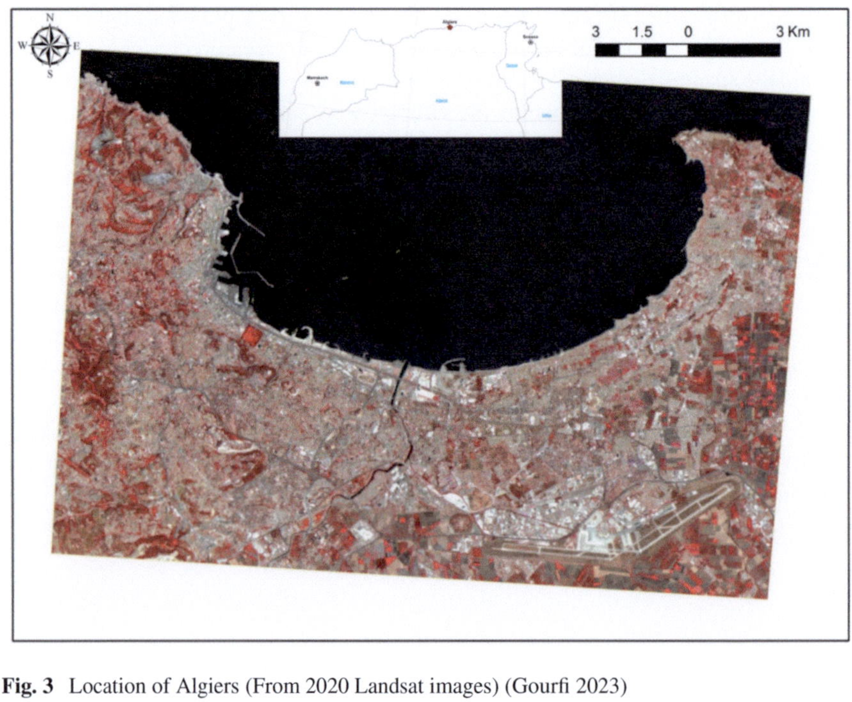

Fig. 3 Location of Algiers (From 2020 Landsat images) (Gourfi 2023)

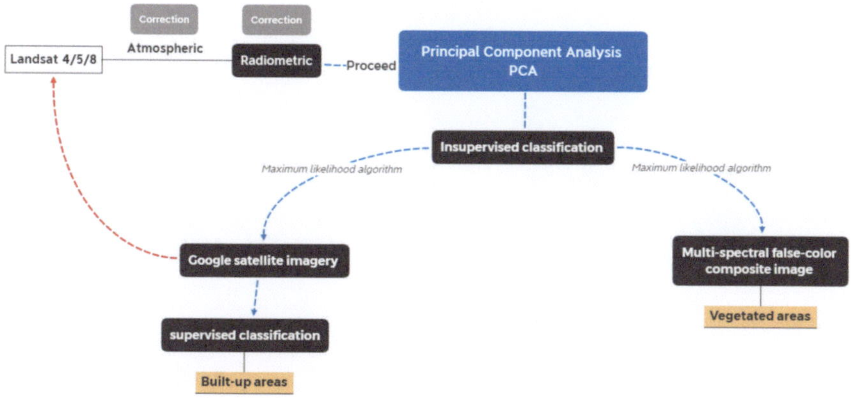

Fig. 4 Different steps of images processing (Gourfi 2023)

3 Results and Discussion

3.1 Evolution of Built-Up and Vegetation Areas: An Overall Urban Sprawl from 1980s to 2020s

In Marrakesh we have an increase of the built-up areas with 11.94 km^2 more (43% more) to the detriment of green spaces which have suffered a 45% decrease in surface area (35.69 km^2) from 1985 to 2020 (Figs. 5 and 6).

We can notice once again an important decrease in green spaces (−38%) in Algiers in conjunction with an increase in built up areas from 1984 to 2021. In 2021, green spaces covered 46.12 km^2 for a total surface area of 377.11 km^2 (including built-up areas and bare soils) (Fig. 7).

In Sousse, the scenario is quite different as we have an increase of both green spaces and built-up areas from 1984 to 2021.

The apparent increase in vegetated areas in Sousse between 1984 and 2020 is only relative, and probably linked to the development of new irrigated agricultural zones on the outskirts of the city, and because the many former olive orchards absorbed into the city, some of which have been preserved but transformed into private gardens, are now irrigated even in summer, whereas they were not in 1984. As long as the town does not become too densely populated, private individuals are able to preserve the olive trees that play such an important role in Soussian culture, especially when it comes to single-family houses. In addition, the town's expansion is taking place in a more or less anarchic fashion, and is very much a case of urban sprawl.

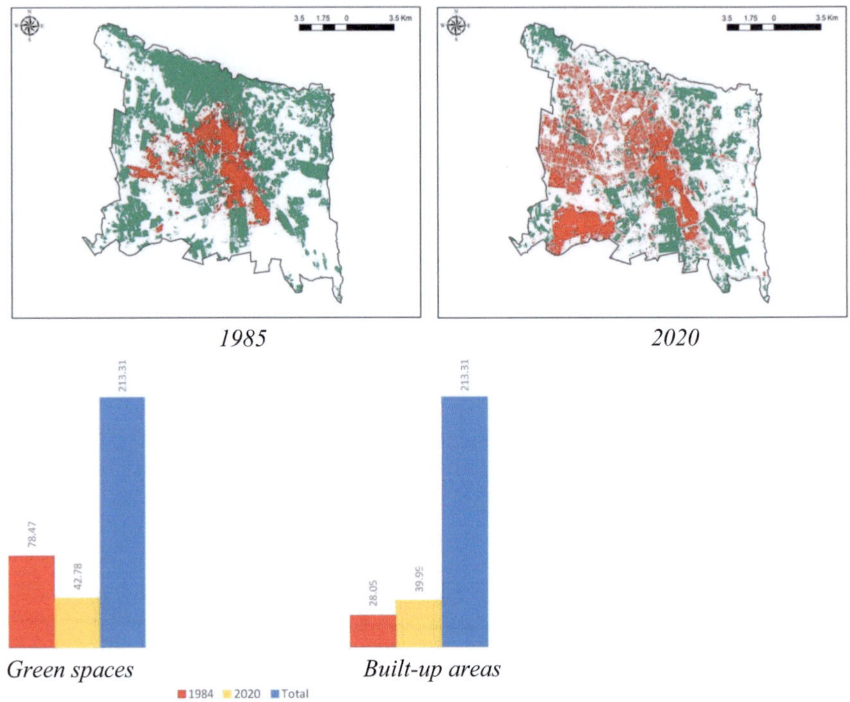

Fig. 5 Evolution of built-up and vegetation areas in Marrakesh from 1985 to 2020 (Gourfi 2023)

3.2 An Overall Downward Trend in Vegetation Cover from 1973 to 2020 the Example of Marrakesh

In Marrakesh, the diachronic study of vegetation cover from the classifications calculated from 1973 to 2020 shows a detailed significant, steady decline over this period (Fig. 8). In 1973, greenery covered 11,389.21 ha, a figure which had dropped to just 4528.90 ha in 2020. This corresponds to an average annual reduction between 1973 and 2020 of −145.96 ha (Fig. 9).

However, behind this overall decline in vegetation cover in Marrakesh between 1973 and 2020 are two distinct periods in reality (Fig. 9). Until 2020, the decline in greenery was swift and fairly steady, but after this date it slowed significantly and even reverses slightly.

Over the first period, between 1973 and 1995, the average annual fall in green spaces was −234.52 ha. During this time, the surface area covered by greenery diminished from 11,389.21 ha (53.39% of the CUM's total surface area) in 1973 to 6229.85 ha (29.21% of the CUM's total surface area) in 1995. Over the second period, with an overall increase of +6.28 ha per year, green spaces dipped from

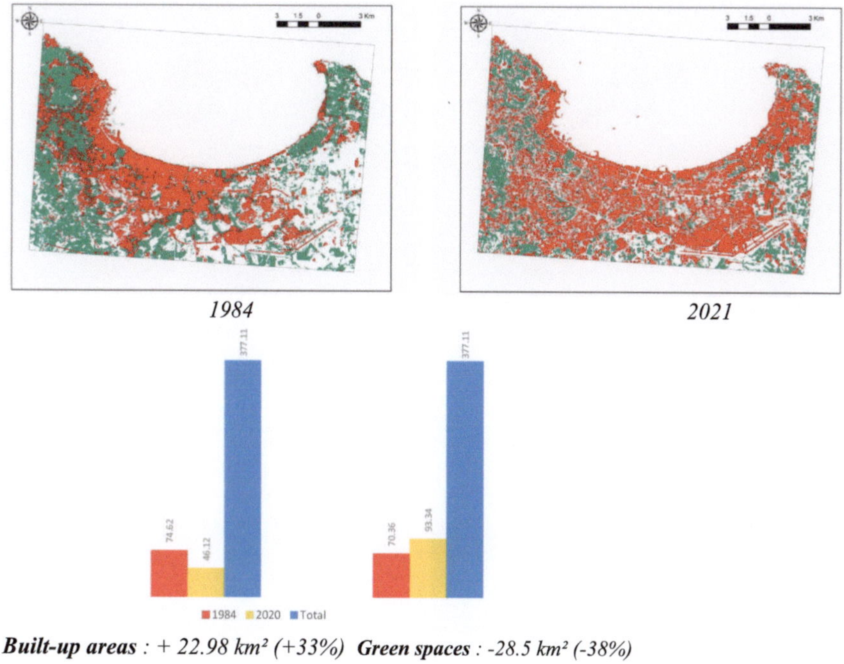

Fig. 6 Evolution of built-up and vegetation areas in Algiers from 1984 to 2021 (Gourfi, 2023)

4403.55 ha (20.64% of the CUM) in 2000 to 4278.00 ha (20.06% of the CUM) in 2020.

This overall change in vegetation cover is not uniform in spatial terms. In reality, during each period, it can be broken down into areas characterized by a decline in vegetation compared with the previous date, stable areas that do not change and areas with a change in land use and development of new plant cover. From 1973 to 2020, although the dominant areas are those where greenery disappeared (with 8674.59 ha between these two periods), some vegetation areas did not change (2720.79 ha), and some new green spaces even appeared (1157.48 ha) (Fig. 10).

The stable green spaces mainly correspond to the heritage gardens (Agdal and Menara) as well as farmland, which will still be there in 2020, forming a more or less protected, if discontinuous, green belt around the city. The dominant areas characterized by disappearing vegetation, meanwhile, correspond for the most part to areas that were urbanized during this period. The new vegetation areas are mostly golf courses or the expansive grounds of hotels, tourism complexes and/or luxury real estate, which have developed in leaps and bounds since 1973.

This overall trend can be analyzed in more detail by spatializing information using a grid of 100 ha hexagonal polygons, thus focusing the analysis on the scale of the CUM's neighborhoods.

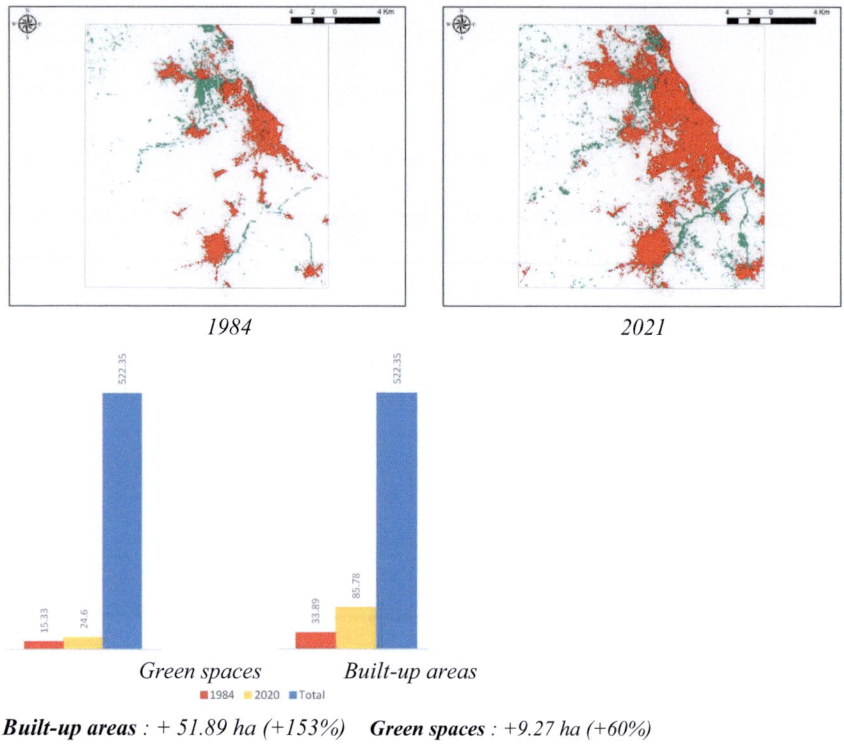

Fig. 7 Evolution of built-up and vegetation areas in Sousse from 1984 to 2021 (Gourfi, 2023)

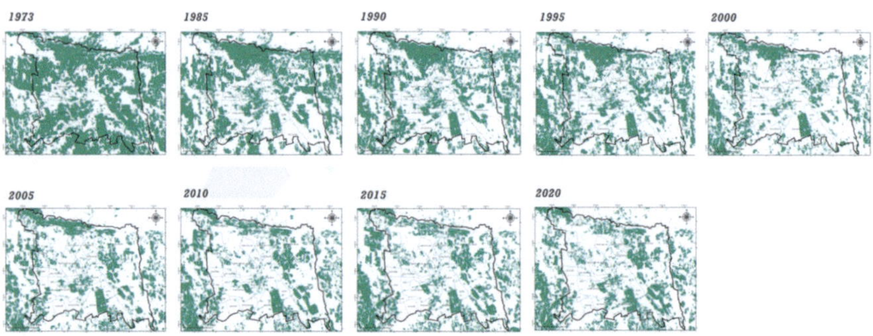

Fig. 8 Vegetation cover from 1973 to 2020 based on the supervised classification of the PCA components (Gourfi 2023)

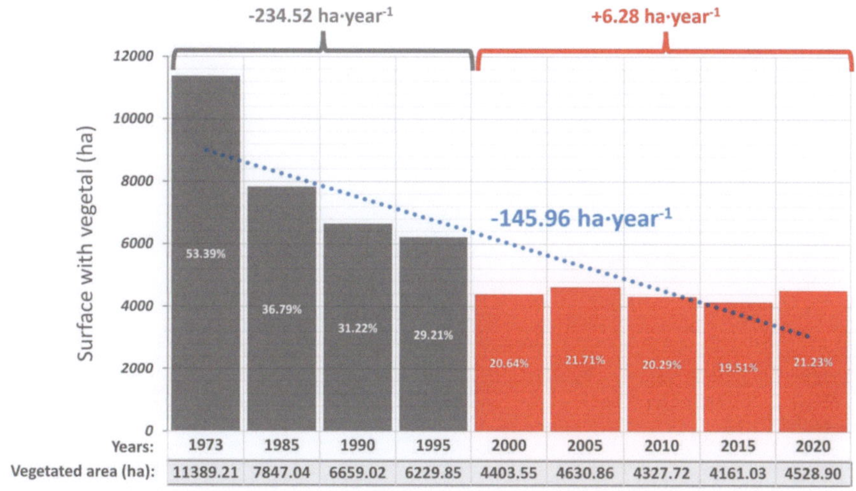

Fig. 9 Variation in vegetation cover in Marrakesh from 1973 to 2020 (Gourfi 2023)

The box plot, calculated at each date based on all of the grid's polygons (Fig. 11), shows that, beyond the overall steady decline in vegetation cover, its spatial distribution across the CUM has also changed.

In 1973, the spatial distribution of vegetation areas in the hexagons was much more uniform than during the later periods where a stronger spatial specialization can be observed. In 1973, the standard deviation is very high (Fig. 11), which tells us that, in each polygon, it is possible to have both built-up areas and bare soil areas, and that there was vegetation pretty much everywhere. After 1985, and after 2000 especially, the standard deviations became lower and land use more uniform. Greenery became increasingly concentrated within clearly defined areas.

3.3 *Variations Associated with Public Policy-Making*

The changing trends observed in terms of Marrakesh's vegetation and the differences observed between its neighborhoods can be attributed to a variety of factors.

Given the city's arid context, the overall decline in vegetation cover between 1973 and 2000 could be put down to diminishing rainfall—particularly given the droughts of the 1970s and 1980s—and the levelling off of this decline, and subsequent regreening in some cases, to an increase in rainfall. And yet the growth in green spaces, calculated on the basis of NDVI data, with average annual rainfall across the CUM over this period (Fig. 12), shows that this is simply not the case. For the most part, green spaces are watered at regular intervals and do not depend on available rainfall.

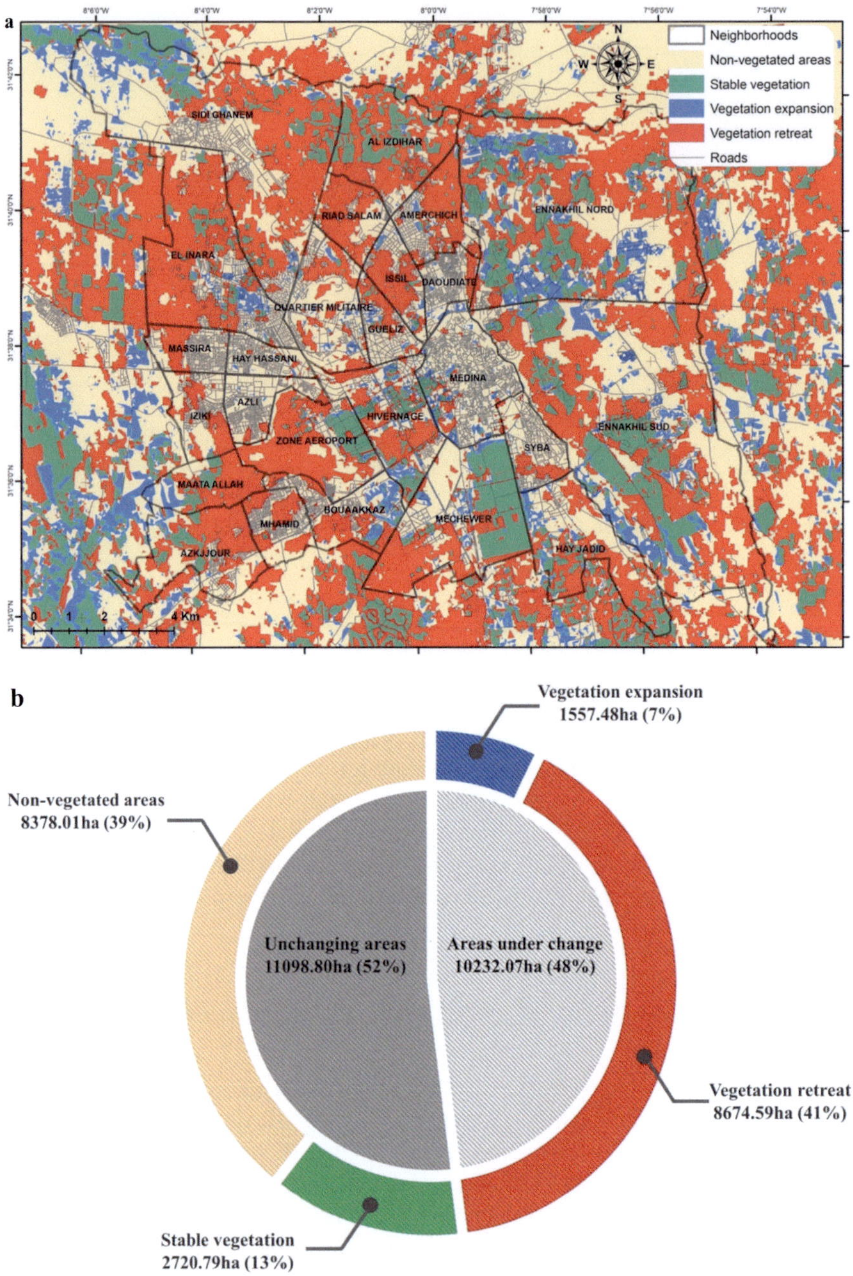

Fig. 10 **a** Variation in vegetation cover in Marrakesh from 1973 to 2020 (Gourfi 2023). **b** Changes in vegetation cover in Marrakesh from 1973 to 2020 (% and surface areas) (Gourfi 2023)

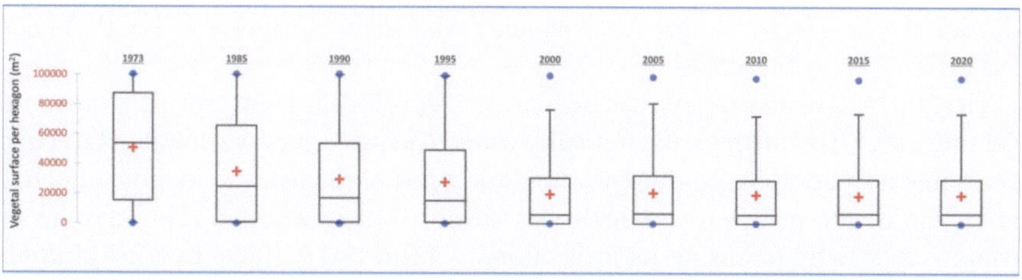

Fig. 11 Box plot showing the standard deviation and the average vegetation surface area per hexagon at each date for the whole of the CUM (Gourfi 2023)

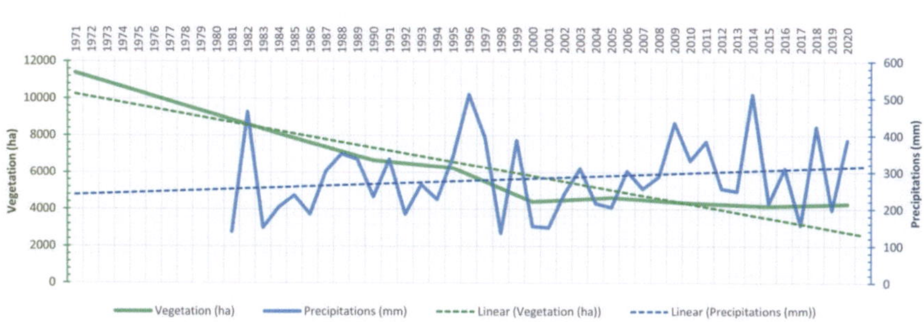

Fig. 12 Cross-linking of areas of vegetation (from NDVI) with average annual rainfall data from 1971 to 2020 for the region of Marrakesh (Gourfi 2023)

The reasons for such variations therefore lie in the development of the Marrakesh Urban Community (CUM) and the public policies setting this in motion.

The decline in green spaces from 1973 to 2000 stems, of course, from urbanization, in connection with demographic growth and positive natural balance and inward migration. The increase in built-up areas and decline in vegetation have kept pace with this swift growth.

The middle classes' loss of interest in the public green spaces laid out during the colonial period and their move away from them, leaving them for the working classes without investing in their upkeep, also explain why these spaces have deteriorated so sharply and why vegetation cover has fallen.

Various factors then explain why the decline in green spaces slowed before reversing slightly from the 2000s across the CUM. A range of projects of varying sizes and scales have promoted the protection of existing green spaces, reforestation (olive and palm groves especially) or the layout of new areas of vegetation.

Note, in particular, the creation in 2001 of the Mohammed VI Foundation for Environmental Protection, which is a nationally- and regionally- acclaimed leader in the protection and restoration of green spaces and sites of biological and ecological interest (SIBE) in Morocco. It can be considered the impetus behind the momentum to protect and restore green spaces in Marrakesh. One of the foundation's projects bears specifically on the safeguarding and development of the Marrakesh palm grove. The foundation also leads projects to restore historic parks and gardens across several

Moroccan cities, including the Arsat Moulay Abdeslam garden (or Cyber Park) and Ghabat Chabab in Marrakesh.

The 2007–2018 programme to safeguard and develop the palm grove thus aimed at planting 580,770 palm trees, theoretically watered using treated wastewater. Over and above this reforestation programme, the project has paved the way to more effective protection of the palm grove which was steadily being whittled away despite its listing, supposedly to ensure its protection, as historical heritage by a royal *dahir* (decree) dated 1929. Its decline is tied in with the pressure of urbanization from colonial times, since the Gueliz neighbourhood had been built to the detriment of the palm grove [17], right up until the present day with tourism infrastructure (golf courses, complexes), real estate complexes and luxury villas encroaching on it. This process has been happening amid competition for water where palm trees are no longer the priority. The centuries-old water supply from the *khettaras* (underground galleries that harvest groundwater and deliver it to the surface) has been cut off, and this has all been compounded by the climate crises of the latter half of the twentieth century.

More broadly, this royal foundation and the projects it has undertaken have been a driving force in shaping a global context in favour of the protection and development of greenery in the city.

At the local level, the unveiling of a Local Agenda 21 for Marrakesh in 2002, which advocates the "Protection of natural heritage from any urban planning", is a contributing factor to this momentum in favour of greenery in the CUM. Conducted in partnership with the Ministry of Regional Development, Water Resources and the Environment (MATEE), the UNDP and municipality of Marrakesh, this champions a project to safeguard and protect historic gardens and theoretically to restore and protect the last *khettaras* still in operation to prevent them from disappearing. From the city's environmental profile, produced back in 2003 for this Agenda 21, the participatory workshops' method in 2004, and the action plan and initial achievements from 2005, three priority areas stand out: sustainable management of water resources, heritage preservation and tourism promotion, and access to urban services and social integration. These three areas have given rise to a plan of proposed actions including, in particular, the listing of the Northwest palm grove, covering 250 ha, as a Site of Biological and Ecological Interest (SIBE), and other measures aimed at creating green spaces [4].

The "Vivre ma ville" initiative, set up by the Mayoress of the CUM in 2009, includes a section dedicated to green space and has also contributed to the regreening of the city [4]. The goal was to make Marrakesh a green, clean and attractive city by implementing a range of projects across various sites including Ghabat Chabab, Jnane Sidi Bel Abbas, the Northwest palm grove, Ryad Laârouss, Borj Zaitoune, two sites known as Unit 3 and Unit 5, the oasis Hassan 2, and the historic gardens of Menara and Agdal. However, despite the stated ambition, only the projects coinciding with those of the Mohammed VI Foundation could be conducted. The others were either not seen all the way through—or embarked on at allowing to land issues, no preliminary analysis and political infighting, the elected official's term in office having come to an end before the project's completion.

Beyond these examples of restoration and protection of iconic green spaces, it is also possible to note changes in real estate developers' practices, not least in the context of a "programme-contract" on social housing, with increased tax exemptions in their favour, in connection with a decision by King Mohammed VI to make so-called "social" housing a "national priority" (20 August 2002 speech) to address the housing crisis of the early 2000s. This programme-contract includes planning guidelines for green spaces which specify that the developer must plant a corresponding number of trees in light of the number of homes built, at least 3 m high, and undertake to look after them for one year or, if the project property base does not enable the corresponding number of trees to be planted, the remainder must be planted in an area designated by the municipality concerned by the project. However as real estate developers often don't respect these engagements, the green open spaces are frequently created and maintained by the inhabitants themselves.

This post-2000s period, together with the raising of the city's international profile and reputation since the 1990s [28], also coincides with a thriving tourism scene driven by the showcasing of the city [28] where greenery takes pride of place, in addition to the development of the events and recreation sectors [28]. Myriad luxury tourist facilities have been built over this period, setting greater store by vegetation, in gardens and golf courses, because such green space boosts appeal. Today, Marrakesh boasts twelve 18-hole golf courses, versus just two before 2000 [4].

Against this backdrop, luxury real estate properties also soared in the early 2000s, with developers giving precedence to "attractive natural" elements to enhance the sales value of their projects.

Marrakesh's residential and tourism real estate market grew exponentially across the board in the 2000s, with a glut of very-high-end properties on offer [4].

In the wake of the global economic crisis, the downturn in the real estate market from 2008 perhaps explains, on the other hand, the slowing decline in rural farmland, with mid-range real estate projects on the city's outskirts called off.

4 Conclusion

The example of Marrakesh analysed in detail here is representative of the trends occurring in other cities in north Africa, among which Algiers and Sousse.

The observed variation in green spaces since the 2000s could strike as reassuring given Marrakesh's frenzied pace of urbanization over recent decades.

However, on closer inspection of the slowing decline in vegetation cover in the city from 2000 (a sign of renewed interest in greenery in Marrakesh), the percentage of vegetation per capita continues its relative downward trend after this date, since the population has carried on growing (from 823 154 in 2004 to 928 850 in 2014 and will reach 1 175 770 inhabitants in 2030). The proportion of green space per capita shows that Marrakesh is a long way from being able to call itself a green city, since it falls short of the minimum 10 m^2. per capita recommended by UN-Habitat and by the green plan guide commissioned by Morocco's Ministry of National Planning,

Urban Planning, Housing and City Policy (a guide which has not been legislated and is not enforceable as a result).

The neighborhood-based analysis shows an imbalance at city level with stark inequalities. The city's overall level of greenery stems chiefly from the green spaces in some affluent neighborhoods (Hivernage, Ennakhil Nord and Sud) or in the royal grounds (Mechwer), which are, therefore, private green spaces. The level of greenery in working-class neighborhoods is very low that do not exceed 3 m^2 per capita if we exclude the remaining tracts of farmland on the outskirts of the most recently urbanized neighborhoods.

Accordingly, for the five *arrondissements*, the average area per capita is 10.41 m^2 for Gueliz; 2.64 m^2 for Medina; 1.17 m^2 for Menara; 0.98 m^2 for Sidi Youssef Ben Ali and 0.30 m^2 for Annakhil, resulting in an overall average of 3.1 m^2 per capita.

The image of Marrakesh as a green city—a marketing tool to promote the city—is based upon a few features that are by no means representative, including the handful of tree-lined avenues along the main thoroughfares taken by tourists between the airport, the city center and the touristy areas, as well as some landmark parks and gardens, both public (Arsat Moulay Abdeslam, Menara etc.) and private (Majorelle in particular).

The delisting over recent decades of a neighborhood like Gueliz comes hand-in-hand with a reduction in plant cover and their quality of life.

The socio-spatial segregation initiated during the colonial period by the unfettered urban planning of the time still endures to this day. The physical separation of the "indigenous" city (Medina and Mechwer) from the "new city" (Gueliz, Hivernage and military district) already showcased the former's heritage, with its historical and religious monuments and picturesque old walls, considered to form a considerable touristic capital [13] and representing the "backdrop" of the European city [34]. This urban design had another, health-giving goal (synonymous with modernity, civilization and social progress) by allowing the city to breathe through light, open green spaces with roadside trees (*Citrus aurantium, Ficus*, etc.) that fully chimed with the urban planning rationale of the landscape architect JCN. Forestier, and implemented by the urban architect Prost [37].

Gardens, along with other public spaces, used to symbolize the middle-class practices and designs of Western cities. But once they had been transposed to the Arab world in the new parts of cities, they were abandoned by the same middle classes following independence, shifted to private ownership [20] and forsaken by the local authorities. Left to the poorer working classes, these green spaces, public gardens and roadside trees deteriorated for want of proper upkeep and overuse as there were not enough of them. The slowing decline in vegetation observed post-2000 seems to signal a reversal in this situation and renewed interest in greenery.

However, any replanting observed remains primarily private, in the gardens of affluent villas or luxury developments of the palm grove or the scores of golf courses for national or foreign visitors. Greenery has thus once again become an elitist marker of space. Changes in the uses initially made of spaces (for agriculture, particularly in the palm grove), now repurposed into ornamental sites for enjoyment, first by well-heeled Europeans during colonisation and then by wealthy Moroccans today, have ended up excluding the working classes, for whom they no longer serve any justifiable purpose, especially in the neighborhoods of Ennakhil Nord and Sud.

The repurposing or design of new landscapes, whether urban or rural, forms part of the approaches to controlling territories and their occupants, and "often convey a social or even political message" [30]. "To build a garden is to project an image of oneself, of one's city and one's country; it is also, as during the second Empire, to proudly proclaim the victory of a middle-class social order one seeks to establish" [30].

Nowadays with the growing and hard mineralization of the city, once called a "city garden", the different stakeholders have to reconsider the urban development model of Marrakesh, necessarily involving more of its inhabitants.

Algiers and Sousse sharing some major elements of their history with Marrakech, such as French colonisation and independence, show the same trends in vegetation evolution.

In the three cities, the parks and gardens inherited from the French colonial period were largely neglected after independence, and almost completely ignored by the urban planning and cities management departments, resulting in a lack of care for intra-urban green spaces. For a long time, urban planning departments considered the development of parks and gardens in urban areas as an element in the well-being and quality of the urban environment, to be secondary to the primary functions of traffic, housing and commerce.

However, we have noticed that the same renewed interest in urban greenery in these cities since the 2000s, in the form of new laws and regulations and rehabilitation projects of French colonial gardens and parks.

5 Recommendations

Remote sensing is a tool that can be used to monitor urban vegetation in great detail and understand its spatial distribution and evolutionary dynamics.

Acknowledgements This work was supported by a Partenariat Hubert Curien France-Maghreb program (38MAG20 projet 43632PH), the FMSH-CASS (programme Atlas 2021) and FMSH (programme "Réseaux internationaux en SHS" 2024).

References

1. Ajmi R, Belfekih Boussema S, Essasi S, Taïbi AN, Khebour Allouche F (2024) A qualitative urban green spaces assessment for a sustainable management, case study of Sousse City (Tunisia). In: Ksibi et al. (eds) Recent advances in environmental science from the euro-mediterranean and surrounding regions (3rd edn). EMCEI 2021. Advances in Science, Technology & Innovation. Springer, Cham. https://doi.org/10.1007/978-3-031-43922-3_138

2. Ajmi R, Khebour Allouche F, Taïbi AN, Bel Fekih Boussema S (2023) Developing a qualitative urban green spaces index applied to a Mediterranean City. Urban Sci 7(4):115. https://doi.org/10.3390/urbansci7040115
3. Barety L (1932) La France au Maroc. Félix Carbonnel, Paris
4. Barthel PA (2013) Villes pour une Nouvelle Génération, Rapport sur Marrakech. Le PCD : impact du cadre national et régional d'urbanisation, évaluation et recommandations, Rapport d'étude Version 22 octobre 2013, ORLD BANK_CMI
5. Benabdeljalil M, Boujmal Z (2008) Guide de l'élaboration des plans verts urbains au Maroc. Ministère de l'habitat et de l'urbanisme au Maroc
6. Benchaâbane A (2014) Marrakech Cité-Jardin : grandeur, décadence et renaissance. Borkane, France
7. Bennani M (2017) Villes-paysages du Maroc, Ed. La Decouverte, Paris
8. Clergeau P (2020) Urbanisme et biodiversité. Vers un paysage vivant structurant le projet urbain, Ed. Apogée, Rennes
9. Dana RH (2002) Mental health services for African Americans: a cultural/racial perspective. Cult Divers Ethn Minor Psychol 8(1):3–18. https://doi.org/10.1037/1099-9809.8.1.3
10. Darrel Jenerette G, Harlan SL, Stefanov WL, Martin CA (2011) Ecosystem services and urban heat riskscape moderation: water, green spaces, and social inequality in Phoenix, USA. Ecol Appl 21(7):2637–2651. https://doi.org/10.1890/10-1493.1
11. Declet-Barreto J, Brazel AJ, Martin CA, Chow WTL, Harlan SL (2013) Creating the park cool island in an inner-city neighborhood: heat mitigation strategy for Phoenix, AZ. Urban Ecosyst 16:617–635. https://doi.org/10.1007/s11252-012-0278-8
12. Dópido I, Villa A, Plaza A, Gamba P (2012) A quantitative and comparative assessment of unmixing-based feature extraction techniques for hyperspectral image classification. IEEE J Sel Top Appl Earth Obs Remote Sens 5:421–435. https://doi.org/10.1109/JSTARS.2011.2176721
13. Du Vivier De Streel ME (1932) Introduction. In: Royer J (ed) L'urbanisme aux colonies et dans les pays tropicaux. Communications et rapports du congrès international de l'urbanisme aux colonies et rapports du congrès international de l'urbanisme aux colonies et dans les pays de latitude intertropicale (Paris – Vincennes 10–15 oct. 1931), Tome 1, Delayance, La Charité sur Loire, pp 1–4
14. El Faiz M, Bendaoud R (2000) Jardins de Marrakech. Actes Sud, Arles
15. El Faiz A, Dounas H, Meddich A, Hafidi M, Ouhammou A (2016) Biodiversité des espaces verts publics de la commune urbaine de Marrakech (Maroc). Acta Bot Malacitana 41:83–100
16. El Faïz M (2002) Marrakech, patrimoine en péril. Actes Sud/Eddif, Arles
17. El Hannani M, Taïbi AN, Brabra N, Giffon S (2017) Les enjeux du végétal dans une ville du « Sud ». Le cas de la ville de Marrakech et la fin d'un modèle de « cité-jardin ». Proj. paysage 16. https://doi.org/10.4000/paysage.5682
18. Forestier J-CN (1906) Grandes Villes et Systèmes de parcs. Hachette et Cie, Norma/IFA, Paris
19. Gillot G (2014) La ville nouvelle coloniale au Maroc: moderne, salubre, verte, vaste. In: Leimdorfer F (ed) Dire les villes nouvelles, 5, Editions de la Maison des sciences de l'homme, Paris, pp 71–96
20. Gillot G (2002) Espaces populaires, pratiques intimes : les jardins publics au Caire, à Rabat et à Damas. Geocarrefour 77:267–274
21. Global Visualization Viewer (2021) Earth resources observation and science center [WWW Document] http://glovis.usgs.gov
22. Gourfi A, Taïbi AN, Salhi S, El Hannani M, Boujrouf S (2022) The surface Urban Heat Island and key mitigation factors in arid climate cities, case of Marrakesh, Morocco. Remote Sens 14(16):3935. https://doi.org/10.3390/rs14163935
23. Grimm NB, Foster D, Groffman P, Grove JM, Hopkinson CS, Nadelhoffer KJ, Pataki DE, Peters DPC (2008) The changing landscape: ecosystem responses to urbanization and pollution across climatic and societal gradients. Front Ecol Environ 6:264–272. https://doi.org/10.1890/070147
24. Guo L, Xi X, Yang W, Liang L (2021) Monitoring land use/cover change using remotely sensed data in guangzhou of China. Sustainability 13(5):2944. https://doi.org/10.3390/su13052944

25. HCP (2014) Recensement Général de la Population et de l'Habitat de 2014, Haut Commissariat au Plan du Maroc
26. Huang B, Zhang H, Yu L (2012) Improving landsat ETM+ urban area mapping via spatial and angular fusion with MISR multi-angle observations. IEEE J Sel Top Appl Earth Obs Remote Sens 5(1):101–109. https://doi.org/10.1109/JSTARS.2011.2177247
27. Kim G (2016) Assessing urban forest structure, ecosystem services, and economic benefits on vacant land. Sustainability 8(7):679. https://doi.org/10.3390/su8070679
28. Kurzac-Souali AC (2011) Marrakech, insertion mondiale et dynamiques socio-spatiales locales. Méditerranée 116:123–132. https://doi.org/10.4000/mediterranee.5441
29. Lee KE, Williams KJH, Sargent LD, Williams NSG, Johnson KA (2015) 40-second green roof views sustain attention: the role of micro-breaks in attention restoration. J Environ Psychol 42:182–189. https://doi.org/10.1016/j.jenvp.2015.04.003
30. Mathis CF, Pépy EA (2017) La ville végétale : Une histoire de la nature en milieu urbain (France, XVIIe-XXIe siècle). Champ Vallon, Clamecy
31. McGrane SJ (2016) Impacts of urbanisation on hydrological and water quality dynamics, and urban water management: a review. Hydrol Sci J 61:2295–2311. https://doi.org/10.1080/02626667.2015.1128084
32. Menjily-de Corny I (1991) Jardins du Maroc. Pierre Zech, Paris
33. Mutlu M, Popescu SC, Stripling C, Spencer T (2008) Mapping surface fuel models using lidar and multispectral data fusion for fire behavior. Remote Sens Environ 112:274–285. https://doi.org/10.1016/j.rse.2007.05.005
34. Prost H (1932) Le développement de l'urbanisme dans le protectorat du Maroc de 1914 à 1923. In : Royer J (ed) L'urbanisme aux colonies et dans les pays tropicaux. Communications et rapports du congrès international de l'urbanisme aux colonies et rapports du congrès international de l'urbanisme aux colonies et dans les pays de latitude intertropicale (Paris – Vincennes 10–15 oct. 1931), Tome 1, Delayance, La Charité sur Loire, pp 59–80
35. Robinson SL, Lundholm JT (2012) Ecosystem services provided by urban spontaneous vegetation. Urban Ecosyst. https://doi.org/10.1007/s11252-012-0225-8
36. Seitz J, Escobedo F (2011) Urban forests in Florida: trees control stormwater runoff and improve water quality 1 how do trees affect the water. J Arboric 1–4
37. Taïbi AN, El Hannani M (2019) Le paysage colonial en Afrique. Le végétal dans l'espace public des villes coloniales de Marrakech, Alger, Antananarivo et Toliara. Enjeux socio-environnementaux d'un « patrimoine » vert dans un contexte urbain en mutation. Africana Stud 32:27–38
38. United Nations (2018) Sustainable development goal 6 synthesis report 2018 on water and sanitation, United Nations
39. Vigneshwaran S, Vasantha Kumar S (2018) Extraction of built-up area using high resolution Sentinel-2A and Google satellite imagery. Int Arch Photogramm Remote Sens Spat Inf Sci - ISPRS Arch 42:165–169. https://doi.org/10.5194/isprs-archives-XLII-4-W9-165-2018
40. Zitkovic M (2008) Managing green spaces for urban biodiversity. Countdown 2010 Fact Sheets

Design Urban Park. Role of Geoinformatics in Urban Planning

Safa Bel Fekih Boussema, Faiza Khebour Allouche, and Khaled Kharbout

Abstract Nowadays, fast and unplanned urbanization growth has an impact on territorial development. Which have led to significant issues with societal disputes, human growth, economic activities that harm natural areas, biodiversity loss, and environmental degradation. Urban green spaces, especially urban parks, have a favorable impact on the physical environment of urban agglomerations as well as the psychology of their residents, and they have been shown to enhance quality of life through their diverse recreational amenities. This is the case of the Bouficha's urban park design in Sousse, Tunisia. The methodology developed in this research comprises three steps. First, the evolution of landscapes and environments in the study area was examined using Google Earth images. Then, a survey was performed to find the concept that would drive the development project. Third, the urban park's design reflected the unique qualities of the region. Solutions that use geoinformatics in urban planning can have an impact on the development of these areas' designs and planning.

Keywords Google Earth · Urban Palnning · Park · Design · Bouficha

1 Introduction

The urban park is a complex entity, and its meaning varies depending on the actors involved (users, planners, gardeners, geographers, politicians, technical services personnel, ecologists, etc.). They are vital ecological recreational places. Many countries such as, Italy, France, United States, China, Japan, etc. have conducted large-scale urban park construction activities. Due to its beneficial effects on urban

S. B. F. Boussema (✉) · F. K. Allouche · K. Kharbout
Department of Engineering and Landscape Planning, High Institute of Agronomic Science of Chott Mariem, Sousse University, Chott Meriem Sousse, Tunisia
e-mail: belfekihsafa@gmail.com

S. B. F. Boussema · F. K. Allouche
Lr GREEN TEAM (LR17AGR01), National Agronomic Institute of Tunis, Carthage University, Tunis Mahrajène, Tunisia

F. Khebour Allouche (ed.), *Environmental Remote Sensing and GIS in MENA Region*, Earth and Environmental Sciences Library,
https://doi.org/10.1007/978-3-031-86574-9_11

greening, public outdoor enjoyment and education, city image restoration, and sustainable urban development, urban parks have become an essential component of city life [34]. From [1], urban green areas have many positive effects on the environment and human health in addition to their aesthetic value. They have also been shown to aid in stress management by fostering social support. Therefore, it is interesting to consider some definitions related to this notion in chronological order. In 1970, Courajoud described that "the image of the park is that of geometry of the mountain, passed through the filter of the urban". So, the urban park is a green lung located in the center of the city; its basic design is the same as the space connected to the collective habitat; and it welcomes a relatively large audience. Nevertheless, this park is enclosed by portals with limited opening hours, providing a range of activities to visitors. According to [5], the use of urban materials can recreate from scratch a framework that, by reference, gives the city the same emotional abilities as those encountered in nature. So, the park is designed as a rationalized geometry while evoking the shapes of virgin nature on the horizon. Boutefeu [10] affirmed that the park is a green space in the city that is open to the public, shared by many citizens, whether it is enclosed or not, with a greater area than a square, measuring between 2 and 3 ha. The park is designed for recreational purposes, consisting of recreational lawns, ornamental trees, and wooded areas. It provides green spaces for residents and visitors to the municipality. Then, in 2007, Clergeau added that urban environments are impervious to species' mobility, and public parks are islands of habitat, more or less natural and isolated. To eliminate this isolation, corridors must be established within the urban matrix.

Urban parks are the image and identity of the city; in addition to urban planning functions, they positively affect the physical environment of urban agglomerations and the psychology of their inhabitants. According to [1], they serve as a natural gathering place for inhabitants, facilitating both social contact and community integration. Furthermore, they stimulate physical activity and mental and psychological relaxation, providing oxygen for breathing and purifying air contaminants. They play several roles, such as:

(i) *Scientific and educational role*: According to [3], the garden is ideal for children to explore nature and try gardening,it helps develop many educational potentials. Thereafter, it is a good place to discover nature and carry out various environmental awareness and education activities with schoolchildren As a result, children will be more attentive to environmental issues and better understand the importance of green spaces in human life.
(ii) *Antierosive role*: Green screens mainly play a role in preventing soil erosion. Bergaoui et al. [4] studied the erosion process of an experimental watershed in Tunisia during 1987–1993 and proved that environmental components play an important role in the process of soil erosion and loss, mainly vegetation cover, which acts as a hydraulic brake, increasing soil roughness. Trees can also weaken the force of wind and weather, protecting the soil. Plants store circulating water, and it is then returned to the atmosphere through a continuous cycle of absorption and evapotranspiration.

(iii) *Hygienic role*: Green spaces play an important role in cleaning the atmosphere; they help filter foreign bodies and dust from various impurities present in the water. Green spaces, especially forest areas, can act as a barrier and filter dust into the air. Gathy [19] highlighted this role, and he give these figures per hectare of forest, where Spruce filters 32 tons of dust and Pine filters 35 tons of dust.

(iv) *Human well-being*: Studies have shown that vegetation and nature improve our spontaneous attention, relax our sensory system, and infuse new energy into us. Thus, visiting the green space is synonymous with relaxation; it makes us more focused because we just have to pay attention spontaneously. In Sweden, Grahn [23] has conducted extensive research on the importance of parks for different groups of people. Studies have shown that outdoor activities have a real curative effect on patients, hospitals, nursing homes, and residents of retirement homes.

In Tunisia, the Ministry of the Environment launched at the beginning of 1990s a National Program of Urban Parks to shape a "green policy" [8]. It was formulated to create hundred urban parks in order to: (i) protect the natural space, (ii) increase the proportion of green spaces; (iii) offer citizens opportunities for rest, relaxation, education, and culture, and (iv) make them aware of the virtues of nature. In 2009, 36 municipal parks were established in 24 governorates [26]. In addition, according to [8], the Sousse Governorate was benefited by the construction of a single urban park in 2009.

The urban park of Hmedet Douik is located in the Sousse city and covers 10 hectares. It includes reception, a playground for children, sanitary blocks, a refreshment bar, a restaurant, kiosks, a fitness trail, and two tennis courts. Thus, municipalities in Tunisia are becoming increasingly interested in creating and maintaining urban parks to provide locations for people to gather, support their social, intellectual, and cultural development, and be an identity for the city. In the instance of the commune of Bouficha, which is situated in the governorate of Sousse, the local decision-makers have shown an interest in the creation of an urban park.

Currently, Geographical Information System (SIG) is crucial for managing natural resources (vegetation, water, and soil) and guiding decision-making processes in their development plans [6]. Maps, satellite images, and aerial photographs are essential goods and analytical tools for a variety of scholarly and professional endeavors. Thus, the application of GIS in urban planning, particularly the establishment of urban green spaces, aids and directs planners in the orderly development of settlements and infrastructural facilities both within and outside of urban areas [33]. Remotely sensed data was utilized in most cases when examining changes in Land Use and Land Cover (LULC) patterns, from [18]. Previous studies has shown that the availability of satellite imagery, particularly remotely sensed data and Google Earth (GE) images, helps to transform these data into meaningful information about the type and rate of changes in urban landscapes and human settlements [13, 20, 21, 30, 31].

This research aims: (i) to study the evolution of landscapes and environments in the study area through a multi-annual analysis using GE time series images, (ii) to integrate citizens in proposing ideas and analyzing their relationship with nature through a survey to find the concept that will drive development project, and (iii) to design the urban park that meets the characteristics of the region and its identity.

2 Materials and Methods

2.1 Study Area

Bouficha is a city located in the eastern center of Tunisia and north of the governorate of Sousse. Nabeul Governorate borders it to the north, Zaghouan Governorate to the west, and Hammamet Golf Course to the east. The study area is 72 km from the city of Sousse and 28.5 km from the city of Hammamet (Fig. 1).

Bouficha occupies 1000 ha of land, and according to data from the National Institute of Statistics in 2020, the delegation has a population of 27.953 inhabitants [28]. It is characterized by a Mediterranean climate. The highest monthly average temperature recorded in July was about 34.3 °C, while the lowest temperature was recorded in December at −7.7 °C. The summer season is known for being dry due to an annual precipitation rate of 240 mm [27].

The map (Fig. 2) shows that the maximum altitude of the Bouficha region reaches 272 m and is located mainly in the peaks of reliefs. The average altitude is 60–120

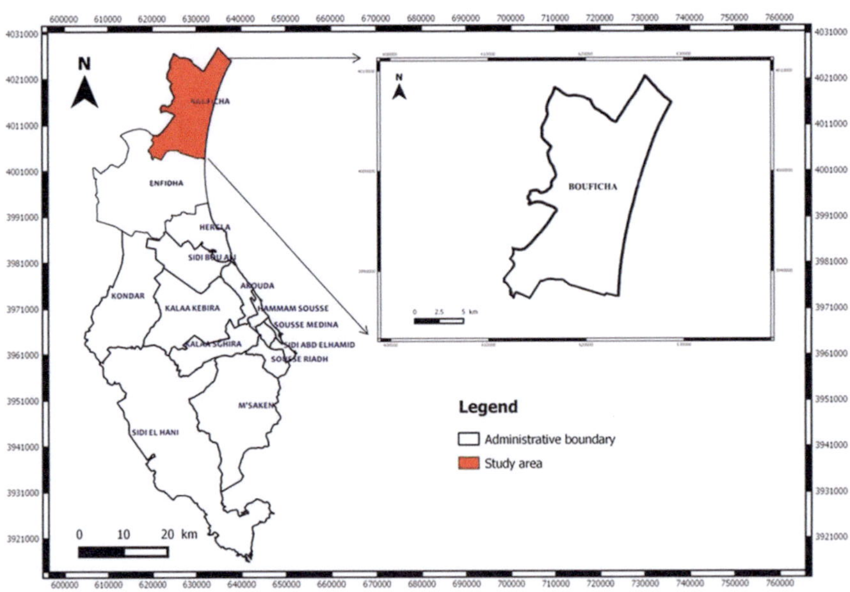

Fig. 1 Localisation map of Bouficha [6]

Fig. 2 Digital elevation model map of Bouficha [6]

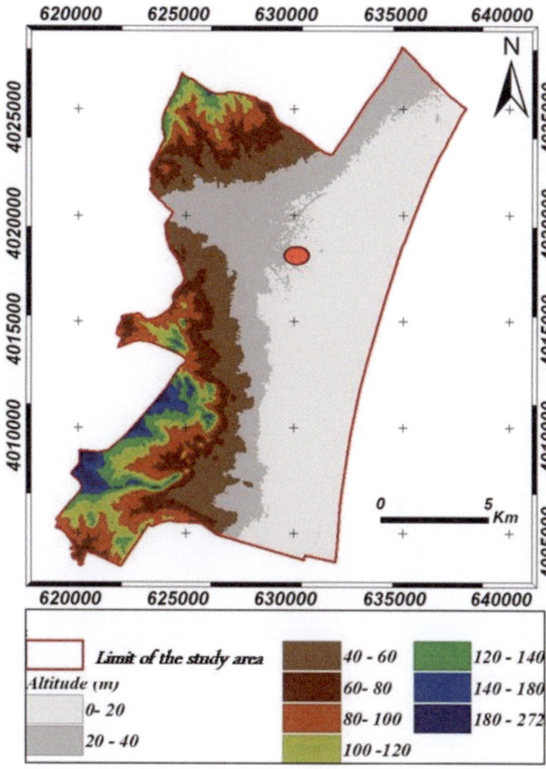

m, with average slopes of 5–10%, while the low altitudes are 20–40 m, located in the center and in the coastal part of Bouficha. This indicates a small slope of 5%.

Agricultural land dominates most of the area. In particular, Fig. 3 illustrates that cereal farming is the dominant crop, which explains the significant production of 8925 tones of cereals with the regions of Enfidha and Kalaa Kobra in 2018 [16]. In addition, olive fields occupy large areas in the region, especially in the middle of the delegation on the east–west axis. On the north side, vegetable crops and vine farming stand out.

2.2 Methodology

Land Use Land Cover Study Between 2004 and 2019

Six GE time series images dating from 2004, 2012, 2014, 2016, 2017, and 2019 were processed with Google Earth Pro © software to extract the evolution and assess LULC changes of the study area.

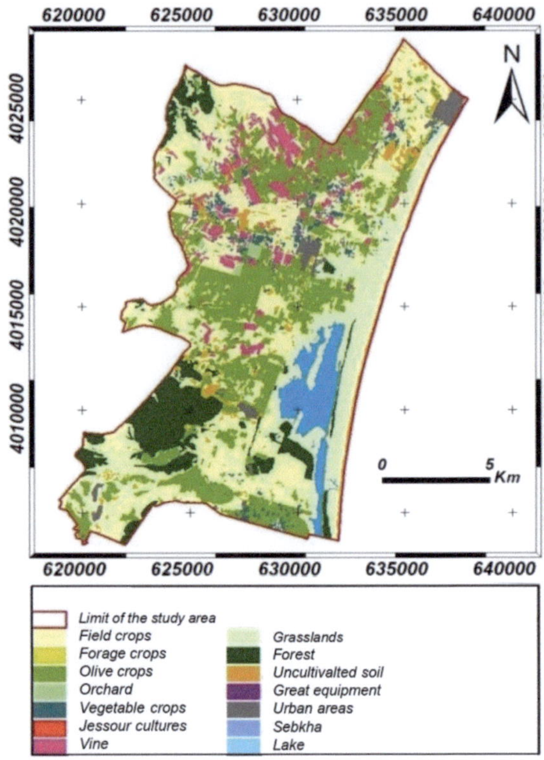

Fig. 3 Land use map of Bouficha [6]

Survey among Bouficha's Residents

A semi-directive survey was completed by 72 residents of Bouficha. The survey is broken into three parts: (i) gathering basic information from respondents; (ii) analyzing the link between residents and green space; and (iii) outlining park development suggestions.

Landscaping Design

The urban park in Bouficha's landscaping design involved creating 2D plans and 3D sketches. Thus, the design tool used is AutoCAD© for the creation of 2D plans and Lumion© for 3D modeling.

3 Results

According to Fig. 4, Bouficha's green areas are mostly represented by the football field in the city center, the garden "Souk" in the northwest near the station and the newly created weekly market, and the gardens "Hay Riyad" in the city's south

Fig. 4 Spatial distribution of green spaces in Bouficha's city in 2019 [6]

end. The volume and distribution of green space in the city are extremely restricted, prompting municipal authorities to declare their desire to establish an urban park in the north. The park's boundaries are as follows: the Oued Rmal in the north, followed by olive lands; a railway in the west, followed by agricultural land for cereal cultivation; a cemetery in the east, followed by the national road RN1 and a mosaic of cultures; and the districts of Bouficha in the south (Fig. 5).

3.1 LULC Analysis Between 2004 and 2019

The study area for the creation of the urban park has undergone many transformations. In 2004, all the ploughed land was used for cereal cultivation. In the Southwest, there is a waste area (public landfill). Then, in 2012, the waste area was removed, and a weekly market was established. In 2014, *Crataegus monogyna* appeared along the RN1 road and near the Oued Rmal. In addition, the presence of anarchic habitat zone in the bare soil is observed on the northwest side. In 2016, the municipal authorities removed this area in order to create plots for olive cultivation. There is also the presence of waste behind the weekly market. In 2017, there was an increase in the extent of waste and the emergence of *Crataegus monogyna* on the southwest side. In 2019, the study area covers an area of 23.8 ha. The majority of the area is devoted to cereals, occupying 12.93 ha. The bare soil has 6.95 ha with an area of olive cultivation of 2.44 ha, a waste area of 1.1 ha, and a habitat area of 0.38 ha (Fig. 6).

Fig. 5 Urban park boundaries [6]

3.2 The Survey's Findings Among Bouficha's Residents

3.2.1 Green Spaces as an Essential and Sufficient Component in the Modern City of Bouficha

Green spaces have a significant presence in urban areas to enhance our connection with nature, which can improve human well-being. For this reason, all people surveyed believe that green spaces are essential to the modern city (Fig. 7a). As for the number of green spaces in the city of Bouficha, 94% of respondents say it is insufficient, while 2% consider it sufficient (Fig. 7b).

3.2.2 Roles of Urban Green Spaces

As urban green spaces are a major issue for the living conditions of city dwellers, they play important roles and serve multiple functions. Most respondents define the role of a green space as a leisure role (34%), 22% as an aesthetic role, 18% as a place of relaxation, 15% as an image of nature, 10% as a sports space (to play football), 7% as a shading tool, and 4% as a pollution-fighting element (Fig. 8).

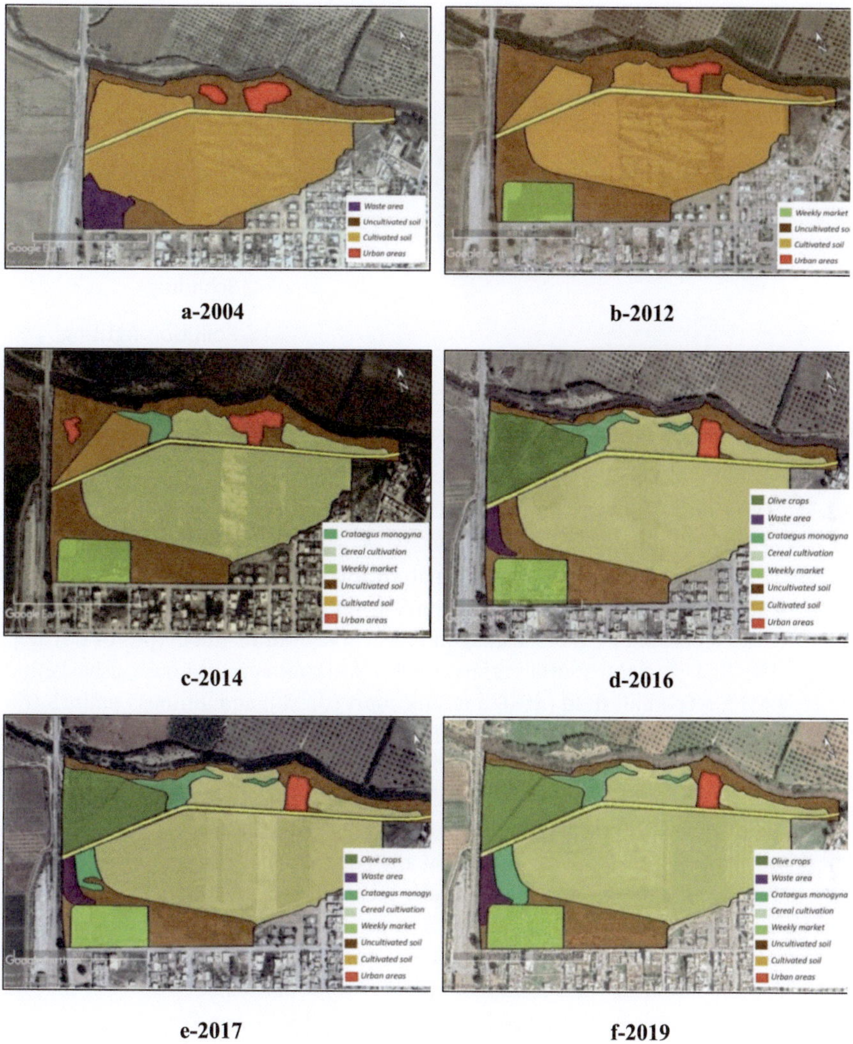

Fig. 6 LULC evolution between 2004 and 2019 [6]

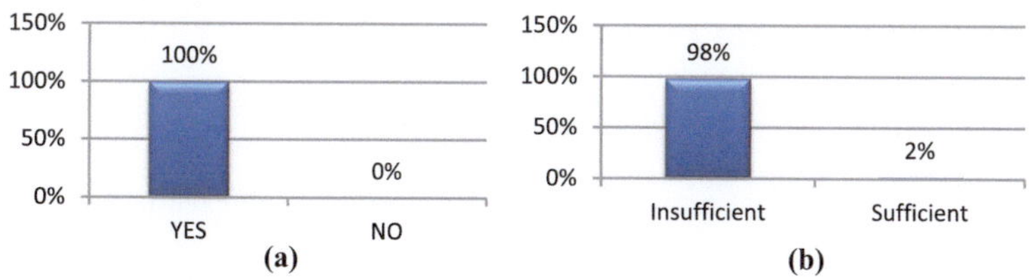

Fig. 7 Green spaces as an essential (**a**) and sufficient component (**b**) [6]

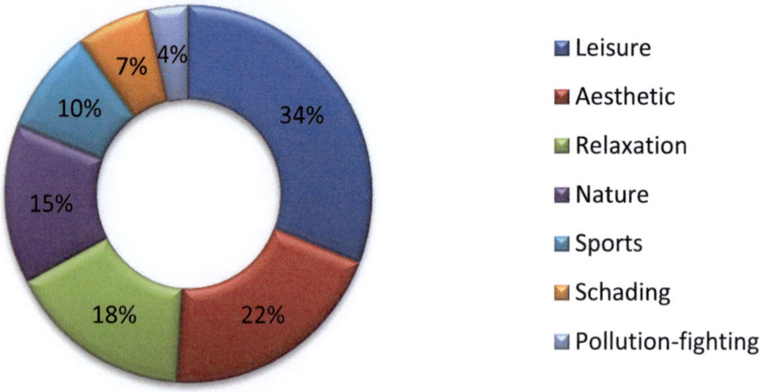

Fig. 8 Roles of urban green spaces [6]

3.2.3 Types and Reasons for Desired Developments

Access to green spaces in the city and to nature allows residents to engage in various outdoor activities; this is primarily dependent on the many facilities available in these areas. According to Table 1, respondents chose four distinct types of developments. The first is of the sports variety, with a workout space (4.6%), a basketball court (4.6%), a football field (5%), a tennis court (0.7%), and a fitness path (8.6%) represented. The second proposed development is of an entertainment and recreational nature, including a snack (10.2%), relaxation space (13.5%), playgrounds (25.5%), rest area (11.7%), and camping area (2%). A lake view (6.3%), sculpture (0.9%), and a lawn (2.9%) are examples of aesthetic and ornamental arrangements. The fourth type of development is cultural and educational, which is represented by a shared or community garden (1.9%) and a theater (0.9%).

The respondents want to see these various sorts of developments planned in Bouficha's future green areas. This is due to well-defined factors, such as family gatherings on weekends and holidays (48%), sports participation (36%), and the desire for Bouficha to be designated as a green city (16%), Fig. 9.

3.3 Bouficha's Urban Park Design

Project Concept

The development concept stems from the combination of the recreational system proposed by the city's residents thanks to the semi-directive surveys carried out and the ecosystem of the region. Moreover, the site will also arouse the interest of visitors from other regions because the objective is to create an eco-village integrated into the urban park and surrounded by various useful entities.

Table 1 Responses of the people surveyed for the development of the urban park

Response	Percentage (%)	Type of development
Snack	10.9	Entertainment and recreational
Relaxation space	13.5	
Playgrounds	25.5	
Rest area	11.7	
Camping area	2	
Fitness path	8.6	Sports
Workoutspace	4.6	
Basketball court	4.6	
Football field	5	
Tennis court	0.7	
Theater	0.9	Cultural and educational
Shared or community garden	1.9	
Sculpture	0.9	Aesthetic and ornamental
Lawn	2.9	
Lake view	6.3	

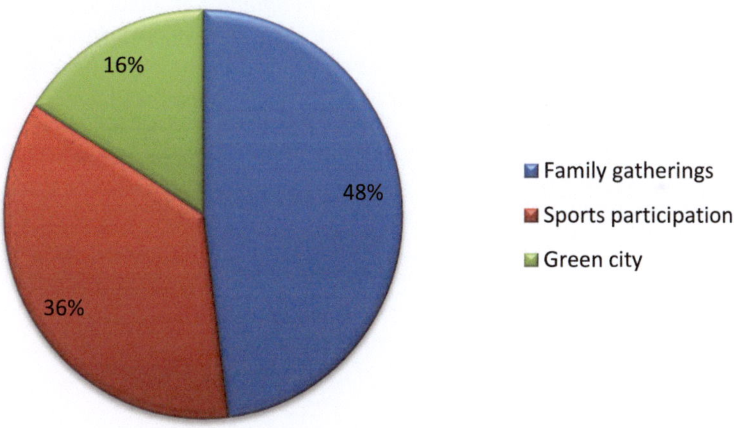

Fig. 9 Reasons for desired developments [6]

Project Components

As shown in Fig. 10, the park has two main entrances: the first leads directly to a station where visitors can rent bicycles to move, to contemplate the space, and to benefit from its services. However, the second leads directly to an eco-museum. In addition, the ecological village consists of a restaurant, a reception, a meeting space, and houses that are built with stones and clapboard ceilings. Sports complex is located in the western part of the site, consisting of a fitness trail, two petanque areas, a football field, two mini-football fields, four basketball courts, three tennis courts, four changing rooms, public toilets, a drugstore, and street workout spaces. On the east side, the leisure complex consists of a snack bar, an open-air theatre, a reading area and a children's playground.

Project Description

The portal of the main entrance to the park is in the shape of the Roman portal of the archaeological site "Pheradi Majus" in order to give the place the historical imprint of the city (Fig. 11). Near each main entrance and within the park, there is parking for vehicles (Fig. 12). Between the main entrance and the sports complex, there is a bike rental station for visitors to tour the park and explore the cycle path with a total distance of 2792 m (Fig. 13). In the middle of the two entrances, a rest area is fenced with a plant screen consisting of *Washingtonia robusta* and equipped with wooden benches under the trees where visitors can sit (Fig. 14). In the middle of the park extends the eco-village on an area of 1.34 ha. It consists of a reception area for visitors at the entrance, a restaurant, 36 ecological houses, and a meeting space equipped with picnic tables and chairs in the center (Fig. 15). On the west side of the eco-village, there is a shared garden consisting of 30 small plots whose dimensions

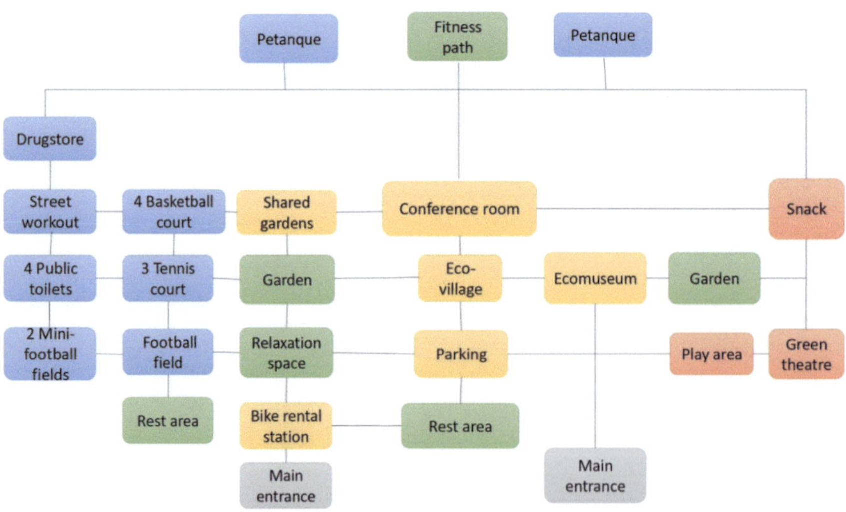

Fig. 10 Conceptual diagram of the Bouficha urban park [6]

Fig. 11 Main portal of the park in 3D [6]

are five meters long and 2.5 m wide (Fig. 16). To the west of the park, there are two football fields, each 40 m long and 24 m wide, and a football field 90 m long and 45 m wide, as well as a dressing room for players (Fig. 17). To the east of the park, the green theatre is located between the play area and the snack bar. It is surrounded by eight outdoor reading kiosks (Fig. 18).

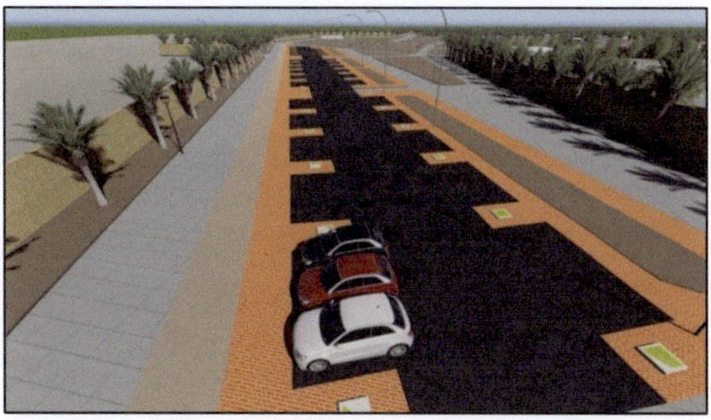

Fig. 12 Parking lot in the park in 3D [6]

Fig. 13 Bike rental station in 2D [6]

Fig. 14 Rest are a next to the main entrance in 3D [6]

Fig. 15 Eco-village in 2D [6]

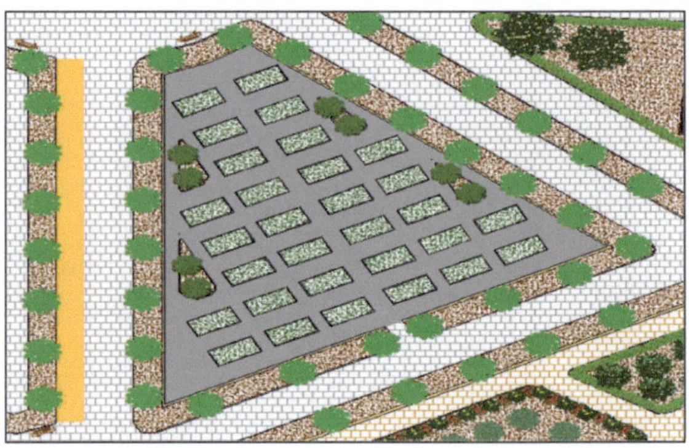

Fig. 16 Shared garden in 3D [6]

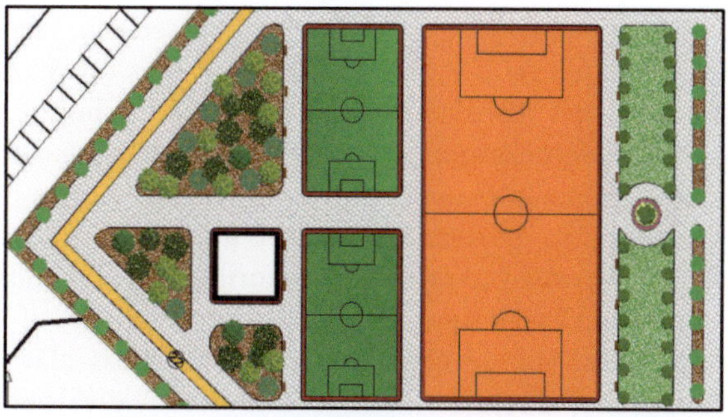

Fig. 17 Football fields in 2D [6]

Fig. 18 Green theatre and reading spaces in 3D [6]

4 Discussion

The spatial distribution of green spaces in Boufícha's city shows that their amount and distribution are severely limited. Similar findings were noted for Sousse City, where [8] confirmed that an unbalanced distribution of green areas across the city exacerbates North–South division and affects social cohesion and balance between districts. Therefore, it is advised to establish additional green areas, such as via managing the urban park in Boufícha's northern. GE images were helpful in detecting the LULC changes that has taken place from the year 2004 to 2019 in the study area. The city's population was surveyed to learn about their objectives and suggestions for park design, as well as better to comprehend the relationship between the people and nature. Thus, it enabled the implementation of environmental governance on a local level. Thus, the survey's results served as the foundation for the urban park development concept, which is significant for the area's growth and the enhancement of the local population's quality of life. Consequently, it is a component of Boufícha's sustainable development initiative.

Thanks to progress in GIS and remote sensing, satellite data is regarded as critical data required by planners and policymakers for the performance of various urban planning and management tools [25]. According to [18], accurate knowledge of LULC is required to implement different planning, land use, and urban development plans. Previous research employed GE for various purposes, namely forest cover mapping [24], population [29, 32], urbanization [7, 22]; biodiversity conservation [6, 8], and urban ecosystem monitoring [2, 12]. Landscape potential, such as green areas, contributes significantly to improve quality of life [11]. Similarly, according to [11] and [1], developing urban green areas play a crucial role in maintaining their ecological roles. Recent studies on ecological networks and green infrastructure found significantly positive effects on green areas, human well-being and sustainable development [9, 17].

5 Conclusion

The Bouficha area, located between the two tourist cities of Sousse and Hammamet, benefits from a strategic location with great geographical variety. Thus, the development concept offered is critical to the growth of this area and the betterment of the local population's living conditions. The many entities presented and the richness and diversity of their roles and services makes this development a multifunctional site and a focal point that unites the various areas that surround it and trace certain continuity with the neighboring regions. In conclusion, the use of GE in urban planning has many advantages. For example, it offers the most recent satellite images with a spatial resolution of less than one meter [7]. Additionally, from [33], it enhances the standard of public service and the spatial analysis process in urban planning organizations. It also causes a rise in the community's territorial consciousness and spatial thinking (local governance).

6 Recommendations

1. The implementation of urban parks strengthens green urban infrastructure design in the city.
2. The use of Google Earth images enhances the standard of public service and the spatial analysis process in urban planning organizations.
3. Solutions that use geoinformatics in urban planning can have an impact on the development of these areas' designs and planning.
4. Geographical Information System is crucial for managing natural resources (vegetation, water, and soil) and guiding decision-making processes in their development plans.
5. It also causes a rise in territorial consciousness and spatial thinking in the community (local governance).

References

1. Ajmi R, Khebour Allouche F, Nuscia Taîbi A, BoussemaBelFekih S (2023) Developing a qualitative urban green spaces index applied to a Mediterranean City. Urban Sci 7(4):115. https://doi.org/10.3390/urbansci7040115
2. Allouche FK, Ajmi Ismail R, Boussema Bel Fekih S, Serban G (2023) Ecosystem behaviors face to climatic changes and anthropogenic actions. Case study: a north-eastern urban wetland, Tunisia. Forum Geogr 11(2):172–185. https://doi.org/10.5775/fg.2022.217.d
3. Basset F, Baudelet L, Le Roy A (2002) Shared gardens. Utopia, ecology, practical advice, Terre Vivante, Paris, p 144
4. Bergaoui M, Camus H, Nouvelot JF (1998) Test of modeling of solid transport on the micro watersheds of Tebaga (central Tunisia). Drought 9(1):51–57
5. Blanchon B (1997) Landscapers in France since 1945. In: Le Moniteur (ed) Modern public spaces. Paris
6. Boussema SBF, Allouche FK, Ajmi R, Chaabane B, Gad A (2023) Assessing and monitoring the effects of land cover changes in biodiversity. Case study: Mediterranean coastal region, Sousse, Tunisia. Egypt J Remote Sens Space Sci. https://www.sciencedirect.com/science/article/pii/S1110982323000029
7. Boussema BFS, Allouche FK, Bekaoui A, Khalifa Y, M'Sadak H (2020) Using Google Earth™ and geographical information system data as method to detect urban sprawl and green spaces for better well being: case of a coastal landscape. Int J Res–GRANTHAALAYAH 8(9):266–276. https://doi.org/10.29121/granthaalayah.v8.i9.2020.1524
8. Boussema S, Allouche FK, Saidaine I, Mechmech Y, Bettaieb T (2018) State of the art of greenway concept application in Tunisian green policy: a case study of an urban landsape in Sousse city. Int J Environ Geoinform (IJEJEO) 5(1):36–50. http://dergipark.gov.tr/download/article-file/452537
9. Boussema BFS, Souidi Z, Chaabane B, Hammami S, Ajmi R, Aloui B, Khebour Allouche F, Boukharouba K (2024) Green infrastructure, a sustainable development tool. Urban Art Bio 2(3):26–31. https://www.ajpsdz.org/rv/index.php/uab/article/view/110
10. Boutefeu E (2005) Social demand for nature in the city: survey of residents of the Lyon metropolitan area, Puca – Certu 154:81
11. Cetin M (2015) Using GIS analysis to assess urban green space in terms of accessibility: case study in Kutahya. Int J Sustain Develop World Ecol 22(5):420–424. https://doi.org/10.1080/13504509.2015.1061066
12. Chaabane B, BoussemaBelFekih S, Khebour Allouche F (2021) Mapping and monitoring of Large-Sfax wetlands (center-east of Tunisia) using radiometric indexes and GIS tools. Algerian J Eng Technol 5:64–71. https://doi.org/10.5281/zenodo.5749971
13. Chen X, Nordhaus W (2015) A test of the new VIIRS lights data set: population and economic output in Africa. Remote Sens 7:4937–4947. https://doi.org/10.3390/rs70404937
14. Clergeau P (2007) An ecology of the urban landscape. Rennes, Apogée, p 136
15. Courajoud M (1970) Green Spaces 25:32
16. CRDA (2020) Agriculture report of Sousse. Regional commission for agricultural development of Sousse, p 42
17. Fang X, Li J, Ma Q (2023) Integrating green infrastructure, ecosystem services and naturebased solutions for urban sustainability: a comprehensive literature review. Sustain Cities Sociandy 98:104843. https://doi.org/10.1016/j.scs.2023.104843
18. Gadrani F et al. (2018) Assessment of landuse/landcover (LULC) change of Tbilisi and surrounding area using remote sensing (RS) and GIS. Ann Agrarian Sci 16:163–169
19. Gathy P (1962) The protective forest. Bull Soc R Belg 69(3):20
20. Gaughan AE, Stevens FR, Linard C, Jia P, Tatem AJ (2013) High resolution population distribution maps for Southeast Asia in 2010 and 2015. PLoS ONE 8:e55882. https://doi.org/10.1371/journal.pone.0055882

21. Goldblatt R, Deininger K, Hanson G (2018) Utilizing publicly available satellite data for urban research: mapping built-up land cover and land use in Ho Chi Minh City, Vietnam. Develop Eng. https://doi.org/10.1016/j.deveng.2018.03.001
22. Goldblatt R, You W, Hanson G, Khandelwal AK (2016) Detecting the boundaries of urban areas in India: a dataset for pixel-based image classification in Google Earth engine. Remote Sens 8:634. https://doi.org/10.3390/rs8080634
23. Grahn P, Hunt JD (1996) The art of the garden and its history. O. Jacob, Paris
24. Hansen MC, Potapov PV, Moore R, Hancher M, Turubanova SA, Tyukavina A, Thau D, Stehman SV, Goetz SJ, Loveland TR, Kommareddy (2013) A high-resolution global maps of 21st-century forest cover change. Science 342:850–853
25. Malarvizhi et al. (2016) Use of high resolution google earth satellite imagery in landuse map preparation for urban related applications. Procedia Technol 24:1835–1842
26. MET Ministry of the Environment Tunisia (2010) Regional indicators of improvement in living conditions, p 5
27. NIM (2020) Annual report from the National Institute of Meteorology Tunisia, p 31
28. NIS (2020) Monthly Statistical Bulletin (2020) National Institute of Statistics Tunisia, p 15
29. Patel NN, Angiuli E, Gamba P, Gaughan A, Lisini G, Stevens FR, Tatem AJ, Trianni G (2015) Multitemporal settlement and population mapping from Landsat using Google Earth Engine. Int J Appl Earth Obs Geoinf 35:199–208
30. Potere D, Schneider A, Angel S, Civco DL (2009) Mapping urban areas on a global scale: which ofthe eight maps now available is more accurate? Int J Remote Sens 30:6531–6558. https://doi.org/10.1080/01431160903121134
31. Seto KC, Fragkias M, Güneralp B, Reilly MK (2011) A meta-analysis of global urban land expansion. PLOS One 6:e23777. https://doi.org/10.1371/journal.pone.0023777
32. Trianni G, Lisini G, Angiuli E, Moreno EA, Dondi P, Gaggia A, Gamba P (2015) Scaling up to national/regional urban extent mapping using Landsat data. IEEE J Sel Top Appl Earth Obs Remote Sens 8:3710–3719
33. Xhafa S, Kosovrasti A (2015) Geographic information systems (GIS) in urban planning. Eur J Interdiscip Stud 1(1). https://revistia.org/files/articles/ejis_v1_i1_15/Sonila_Xhafa.pdf
34. Zhang S, Liu J, Song C, Chan CH, Pei T, Wenting Y, Xin Z (2021) Spatial-temporal distribution characteristics and evolution mechanism of urban parks in Beijing, China. Urban For Urban Greening 64:127265. https://doi.org/10.1016/j.ufug.2021.127265

Impact of Human Activities on the Bay of Algiers (Algeria)

Halima Slimani, Sahraoui Bensaid, Amani Bakalem, and Maya Flici

Abstract Currently, two-thirds of the Algerian population lives in the northern part of the country representing 4% of the entire territory. To provide needs of this population, the extension of the urban area of coastal has constantly expanding. As the capital and central core of the economy, Algiers is home to around 3 million people. To structure the inevitable expansion of Algiers, the government has put in place urban and coastal planning policies and enacted a set of laws regulating these urban planning. In order to illustrate the importance of urban expansion to the detriment of natural areas, a diachronic study is applied on the bay of Algiers. Geographic Information System (GIS) and imagery are used to assess the expansion of the mineral element and the necessity opening the bay of Algiers and its infrastructures onto the sea. It aims to quantify and illustrate the diversion of coastal land towards buildings. The study shows a constant expansion of the mineral element to the detriment of the plant element. Analysis of a few examples of projects shows that they do not always meet the need for a Mediterranean city and sustainability concepts.

Keywords Bay of Algiers · GIS · Mineral Element · Planning · Plant Element

1 Introduction

Coastal zones are very important natural resources [8] for the variety of ecosystem services they provide to society. But they are also highly vulnerable [3, 7, 13] and sensitive to various human activities, such as urbanization [5, 6] and climate change [1]. Several studies point out the threat to these ecosystems, which can also exacerbate erosion [7, 19], with serious repercussions for coastal populations, infrastructure goods and services provided [1, 6].

H. Slimani (✉) · S. Bensaid · A. Bakalem · M. Flici
Laboratory of Plant Ecology and Environment, University of Sciences and Technology Houari Boumediene, Algiers, Algeria
e-mail: slimani_h@usthb.dz; hslimani@usthb.dz

F. Khebour Allouche (ed.), *Environmental Remote Sensing and GIS in MENA Region*, Earth and Environmental Sciences Library,
https://doi.org/10.1007/978-3-031-86574-9_12

The Mediterranean coastal zone is one of the most fragile and densely urbanized areas in the world (Satta et al. 2017). In this context, the Algerian population lives predominantly in the northern part of the country, in coastal areas, and more than 50 percent lives in cities [15]. To respond to the needs of this population, the urban fabric in coastal cities is expanded. As the capital and the core of economy [21], Algiers hosts nearly 3 million inhabitants according to the Office of National Statistics [17].

To structure the inevitable expansion of Algiers, the State implemented urban and coastal planning policies, as well as a set of laws regulating urban and coastal management (e.g. Law 92–20 on urban planning). However, several works on the Bay of Algiers indicate worrying results and pessimistic projections [20].

In order to identify anthropogenic impacts, in particular the decrease in natural areas and urban expansion, we completed a diachronic survey of Algiers Bay using a Geographic Information System (GIS), satellite imagery and the OSM database (Open Street Map). Three dates have been selected for this study (2001, 2011 and 2022). This has permitted assessment of the expansion of the mineral element often of the detriment of the plant element, and the need to open up the Bay of Algiers and its infrastructures onto the sea. Therefore, the diversion of land use to buildings can be quantify and illustrate. The "mineral" element can be defined as all the built structures, roads and any other anthropic occupation of the land. The "plant" element is all the green surfaces, managed or not, that provide vegetation and an opening in the urban space.

For the Bay of Algiers, in view of its potential, its population and its central place it occupied in the economy and business, the question is: is the city opened up to the sea? Does it have any prospects and ensure sustainability? To this end, we chose three key sites (*El Kettani*, the *Sablettes promenade* and the *Mohammadia business district*) that underwent dramatic change between 2001 and 2022, to take a critical, analytical look at coastal development.

2 Methodology

2.1 Study Area

The Bay of Algiers is part of the Wilaya of Algiers (administrative divisions of the country). It extends over 35 km and is composed of 10 municipalities: El Marsa, Bordj El Bahri, Bordj el Kiffan, Mohammadia, Hussein Dey, Belouizdad, Algiers-centre, Casbah, Bab El Oued and Bologhine (Fig. 1). The wilaya of Algiers has a maritime area of 1,326 km^2 [2].

This study is carried out along a band of coastline 800 m wide, according to the law 90–29 on urban management, which defines the coastline as all islands and islets as well as a strip of land at least eight hundred metres (800 m) wide running alongside the sea.

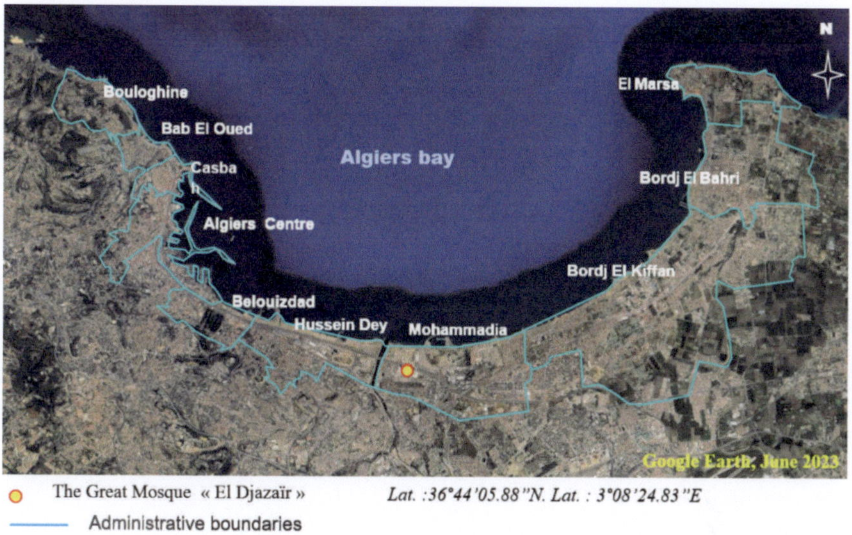

Fig. 1 Municipalities bordering the Algiers Bay (Slimani et al. 2022)

The Bay of Algiers lies on the Mitidja plain characterized by pocket beaches, cliffs and rocky capes which are severely eroded despite the construction structures to limit the decline of the coastline [14].

The variability in dry periods and heavy precipitations also increase erosion. The plain is surrounds on the west by the mount Bouzareah which rases to 407 m.

The coastal dune, largely modified, still exists between Wadi El Harrach and and Bordj El Kiffan. The Bay of Algiers includes the mouths of wadi El Harrach in the centre, wadi El Hamiz in the east and wadi Koriche in the west. These wadis increase flood risks during rainy season.

The study area has a Mediterranean-type climate characterized by two seasons: mild and rainy winter and a long and warm summer.

2.1.1 Urban Environment

Survey show that the urbanisation and artificialization of the coastline rate is about 50% [11] and up to 81.7% [2]. The Algerian urban fabric is complex and dense with a heterogeneous and non-linear development [4]. Consequently, the urban environment of the Bay of Algiers is constantly expanding [22] to the detriment of natural areas. The loss of land is also an alarming problem, particularly land for tourism. At present, half of the tourist land has disappeared towards increased the housing needs of the population. The metropolization of Algiers, even supervised by public authorities, has had negative effect on the quality of life and the urban plan [24]. These include heavy pollution of the urban fabric, urban sprawl, coastal degradation, as well as the misappropriation of lands and unlawful construction. We have one of the largest maritime ports on Algeria, Stretching from the Algiers-Centre to the Belouizdad.

2.1.2 Population and Demography

According to the General Population and Housing Census [17], 63% of Algeria's population lives in the north where 30% in the coastal wilayas.

The population living in the 450,000 km^2 coastal strip has doubled between 1998 and 2008 (last census), rising from 7.5 million to 15 million. That creates an imbalance in population density and accelerate many pressures on the natural coastal areas. Algiers, the major economic centre, is subject to anthropic pressure due to its high population density. At the end of 2020, the wilaya of Algiers have a total population of approximately 3 309, 896 with a density of 4,090 inhabitants per km^2 [2].

2.2 *Method*

In this study, three dates (2000, 2011 and 2022) were chosen to compare changes in land use by decade. For that, supports documents and data are consulted and collected from the National Agency for Land-use Planning and Attractivity and the National Coastal Commission. Data are completed by field observation and measurements.

The GIS (Geographic Information System) is used to collect, organize, manage, analyze, combine, process and present data from a variety of sources [10]. The GIS enabled us to extract the study area, georeference of satellite images, digitize features and produce map (Fig. 2). Several layers are isolated: building, land-use, beaches, roads and other lines, and other land uses.

We used two Landsat satellites images from the period 2001–2011, downloaded from Google Earth Pro. We also used a satellite image downloaded from Yandex via SASplanet (Table 1).

The mineral element (buildings, roads, etc.) and the plant element are isolated, allowing the surfaces occupied by each element to be defined. This approach enabled us to obtain maps of the vegetation and mineral elements for each of the three years selected (2001, 2011, 2022), in order to quantify the expansion of mineralization to the detriment of natural spaces along the bay.

To shed a critical and analytical eye on coastal development, we chose three key sites which have experienced dramatic changes between 2001 and 2022. These sites (El Kettani, Sablette Promenade and Mohammadia Business district) are located in the municipalities of Bab E Oued, Hussein Dey and Mohammadia respectively.

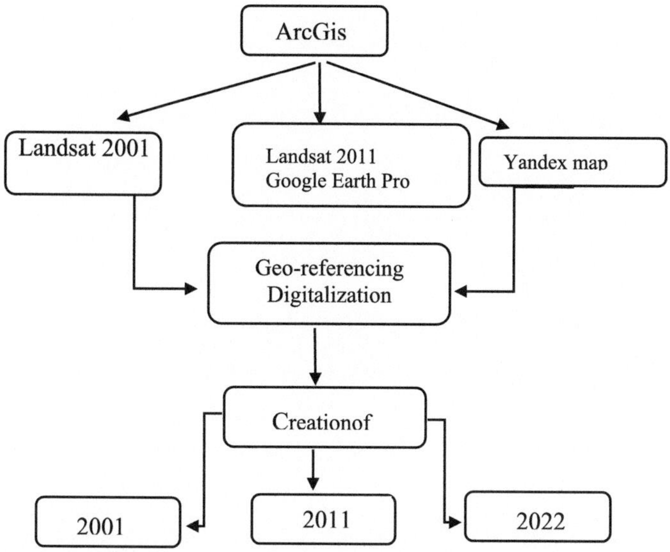

Fig. 2 Overall methodology for the diachronic study (Slimani et al. 2022)

Table 1 Characteristics of the satellite imagery used

Satellite image	Date	Source	Resolution
Bay of Algiers 2022	January 2022	Yandex.maps	77,633 × 41857px Number of panels 304 × 165
Bay of Algiers 2011	June 2011	Google Earth Pro (LandSat)	Maximum quality 4800 × 2292px
Bay of Algiers 2001	August 2001	Google Earth Pro (LandSat)	Maximum quality 4800 × 2292px

3 Results and Discussion

3.1 Assessment of Mineralized and Planted Surfaces

The maps created represent the vegetation element and the mineral element, which includes all built components, roads and other urban structures. The three mineral and plant element maps of Algiers Bay, generated at different dates (2001, 2011 and 2022), show some differences (Figs. 3, 4, 5).

In 2001, the bay presented the largest proportion of vegetated surface (compared to 2011 and 2022), with almost 21% (Table 2) of the total area. The surface area occupied by the mineral element was very large (22.75 km^2) but relatively small compared with the other two dates considered (Table 2).

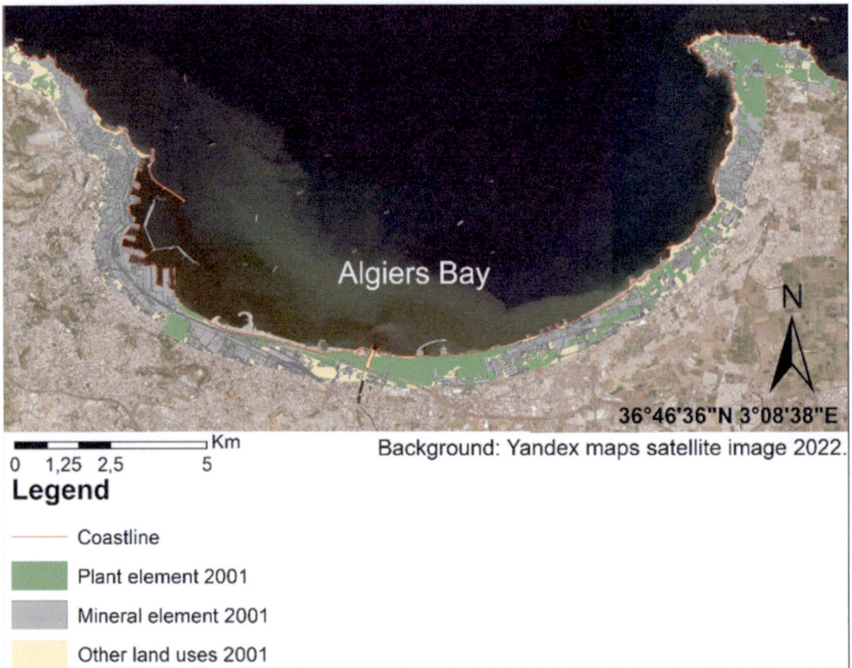

Fig. 3 Map illustrating plant and mineral elements of Algiers bay in 2001(Slimani et al. 2022)

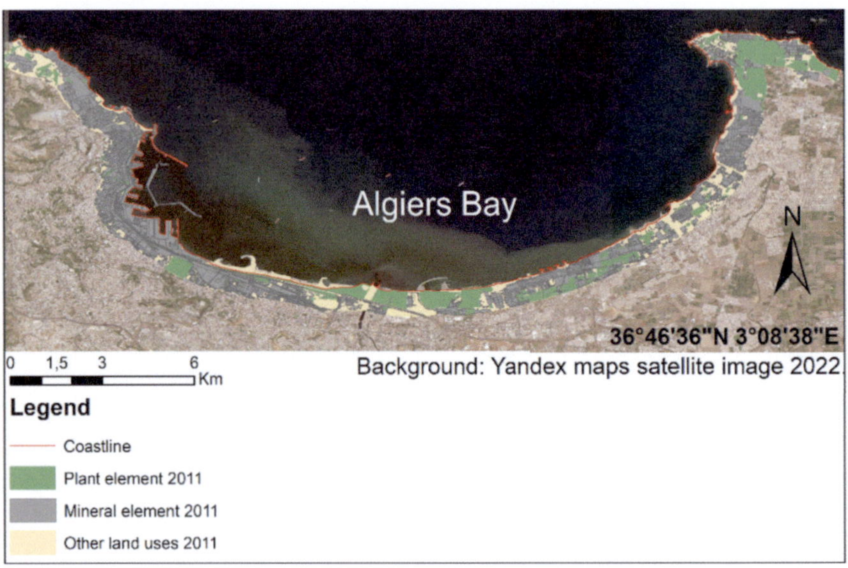

Fig. 4 Map illustrating plant and mineral elements of Algiers Bay in 2011 (Slimani et al. 2022)

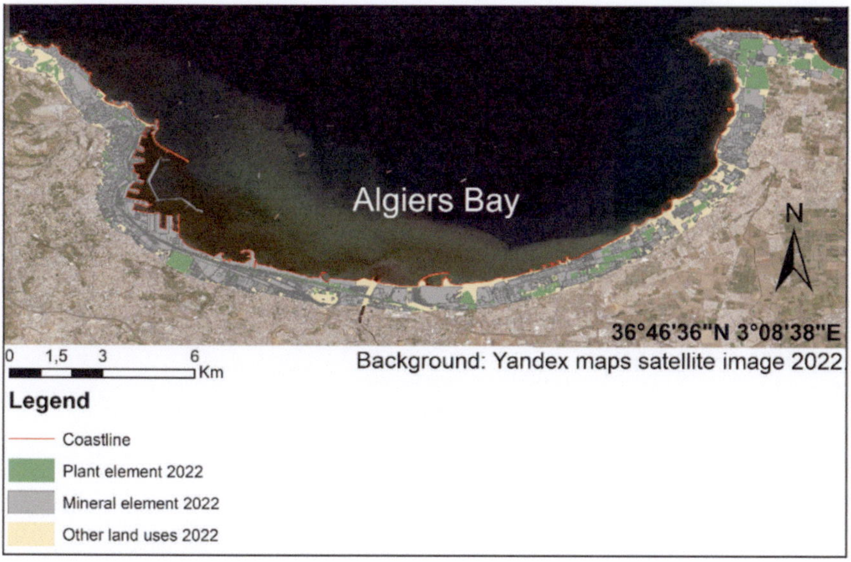

Fig. 5 Map illustrating plant and mineral elements of Algiers Bay in 2022 (Slimani et al. 2022)

Table 2 Surfaces of mineralized and vegetated area in the Bay of Algiers (2001–2022)

	Plant element km^2 (%)	Mineral element km^2 (%)	Other land uses km^2 (%)	Total km^2
2001	6.9 (21)	22.75 (70)	2.83 (9)	32.48
2011	5.79 (17)	25.45 (74)	3.28 (9)	34.52
2022	4.35 (17)	27.5 (73)	5.86 (16)	37.71

The 2011 results show an increase in the surface occupied by the mineral element covering about 74% of the total land area (Table 2, Fig. 4). This year corresponds to a period where several projects undertaken in the Bay of Algiers.

For the year 2022, the urban fabric is very dense, with greater mineralization (Table 2, Fig. 5). Buildings and road infrastructures in the Bay of Algiers occupy most surfaces. In the last few years, this expansion has been due to an artificial increase in coastal areas, rather than the restoration of natural ecosystems such as dune belts.

The "Other surfaces" which cover about 16% of the total land of the Bay Algiers (Table 2), correspond to infrastructures such as cemeteries and construction underway (such as creation of a new marina in El Mohammadia and new jetty in Bordj El Kiffan).

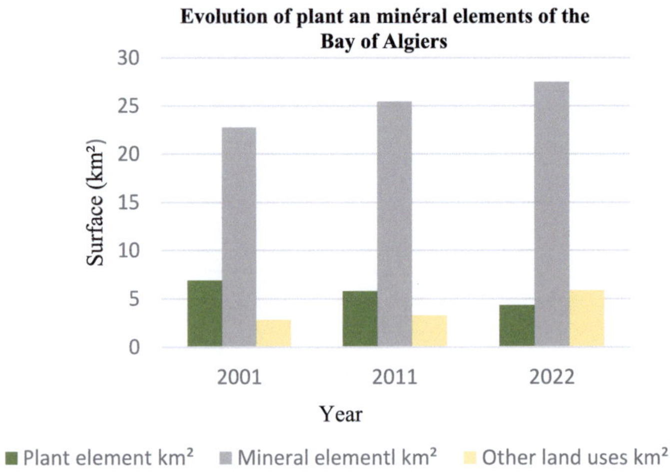

Fig. 6 Evolution of plant and mineral elements of the Bay of Algiers (2001–2022) (Slimani et al. 2022)

3.2 Trends Over 20 Years

The results show a steady increase of mineralization between 2001 and 2022 (Fig. 6). The loss of green space between 2001–2011 and 2011–2022 was 1.11 km^2 and 1.44 km^2 respectively. In total, Algiers Bay has lost 2.55 km^2 of plant areas, or a loss of over 6.5% of the bay's total area in 2022.

This observation is all the more critical in the absence of any real prospect of rehabilitating coastal areas and without any strong, well-thought-out environmental policy.

Moreover, numerous extensions of up to 80 m towards the sea have been built, and we should, therefore, expect an increase in green spaces thanks to these new developments.

3.3 Critical View at the Development of Algiers's Bay

Since independence, Algerians have turned their backs on the sea, because development policies have not been geared towards this goal. However, since the early 2000s, numerous development projects have been carried out. To take a critical look at the development projects carried out since 2002 along the Bay of Algiers, we selected three main sites: the El Kettani site (Figs. 7, 8), Sablette promenade site (Figs. 9, 10) and the business district of Mohammadia (Figs. 11, 12).

Fig. 7 Location of "El Kettani" area, municipality of Bab El Oued (Slimani et al. 2022)

Fig. 8 El Kettani (Slimani et al. 2022)

Fig. 9 Limit of the "Sablette Promenade", municipality of Hussein Dey (Slimani et al. 2022)

Fig. 10 Build in the "Sablette Promenade" (Slimani et al. 2022)

Fig. 11 Limits of the business district of Mohammadia (Slimani et al. 2022)

Fig. 12 Limits of the business district of Mohammadia (Slimani et al. 2022)

El Kettani Site

The El Kettani site is located in the municipality of Bab El Oued (B.E.O.), in the western part of the Bay of Algiers (Fig. 7). The esplanade was made after the B.E.O. floods in 2003. It is indeed open to the sea, even if the primary objective of this development is to allow the water to have an exit route, and find its natural way which was obstructed by dwellings. This planning was not designed for the population or to promote openness to the sea, but it's a relatively positive change. The surface of the beach has also been artificially increased using jetties (Fig. 8). This contemporary urban project, which reinforces B.E.O.'s tourism vocation, meets the objectives of the Algiers 2010/2029 strategic plan and could diversify economic income. However, the project could be improved by correcting the imbalance between the artificial and the natural through the creation of green spaces.

The Sablette Promenade Site

The 4.5 km-long Sablettes promenade was one of the government's flagship projects for the development of the Bay of Algiers (Fig. 9). The project began in 2012, and was handed over in 2021. One of the main features of this development is that it includes extensions of 80 to 100 m towards the sea. One of the most striking aspects of the Sablettes site is its high degree of mineralization and lack of access (physical and visual) to the sea. In fact, the vast majority of the site is concreted and some of the buildings obstruct the view of the sea (Fig. 10).

Numerous built facilities such as sanitary blocks, restaurants and other refreshment outlets completely obstruct the view of the sea (Fig. 9). Their location is not well thought-out and seems random, based on a "plot of land" logic, so each plot is allocated to one or more shops or services, with no regard for overall appearance, natural ecosystems or proximity to the sea. The appearance is unsightly and heterogeneous, giving off a chaotic image.

When planning a project, planting is crucial. It must constitute the greater part of the space. The species chosen must meet specific criteria, for example providing ecosystem services (the most important service is shading). High mineralization coupled with poor species selection in the grass plots results in almost non-existent shade, forcing visitors to confine themselves to "shelters", far from the sea and behind buildings.

Business District of Mohammadia

A number of projects have already been launched in Algiers, notably in the Mohammadia area. Unfortunately, there is a very high level of the mineral element (Fig. 11), to the detriment of the vegetation and primary ecosystems typical of the area. The site shows large areas of wasteland cleared for building. This is in total contradiction with the provisions of the Littoral law, prohibiting construction less than 100 and 800 m from the shore. This project considerably reduces space and the opening onto the sea.

On a positive note, olive trees, *Olea europaea* L., have been planted as trees in rows in the parking lot (Fig. 12). The olive tree is a typically Mediterranean specie, requiring little or no maintenance or irrigation, especially in these times of water stress and is also aesthetics. To ensure sustainability, and with regard to climate change, it is essential that thinking is in line with the typical challenges facing Algeria and Algiers. For example, the introduction of native dune species such as Oyat (*Ammophila arenaria* (L.) Link), which naturally retains sand dunes and helps strengthen dune belts, reduces the use of introduced species which can create an imbalance and overtake local species.

It would be preferable to install a garden in this area of the bay, to promote native coastal species and improve the setting for visitors.

This project creates a barrier to pedestrian access to the sea. They also break up the view of the sea, and present a mobility problem for the elderly, disabled and others.

Urbanization and inappropriate development are leading to major changes in natural ecosystems [21, 22], particularly in coastal areas. It is apparent that this phenomenon has serious consequences for landscape [12, 23], biodiversity [9, 25] and dune erosion [18] resulting in a significant loss of green space.

Faced with the fragility and continuing deterioration of this strategic area of the territory, and the perceptible specter of irreversible situations [16], it has become urgent to have a clear coastal strategy to reversing the trend. The first step in that this strategy would be to implement the law on the protection and valorization of coastal areas.

4 Conclusion

Our study area focused on the coastline of the Bay of Algiers, an area subject to a Mediterranean climate and very strong demographic growth and economy. From a sociological point of view, the people of Algiers have turned their backs on the sea. This is reflected in coastal development, in the way they invest and construct their space, and in the resulting impacts.

In order to assess the impact of these anthropogenic activities, particularly in terms of opening up to the sea and development perspective, a diachronic study was carried out, based on satellite images and GIS databases. Three dates are selected: 2001, 2011 and 2022. The study area was set at a width of 800 m inland from the coastline.

All data processing resulted in three geographic databases for each year. This provided a surface area estimate to quantify the expansion of the mineral element to the detriment of natural areas over a 21-year period.

From 2001 to 2022, the study area lost 2.55 km^2 of vegetation. This represents 6.76% of the total surface area for the year 2022.

The results show that the importance of the mineral element, represented by buildings, roads, parking lots, etc., continues to increase over time. On the other hand, planted spaces, already reduced since 2001, have not taken a significant place in development projects.

The cartographic results show that coastal development is not in line with the concept of the green and blue belts to the costal and sea. Developments are conceived and built without integrating notions of ecological connectivity: no ecological corridors and no initiatives to restore nature in the city. The city's current dynamic development reinforces the feeling of enclosure and encourages people to turn their backs on the sea. The development of the Bay of Algiers should be reviewed in depth, as it does not meet the needs of a Mediterranean city, an open city or the concepts of sustainability.

This work is an attempt to assess the impacts of anthropogenic activities in their cartographic aspects via the GIS tool. In the future, it will be necessary to extend this assessment to larger areas, preferably the entire Algerian coastal territory.

5 Recommendations

The Bay of Algiers is a highly sought-after area of great potential and attractiveness whose development needs to be reviewed. The bay's coastline needs real, well-thought-out planning, based on concepts of sustainability and openness the city to the sea, if it is to become a modern, attractive and green Mediterranean city.

In terms of rehabilitation, the main difficulty lies in the importance mineralization of surfaces and the dilapidation of coastal land. Algiers has a number of alternatives, in addition to changes to the urban fabric itself (demolition for example). There can be no doubt that, to improve the natural coastal protection functions of coastline, it is necessary to allow vegetation to reinvest the Algerian environment, particularly in areas where the urban fabric is the densest and most complex (the Bab el Oued and Casbah districts, for example).

However, Aware of this, and with a view to reconciling the people of Algiers with the sea, a project has recently launched to develop the Bay of Algiers. The project, part of the Wilaya d'Alger's strategic plan for 2030, covers an area of 110 km^2.

This development focuses on landscape structure and the planning of new neighborhoods in a context of eco-environment and eco-development it would be advisible for the authorities to agree with our recommendations and integrate them into the current development project for the Bay of Algiers.

References

1. Allenbach K, Garonna I, Herold C, Monioudi I, Giuliani G, Lehmann A, Velegrakis AF (2015) Black Sea beaches vulnerability to sea level rise. Environ Sci Policy 46:95–109. https://doi.org/10.1016/j.envsci.2014.07.014
2. ANIREF (2022) Budget statistical yearbook, 1st édn, p 142. https://dgb.mf.gov.dz/wp-content/uploads/2022/06/Yearbook-2020.pdf

3. Arnaud A, Claeys C, Lambert M-L (2023) Vulnerability to coastal erosion: mapping four cases in the French West Indies and mainland France. Mappemonde 136:25. http://journals.openedition.org/mappemonde/9043
4. Bensaid S, Calvão T, Djelid S A (2021) Diversité floristique et espace urbain: cas de l'’algglomération d'Alger. Acta Botanica Malacitana 46:57–79. https://doi.org/10.24310/abm.v46i.12864
5. Čeplová N, Kalusová V, Lososová Z (2017) Effects of settlement size, urban heat island and habitat type on urban plant biodiversity. Landsc Urban Plan 159:15–22. https://doi.org/10.1016/j.landurbplan.2016.11.004
6. Chen Y, Huang B, Zeng H (2022) How does urbanization affect vegetation productivity in the coastal cities of eastern China? Sci Total Environ 811:152356. http://www.ncbi.nlm.nih.gov/pubmed/34921883
7. Cruz-Ramírez CJ, Chávez V, Silva R, Muñoz-Perez JJ, Rivera-Arriaga E (2024) Coastal management: a review of key elements for vulnerability assessment. J Mar Sci Eng 12(3):386. https://doi.org/10.3390/jmse12030386
8. Dugan JE, Hubbard DM (2009) Loss of coastal strand habitat in Southern California: the role of beach grooming. Estuaries Coasts 33(1):67–77. https://doi.org/10.1007/s12237-009-9239-8
9. Forys E, Allen CR (2005) The impacts of sprawl on biodiversity: the ant fauna of the lower Florida Keys. Ecol Soc 101:25. https://www.jstor.org/stable/26267714
10. Gilliot J (2000) Introduction to GIS. Introduction & spatial information DAA AGER agronomic-environment department, p 61. https://fr.scribd.com/document/42200200/01-intro-sig
11. Grimes S (2020) Diagnostic report on the implementation of the national strategy for integrated coastal zone management in Algeria (2030). Ministry of the Environment-GIZ, Report No. 1, p 60. https://faolex.fao.org/docs/pdf/alg215561.pdf
12. Leston LFV, Rodewald AD (2006) Are urban forests ecological traps understory birds? An examination using Northern cardinals. Biol Cons 1(31):566–574. https://doi.org/10.1016/j.biocon.2006.03.003
13. Machard A, Abi Moussa T, Alessandrini J-M, De Oliveira F, Devys M et al (2024) Summer thermal comfort, vulnerability of the building stock to overheating and behaviour adaptation to extreme heat: state of the art. Scientific and technical centre for construction, p 91. https://cstb.hal.science/hal-04695333
14. Maouche S, Morhange C, Meghraoui M (2009) Large boulder accumulation on the Algerian coast evidence tsunami events in the western Mediterranean. Mar Geol 262(1–4):96–104. https://doi.org/10.1016/j.margeo.2009.03.013
15. MATE (2006) State and future of the environment report. Ministry of planning and environment. Algeria 490
16. McKinney ML (2006) Urbanization as major cause of biotic homogenisation. Biol Cons 127:247–260. https://doi.org/10.1016/j.biocon.2005.09.005
17. ONS (2020) Statistical yearbook for the wilaya of Algiers. National statistics office, p 91. https://alger.mta.gov.dz/wp-content/uploads/sites/6/2022/03/ANNUAIRE-STATISTIQUE-DE-LA-WILAYA-DALGER-2020.pdf
18. Otmani H, Belkessa R, Rabehi W, Guerfi M (2019) Degradation of the Algerian coastal dunes between urbanization pressure and consequences on shoreline evolution. Geo-Eco-Marina 25:131–145. https://doi.org/10.5281/zenodo.3609777
19. Peduzzi P, Velegrakis AF, Estrella M, Chatenoux B (2013) IIntegrating the role of ecosytems in disaster risk and vulnerability assessments: lessons from the risk and vulnerability assessment methodology development project (RiVAMP) in Negril Jamaica. In: Renaud FG et al (eds) The role of ecosystems in disaster risk reduction. United Nations University Press, pp 109–139. https://www.researchgate.net/publication/285754276
20. Plan U-M-B (2009) Report on the state of the environment and development in the Mediterranean, Athens, p 204. https://planbleu.org/wp-content/uploads/2020/04/SoED2009-FR_compressed-1.pdf

21. Rabehi W, Guerfi M, Mahi H (2019) The Bay of Algiers, a prized coastal area, between urbanization pressures and territorial governance. Geo-Eco-Marina 25:141–158. https://doi.org/10.5281/zenodo.3609744
22. Rabehi W, Guerfi M, Mahi H, Rojas-Garcia E (2019) Spatiotemporal monitoring of coastal urbanization dynamics: case study of Algiers' Bay, Algeria. J Indian Soc Remote Sens 47(11):1917–1936. https://www.researchgate.net/publication/335620851
23. Torrens PM (2006) Simulating sprawl. Ann Assoc Am Geogr 96(2):248–275. https://doi.org/10.1111/j.1467-8306.2006.00477.x
24. Weber C (2018) Quality of life, quality of city and urban health, p 19. https://hal.science/hal-01888413
25. Weng YC (2007) Spatiotemporal changes of landscape pattern in response to urbanization. Landsc Urban Plan 81:341–353. https://doi.org/10.1016/j.landurbplan.2007.01.009

Using RS and GIS to Map Land Cover Changes in the East of Saudi Arabia: The Al-Ahsa Metropolitan Area

Monaem Nasr and Walid Chouari

Abstract The primary objective of this work is to map the recent evolution of land use and land cover (LULC) in east of Saudi Arabia between 1990 and 2022. To understand this complex theme, an integrative, global, and hierarchical methodological approach, combining multi-source and multi-date data was adopted. This approach involves the creation of a prototype land-use map based on the strict rules of graphic semiology. The main cartographic results present the various land use and land cover changes. The combination of two satellite images from 1990 and 2022 enables a diachronic analysis of the Al-Ahsa Metropolitan Area (AMA), as well as a synthetic description of the organization and spatial distribution of land use and its classes. The application aims to highlight the process of urban sprawl and changes in land occupation and use. It shows the importance of change detection methods for understanding urbanization patterns and assessing their impacts on the oasis and bare soil ecosystems. Post-classification comparison of land-use maps revealed a growth in built-up areas to the detriment of the old oasis center and bare soil. The images of the study area were classified into four classes: agriculture, bare soil, built-up areas, and hills. The Remote Sensing (RS) and Geographic Information System (GIS) approaches are highly effective and offer real added value in visualizing land occupation and land usage. All these documents are indispensable and highly important for rational planning of all types of land occupation and use.

Keywords Land Use/Land Cover Change · LULC · Geographic Information System (GIS) · Remote Sensing (RS) · Al-Ahsa Metropolitan Area (AMA) · Al-Ahsa Oasis

M. Nasr (✉)
GIS and Mapping Sciences, Department of Geography, College of Languages and Humanities, Qassim University, Buraydah, Saudi Arabia
e-mail: mo.nasr@qu.edu.sa

GIS and Mapping Sciences, Department of Geography, Faculty of Arts and Humanities of Sfax, University of Sfax, Sfax, Tunisia

W. Chouari
College of Arts, Department of Geography, King Faisal University, Al-Ahsa, Saudi Arabia
e-mail: wchouari@kfu.edu.sa

F. Khebour Allouche (ed.), *Environmental Remote Sensing and GIS in MENA Region*, Earth and Environmental Sciences Library,
https://doi.org/10.1007/978-3-031-86574-9_13

1 Introduction

Studying and mapping Land Use and Land Cover (LULC) has been the subject of a great deal of theoretical and practical research since the rise of new space technologies like Geographic Information Systems (GIS) and Remote Sensing (RS) in the 1980s. In this context, there is an urgent need to monitor the temporal and spatial patterns of LULC changes and assess how the environment responds to them.

RS is a practical and cost-efficient tool for that, as demonstrated by the growing number of studies on land cover monitoring. However, the limited set of land cover classes used in generalized regional or national scale approaches [9, 21], coarse temporal resolution [34, 35], limited time span [17, 18] and/or the focus on specific topics (like impervious surfaces [23] or cropland [20]) may restrict the use of their derived datasets for finer analysis on specific areas [16]. This happens even when they are conducted over the same area. Since the mid-1980s, with the finesse of the analysis scale and thanks to High and Very High Spatial Resolution sensors such as LANDSAT TM or ETM + (30 m), ASTER (15 m), SPOT 5 (5 m), or IKONOS (5 m multispectral and 1 m panchromatic)., enhancement in sensor resolution has made it possible to increase the production of LULC maps at the plot scale. Furthermore, the rapid development of multi-spatial and multi-temporal RS data has now made it possible to monitor urban LULC changes very efficiently. RS techniques have proven very useful in urban mapping.

The rise of GIS has enabled the creation of databases dedicated to spatialized land use by integrating basic land occupation and use data derived from different sources (satellite images, aerial photos, topographic maps, statistical data, GPS data, and land-use plans) to develop LULC monitoring and prediction models.

These technologies are often complex to implement, as they require knowledge and understanding of the factors conducting to the observed changes and the development of valid reproducible models. Since the study of LULC is closely linked to RS and GIS, this paper focuses on their contribution to land cover mapping and its evolution, using the example of Saudi Arabia in general and the Al-Ahsa Metropolitan Area (AMA) in particular.

With the advent of the first RS satellite (Landsat 1) in 1972, many LULC studies have been undertaken in various areas, including urban areas, agricultural areas, forest areas, and wetlands. Conventional ground-based methods of land use mapping are labor-intensive, time-consuming, and infrequently done. Over time in a rapidly changing environment, the maps are outdated. In recent years, developed satellite RS techniques have been immensely valuable for preparing accurate LULC maps and monitoring changes at regular intervals. Despite the urban environments' spatial and spectral heterogeneity challenges, RS seems to be a suitable source of reliable information about their multiple facets. So, for the analysis of dramatic changes in LULC at global, continental, and local levels and further exploration of the extent of future changes, the current geospatial information on patterns and trends in LULC play an important role [29].

Remotely sensed imagery provides an efficient means of obtaining information on temporal trends and spatial distribution of urban areas needed for understanding, modeling, and projecting land changes [29]. In inaccessible regions, this technique is perhaps the only method for obtaining the required data on a cost- and time-effective basis [28]. Satellite imagery can regularly provide more frequent data collection, unlike aerial photographs, which may provide more geometrically accurate maps of limited coverage and more expense. For example, Nasr [26] worked on LULC mapping and change detection in Tunisia's Cap Bon Peninsula using an RS and GIS technique. He used Landsat data from 1987 and 2014 to conclude that urban areas increased threefold between 1987 and 2014 while wetland and agricultural area sharply decreased. Another study carried out by [26] on land use mapping using RS and GIS techniques in the same area used multiple scales and a hierarchical nomenclature. The mapping results at different scales and using various presentation modes of land uses a global information system. Its three phases create a useful GIS and RS synergy: the design of the LULC using the Hypergraph-Based Data Structure (HBDS) method, RS image processing, and the proper development and implementation of the LULC system.

The understanding of LULC change has moved from simplicity to complexity over the last decades. "The question of global change is currently at the heart of the scientific community's interests, as evidenced by the very large number of contributions relating to this issue" [30]. One of the many aspects of this is the key role played by changes in LULC in the evolution of ecosystems. This role has been demonstrated at both global and local scales [10, 11, 19, 27, 32].

Changes in LULC are the result of a combination of numerous interactions between humans and the environment. The main drivers and impacts of land cover change globally have now been identified, thanks to satellite data and methods that have proved their worth in producing homogeneous, usable information describing changes affecting vast areas [22]. However, the study of LULC changes at the local level also presents major challenges. According to Sparfel [30], "knowledge of these changes of low spatial and temporal amplitude provides relevant elements for environmental diagnosis and forecasting and improves our understanding of the sometimes complex functioning of human–environment systems". However, the production of homogeneous, up-to-date data remains problematic at this scale.

According to the example of the "Land Use and Occupancy Information Systems" by the European Communities in 2001, "land use and occupancy classes are analytical units that make it possible to establish an initial quantitative link between human activities, their impact on the environment and the geographical (spatial) dimension. Information on the evolution of land occupation/use integrates the temporal dimension." This data is of major interest to planners, who evaluate decisions affecting the land, and to the scientific community, who discover the underlying causes and consequences. In general, changes in LULC fall into two broad categories: conversion and modification [31].

According to Sparfel [30], "land use and occupancy and their evolution over time are at the heart of the terrestrial system, which can be broken down into two sub-systems: social (or anthropogenic) and environmental (or ecosystemic)." LULC lies at the interface and results from numerous interactions between social and environmental systems that evolve over time". Studying the evolution of land use provides a better understanding of human relationships with the environment [27, 32]. The terrestrial system influences each sub-system, each of which interacts with the other and, in turn, affects the terrestrial system. Indeed, climatic or biophysical phenomena likely to modify LULC are disrupted or amplified by human actions designed to exploit ecosystems to extract goods and services. Conversely, any change in LULC patterns impacts the processes that control ecosystems, leading to ecosystem modifications and alterations of the goods and services from which humans benefit.

However, the changes from the last fifty years are unprecedented [19, 27]. Tropical deforestation, pasture expansion, agricultural intensification, and urbanization have been identified as the main types of LULC change observable on a global scale [19]. Nevertheless, those changes are highly diverse in nature and scope, with causes and impacts varying in intensity, time, and space. This complexity explains why they continue to elude attempts at theorization [19] and modeling [32].

Following the absence of an evolutionary mapping approach at the Metropolitan scale of Al-Ahsa, this paper aims to:

- Combine multi-source and multi-date data and set up an up-to-date database on land cover in this rapidly changing area.
- Cartographically represent the recent evolution of land use and land cover (LULC) in the Al-Ahsa oasis.
- Describe the organization and spatial distribution of land use and land cover (LULC) in four classes (agriculture, bare soil, built-up areas, and hills).
- Propose a prototype of land use and land cover (LULC) map based on the strict rules of graphic semiology.

2 Research Methodology and Data Collection

2.1 Selection of the Study Area: Al-Ahsa Metropolitan Area

Administratively, Saudi Arabia is divided into six main regions (East, Central, North, North-West, Central-West, and South-West) subdivided into 13 provinces (locally called *Imarats* or *Manatiq*) containing the 136 governorates (locally called *Mouhafadhat*) (Fig. 1).

With a surface area of almost 2.15 thousand km^2, Saudi Arabia is the thirteenth largest country in the world and the second largest in the Arab world. Only a few

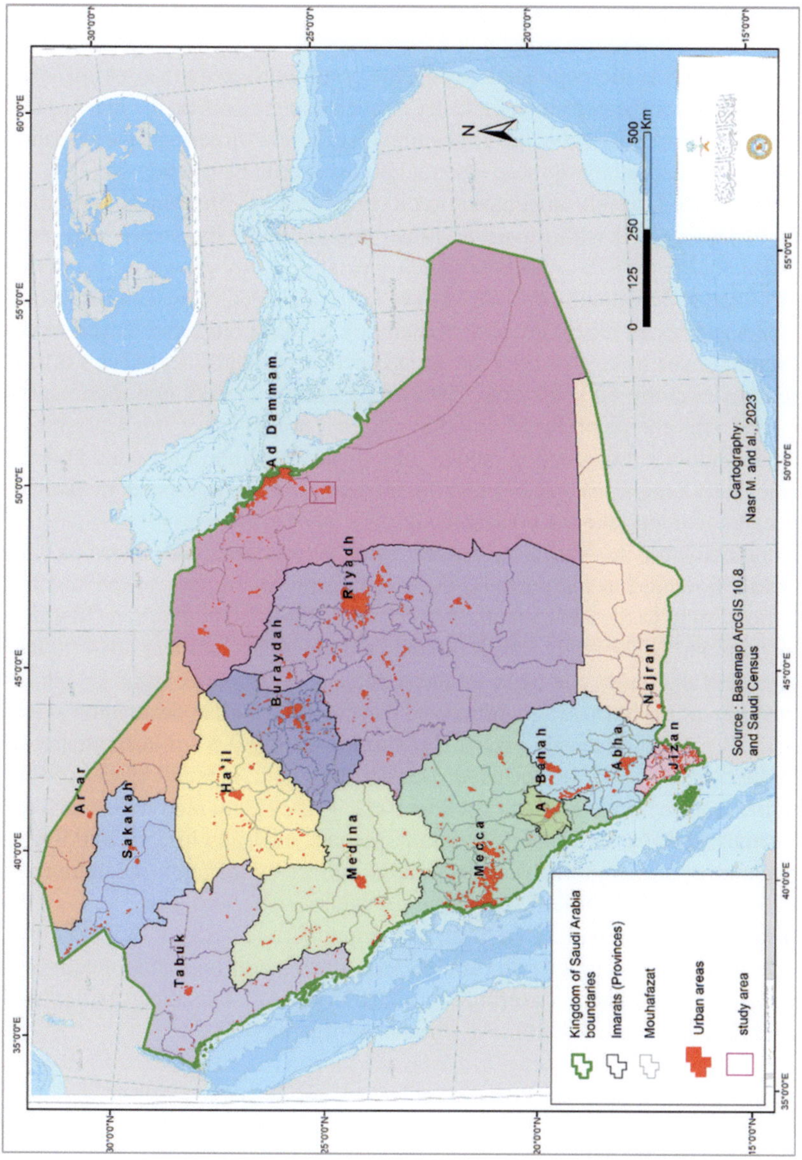

Fig. 1 The various administrative levels of Saudi Arabia (*Source* Central Department of Statistics and Information). Cartography: Nasr et al. 2023

areas of this immense country are populated, particularly the urban areas; about 91% of the total population lives in cities. The five major cities—Riyadh, Jeddah, Mecca, Dammam, and Medina—account for less than 10% of the country's surface area but are home to over 50% of the population. The remaining 50% is scattered over the remaining 90% of Saudi territory [14].

Because of severe natural and climatic conditions, a few areas of the Kingdom of Saudi Arabia (KSA) are populated: towns, the oasis, cultivated areas, or artificial communities built for the petroleum industry or at crossing points on the main roads (Figs. 1 and 2). When portraying the population distribution by cartographic means, and whatever the method used, these essential points should be considered [15].

Saudi Arabia is currently undergoing rapid urbanization. The number of large cities has increased, and with them, the urban population. It has many large and medium-sized cities, various aspects of which specialists study. However, medium-sized cities such as Dammam or Al-Ahsa have been neglected by researchers from a LULC perspective. In recent decades, the city of Al-Ahsa has experienced rapid growth, and land use in its Metropolitan Area is evolving rapidly. Al-Ahsa's urban population has risen by 600,000 from 1990 to 1.3 million in 2022 (an increase of 113.8%). Thus, the choice of this Metropolitan Area is not arbitrary. It is representative of the Kingdom's fast-growing medium-sized cities. In addition, sound knowledge of the study region was taken into consideration. All these aspects motivated the researchers to undertake the present study.

Geographically, the Al-Ahsa Metropolitan Area is located in the eastern part of Saudi Arabia, bordered to the north by the province of Al-Buqayq, to the east by the Al-Jafura desert, to the west, and the south by the Rub' Al-Khali desert (literally the "empty quarter"). The study area is limited by many roads: number 6614 (to the north and east of the oasis) and 6446 (to the south and west of the oasis).

The area (territory) has been characterized by sedentary human occupation since Neolithic times and by the largest oasis in the world, with over 2.5 million palm trees. The oasis landscape, which has evolved over the millennia, presents a way of life typical of the Gulf region of the Arabian Peninsula. This cultural landscape includes gardens, canals, springs, wells, an agricultural drainage lake, and historic buildings. Al-Ahsa's ancient and contemporary landscapes reflect the different evolutionary phases of the oasis and the interaction of cultural and natural heritages.

Al-Ahsa province is a part of the Eastern Region. It covers about 530,000 km^2, 24% of the total area of Saudi Arabia and 68% of the Eastern Region. Its major big cities are Hofuf and Mubarraz, covering about 22,898 km^2 [2]. The other small urban centers include Al-Oyoun, Al-Jafar, Al-Omran, and Juatha (Fig. 3).

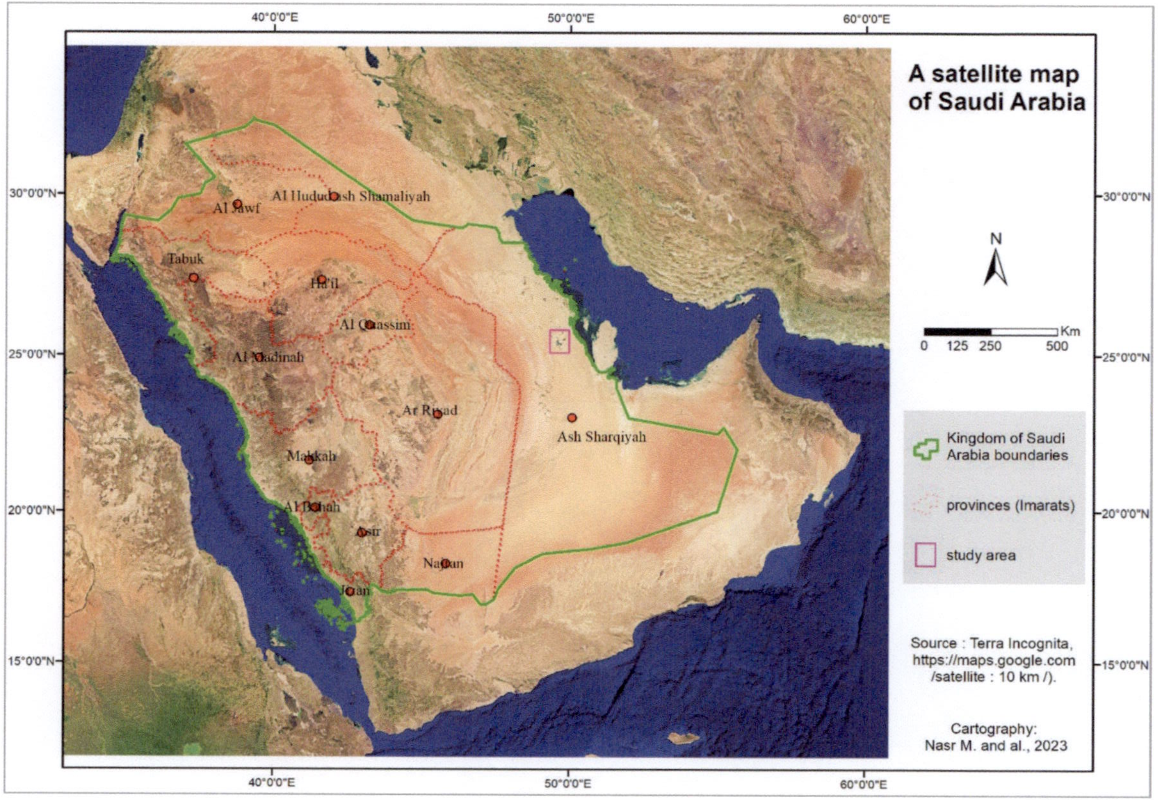

Fig. 2 The greatest desert part of the Saudi territory and the spreading red points corresponding to main cities and localities (*Source* of base map Arc GIS 10.8) Cartography: Nasr et al. [26]

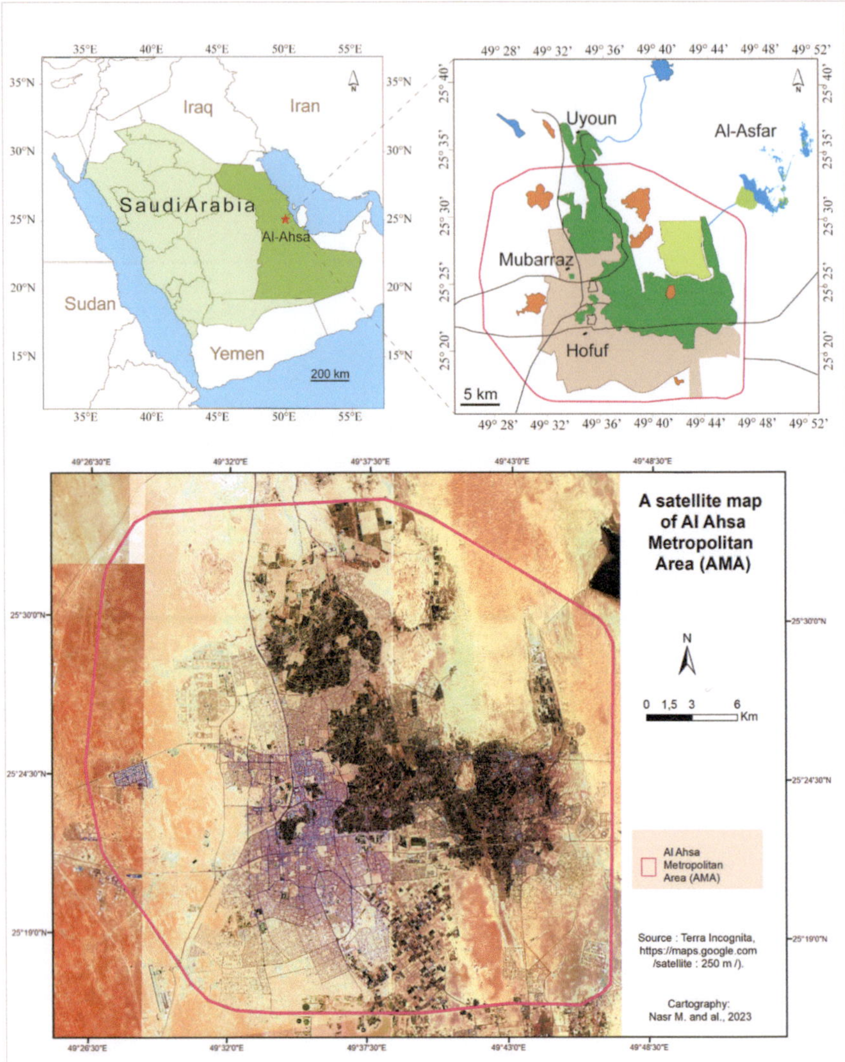

Fig. 3 Location of AMA in Saudi Arabia and a cross scalar of the study area (a satellite map in 2022). Cartography: Nasr et al. [26]

2.2 *Data Pre-Processing*

The present paper was conducted to assess the development of LULC classes around Al-Ahsa Oasis by means of GIS and RS. The selected images were enriched, captured and geo- referenced. Using auxiliary field data, topographic maps, and Terra Incognita imagery, geometric enrichment and data validation were performed. Data acquisition was followed by data processing. The images were geo-referenced and calibrated, which was essential for diachronic analysis. The images were geometrically rectified and projected into the UTM/WGS 84/ZONE 39 North cartographic projection system. The rectified 2022 images were used as a reference for the image-to-image calibration of the previous images. To validate the geometric corrections, the images were overlaid with considerable data on the screen to verify the overlay of notable elements and infrastructures on the site or nearby (road, vegetation, buildings).

A satisfactory overlay of the images allows the extraction of different layers in order to define the landscape units and study their evolution in different periods (1990 and 2022). The Landsat images were downloaded from USGS available at https://earthexplorer.usgs.gov, and used to evaluate the LULC change (Table 1).

Information about LULC transformations is an essential starting point for resource management and vulnerability assessment of the oasis, and a well-designed methodology was used for this study. The LULC maps were produced according to a supervised classification using ENVI 5.7 software. Supervised image classification using the maximum likelihood algorithm was performed. In maximum likelihood, the set of values of the model parameters that maximize the likelihood functions are selected. The present study used it as it is the most popular approach for analyzing RS image data. According to Jensen, it requires a collection of firsthand observations, aerial image interpretation, map assessment, and a field survey to recognize and locate predefined LULC types. Those improved images were used to supervise classification using the maximum likelihood classifier technique, with signatures gathered using the spectral response pattern of each LULC class as a reference. Density slicing was also used to capture the threshold pixel of a few classes (Fig. 4).

In the next step, both classified LULC images were overlapped to determine pixel-based changes in the mainland cover classes. Then, both images were compared in an arc map using the raster calculator for Arc GIS 10.8. After converting the raster LULC images into the vector, changes in LULC classes were evaluated using vector data analysis techniques such as union, intersect, and symmetrical difference. Table 2 elaborates on this research methodology for LULC dynamics detection more broadly.

Table 1 Details of the Landsat images analyzed in the study

Satellites	Spatial resolution (m)	Acquisition period	Bands used	Number of bands	Source
Landsat 5 TM	30	07/1990	2-4-5	07	USGS
Landsat 8 OLI	30–15	08/2022	2-4-5	11	USGS

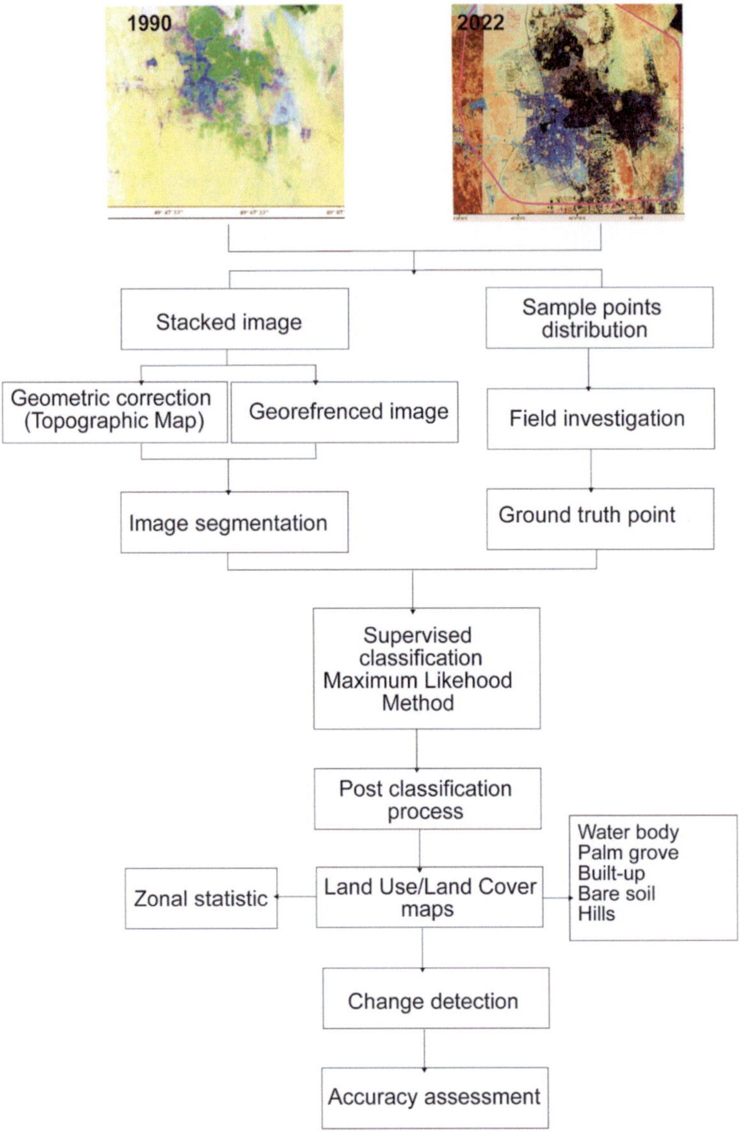

Fig. 4 Methodology of developing LULC maps by using GIS and Landsat data. (conception and realisation Monaem NASR and Walid [11])

Table 2 Details of the Landsat images analyzed in the study

LULC classes	Description
Built-up	Area covered by urban settlements, roads, industries, infrastructure
Agriculture	Area covered by date palm farms (Oasis)
Bare soil	Area Open ground /and areas with no vegetation cover
Hills	Areas covered by raised knolls in relation to surrounding land
Water bodies	Area completely covered by water such as sabkhas and lakes

3 Results and Comments

3.1 Factors of LULC Distribution of Al-Ahsa Metropolitan Area

In recent decades, Saudi Arabia has been no exception. As a result of the socio-economic development that prevailed throughout the Kingdom, many urban and preurban centers have experienced rapid population growth. According to the United Nations Development Program [33], the proportion of the population living in urban areas in Saudi Arabia was 58.4% in 1975, 86.6% in 2001, and 91% in 2016. This rapid growth has increased the size of cities and mirrored a change in land use, particularly in periurban areas [1]. A comparison of the state of LULC was carried out in the Al-Ahsa Oasis, which is considered the largest agricultural oasis in Saudi Arabia and probably the largest irrigated oasis in the world [5]. This L-shaped oasis is located about 70 km from the west coast of the Arabian Gulf. It lies about 170 m above mean sea level to the west and 120 m to the east. There are 10 towns and 60 villages within the oasis [6, 25].

The economic history of the Al-Ahsa Oasis is closely linked to agricultural practices, as the oasis produces a considerable proportion of Saudi Arabia's dates. On the other hand, due to the early discovery of oil in the Eastern Region in 1930, the Al-Ahsa Oasis underwent a considerable increase in population and a change in LULC [1, 6], including a notable urban sprawl at the expense of the ancient oasis core. Since Saudi Arabia's transition to an oil-based economy, the authority has constructed basic infrastructure. This attracted large numbers of workers from inside and outside the Kingdom, which accelerated demand for housing and services. This led the local authorities to plan several new housing projects to meet the urgent demand for housing.

Resource availability, mainly fertile soil and groundwater, is critical in early establishment. Improved economic conditions, easy access to funds, and a good infrastructure network mainly drive the growth. Economically, the region is one of the Kingdom's leading agricultural areas, home to a wide range of economic activities, including agriculture, commerce, tourism, and industry.

In recent decades, it has experienced rapid economic development and has become one of the growth centers in the Eastern Region, underpinning regional development.

Due to its rapid urban development over the last few decades, land use patterns in the region have changed, by invasion or replacement. For example, the development of economic activities and the provision of better services have pushed certain activities to the area's periphery. Secondly, a new type of land use, commercial activity, has emerged. In addition, the creation of King Faisal University in 1975 in its south and Al-Ahsa airport in the southwest played a role in accelerating urban growth. The region's rapid urban expansion has led to the emergence of shopping centers to service the new districts. Many shopping malls, such as Al-Ahsa Mall and Al-Othaim Mall, prefer to set up on the outskirts of cities and along main roads far from the centers. This could be due to the urban center's congestion, noise pollution, and rising rents and land prices.

Local authorities have tried to increase agricultural efficiency by increasing cultivated areas [4]. However, inadequate planning, poor agricultural management, and urban growth have led to radical changes in the LULC. Until recently, many neighborhoods were dotted with farms and green spaces. Now that urban sprawl is invading farmland to the east, many farms have been converted to residential or commercial use. Declining farm incomes and water scarcity are the factors behind this conversion. A study to describe the spatial variations of the LULC in the oasis is needed.

3.2 *LULC Change in Al-Ahsa Oasis*

This section highlights the trends and causes of the unprecedented urban growth in recent decades. In a study by Abdullah et al., LULC analysis of three major cities in Saudi Arabia (Riyadh, Jeddah, and Mecca) revealed that the urban area represented the most altered surface cover and that most of the surface transformed into the urban area came from bare soil between 1985 and 2014. Although a massive decrease in farmland was observed in their analysis, most of the change in farmland was bare soil due to declining water resources. Statistics indicate that the agricultural area of the ancient core of Al-Ahsa Oasis decreased to 33% of its total area in 2009 and to 25% in 2011 [3]. The author also asserted that urban growth would negatively affect the local environment and social life of the inhabitants without proper planning in the future.

To detect changes in LULC, post-classification comparisons (PCCs) have been widely used [12, 13, 24, 36]. They examine changes over time between independently classified LULC data based on two multispectral satellite images acquired in 1990 and 2022. The PCC technique is the simplest method for detecting changes. It depends on comparing independently produced classified images by correctly coding the classification results for the first and second times, and the output results in a change map indicating a complete change matrix. There are two advantages of PCCs. They avoid strict radiometric calibration and minimize the impact of atmospheric, sensor, and environmental differences between multi-temporal images. They also provide a complete matrix of change directions, unlike image differentiation.

In our post-classification process, the image differentiation technique was applied to both images. Change detection statistics provided a detailed picture of the changes between the two classified images, and its statistical report includes a class-by-class image difference. The analysis focuses primarily on changes in the classification of the initial state. Thus, for each initial state class, the study identifies the classes in which the corresponding pixels have changed in the final state image [7].

The change was calculated by applying segmental subtraction for the two classified images representing specific dates. The thematic maps generated after classification are subjected to a PCC operation to visualize and describe the process of urban sprawl by identifying the different changes in the LULC. The confusion matrices generated for both maps showed a satisfactory level of accuracy. The Kappa index showed an acceptable level of accuracy, with values of 85 and 89 for 1990 and 2022, respectively.

Change detection led to the production of a spatiotemporal map illustrating the evolution of the Al-Ahsa Oasis between 1990 and 2022. A comparison of the two maps shows an apparent growth in urbanized areas (Fig. 4). Spatially, the urban sprawl has taken place in three directions: to the north, by encroaching on areas belonging to the palm grove; to the south, along main roads, notably the Al-Khalij road; and to the east, around urban centers in the eastern part of the palm grove, where a process of urban conurbation has been observed, particularly in Al-Halila and Al-Mekdam. In its central part, the palm grove has gradually shrunk and benefited the built-up area via two processes: firstly, the occupation of degraded interstitial surfaces within the palm grove, and secondly, the encroachment of the cultivated surface (Fig. 5).

Figure 5 shows the extent to which built-up areas have grown in relation to vegetation. The built-up area has increased from 137 km^2 (137,000 ha) in 1990 to 373 km^2 (373,000 ha) in 2022, which is an increase of 236 km^2, an estimated annual change of 7.5 km^2. In its central part, the palm grove has undergone the same process. Its surface area increased from 184,000 ha (184 km^2) in 1990 (Fig. 6) to 211,000 ha (211 km^2) in 2022 (Fig. 7), with a gain of 27,000 ha or 0.84 km^2 per year. The bare soil class has thus lost a large part of its surface area, including 267 km^2 converted by the other classes.

Table 3 summarizes statistics for the areas acquired for each surface cover class during the study period. The calculated values (ha) for each class were also produced as the difference between the initial state "t1" (1990) and the subsequent state "t2" (2022), to reveal the final state for each class. The resulting classification revealed that the urban area class had the highest change in surface cover (27.8%). The bare soil class area change was also very high, followed by the vegetation class. The area of the vegetation class was estimated at 184 km^2 in 1990 and reached 211 km^2 in 2022; a gain of 27 km^2 over the other classes. The urban area, estimated at 137 km^2 in 1990, reached 377 km^2 in 2022 due to urban sprawl, thus exceeding the other classes by almost 240 km^2. This increase reflects the continuing growth of the population and its efforts to settle in green areas, as well as other social and economic considerations [8] (Table 3).

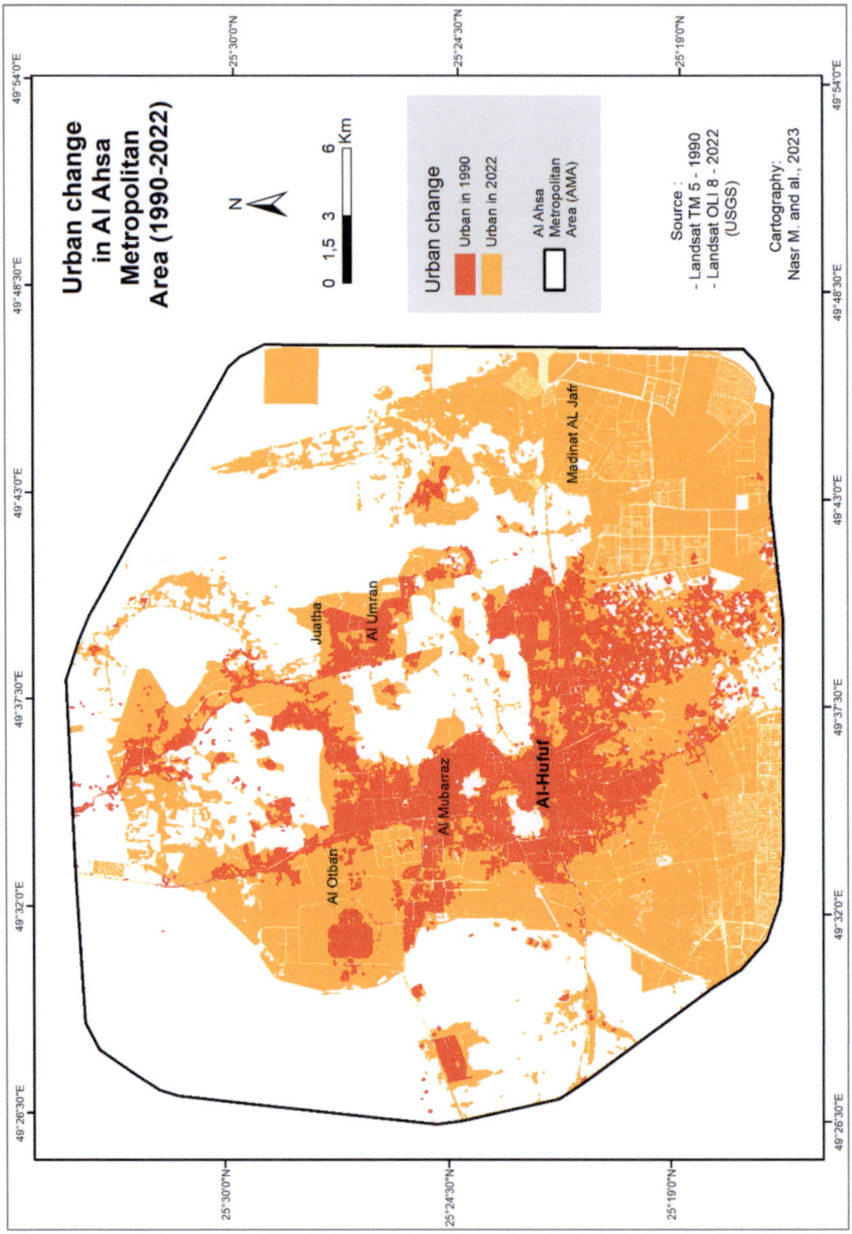

Fig. 5 Urban area change of AMA between 1990 and 2022. Cartography: Nasr et al. [26]

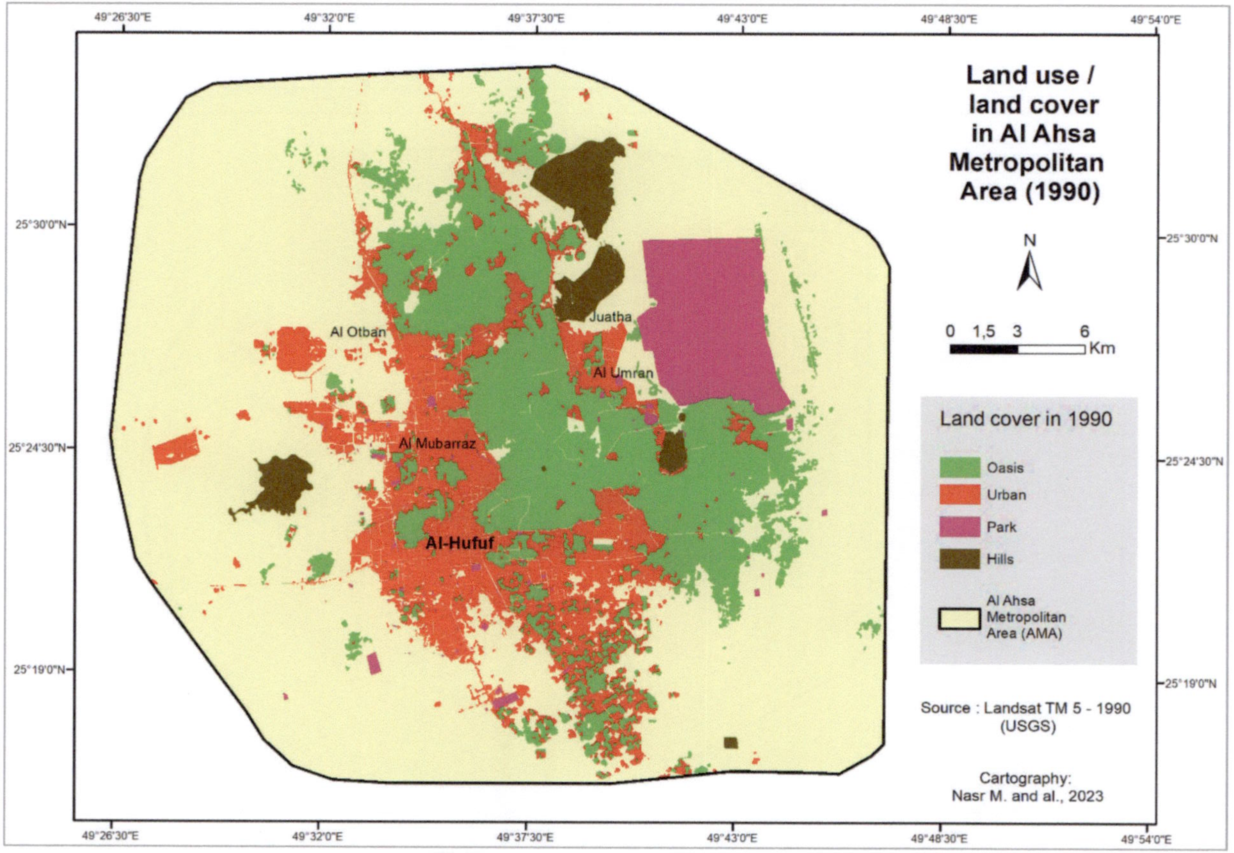

Fig. 6 Changes in the AMA's LULC between 1990 and 2022. Cartography: Nasr et al. [26]

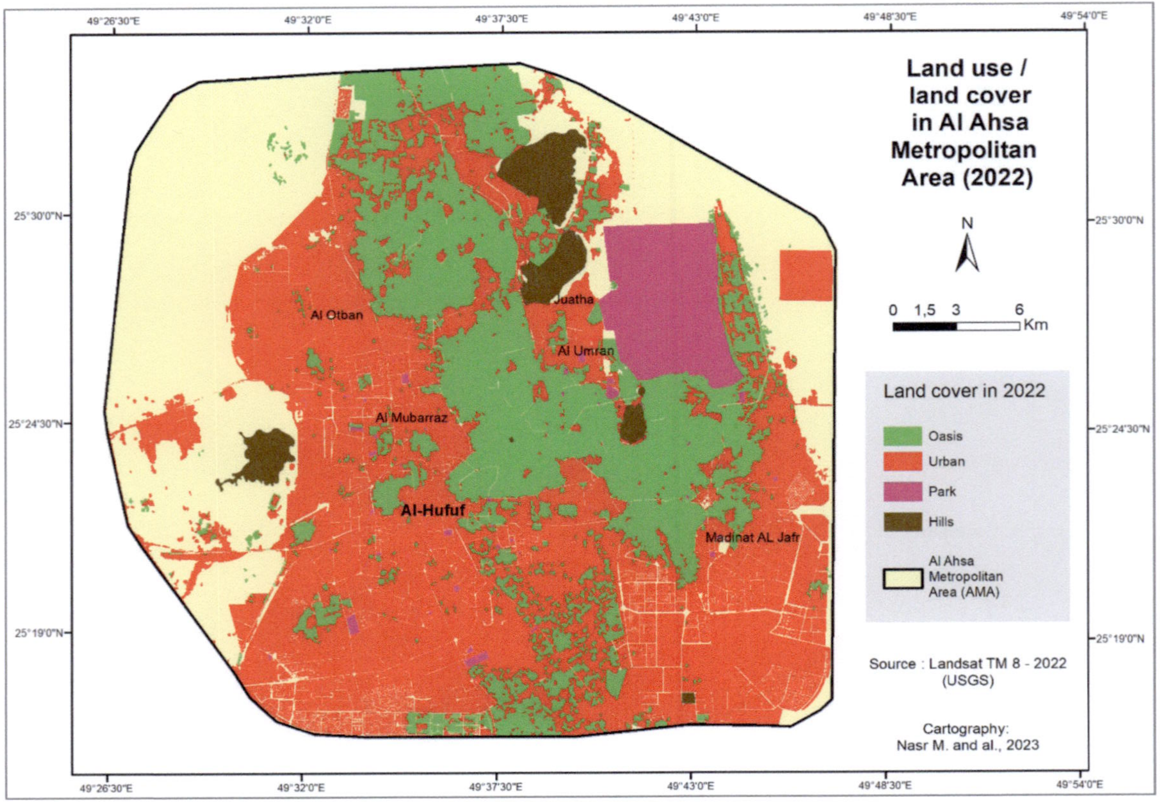

Fig. 7 Changes in the AMA's LULC between 1990 and 2022. Cartography: Nasr et al. [26]

Table 3 Area of LULC classes for the study area from 1990 to 2022

Classes	1990		2022		Area (Ha)	Gain/Loss
	Ha	%	Ha	%		
Hills	25,000	2.9	25,000	2.9	0	Gain
Urban areas	137,000	15.9	377,000	43.7	+240,000	Gain
Vegetation (Oasis)	184,000	21.3	211,000	24.4	+27,000	Gain
Bare soil	516,000	59.8	249,000	28.8	−267,000	Loss
Total	862,000	100	862,000	100	0	

Table 3 shows that the urban class has undergone significant change, with 43.7% of all oasis land occupied by this class in 2022, compared with 15.9% in 1990. A continuous decrease in vegetation cover, estimated at 7 km^2 of the total area of the old oasis core, was observed from 1990 to 2022. Surface gain/loss values (km^2) are given for the urban area and vegetation cover (Figs. 6 and 7). Table 3 also shows that over the study period, almost all urban sprawl occurred on the formerly vacant bare soil class in the oasis and on the periphery of the urban agglomeration with 240 km^2, while only 27 km^2 of the oasis vegetation zone was occupied by the bare soil class. Abdelatti et al. [1], highlighted the threat posed by urban growth in the Al-Ahsa Oasis to the local environment and explained that such urban growth in the area without sound planning in the future would have negative implications for the local environment and social life (Figs. 8 and 9).

4 Discussion

We presented LULC change analysis of Al-Ahsa Metropolitan Area in KSA over the last 32 years. Our findings of increasing urban areas and decreasing ancient oasis in the metropolitan area corroborated the findings of previous studies; the Al-Ahsa Oasis underwent a considerable increase in population and a change in LULC [1, 6], including a notable urban sprawl at the expense of the ancient oasis.

Since 1970, the cities of Hofuf and Mubarraz have experienced rapid population growth due to socio-economic development and the significant expansion of the productive sectors. According to Al-Ahsa [2], the population of the two main cities, covering an area of about 23,000 km^2, rose from 444,970 in 1992 to 768,000 in 2016, and led to increased built-up areas. As a result, most residents of the old districts have moved to new districts on the area's outskirts [25]. Al-Hofuf merged with Al-Mubarraz. Other small urban centers include Al-Oyoun, Al-Jafar, Al-Omran and Juatha. The directions of urban expansion in the study area are almost in all directions, but with high speed toward the north and south. This is due to the presence of palm groves to the east, and the construction of the Saudi National Guard, Arabian-American Oil Company (ARAMCO), a railway line and the Hofuf airport to the west. The total built-up area reached almost 28,000 ha or 280 km^2. The total population of Al-Ahsa was 1,500,000 in 2022.

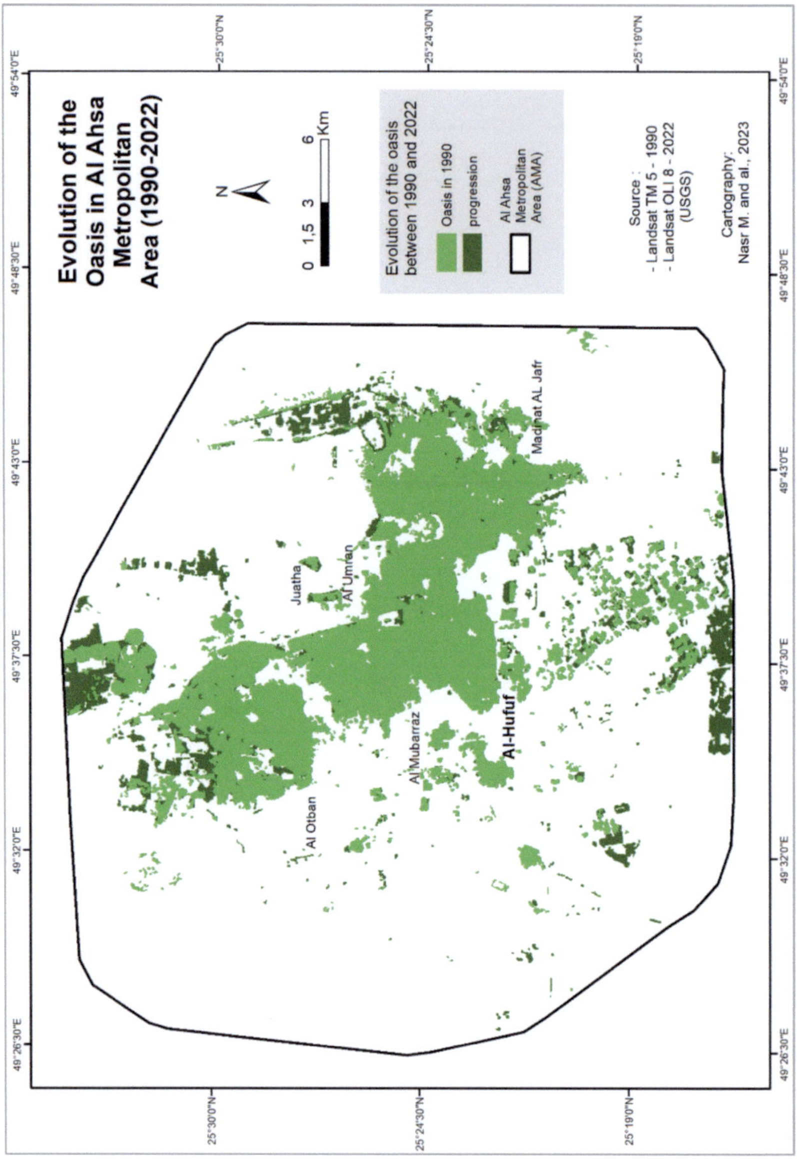

Fig. 8 Evolution of Al-Ahsa Oasis between 1990 and 2022. Cartography: Nasr et al. [26]

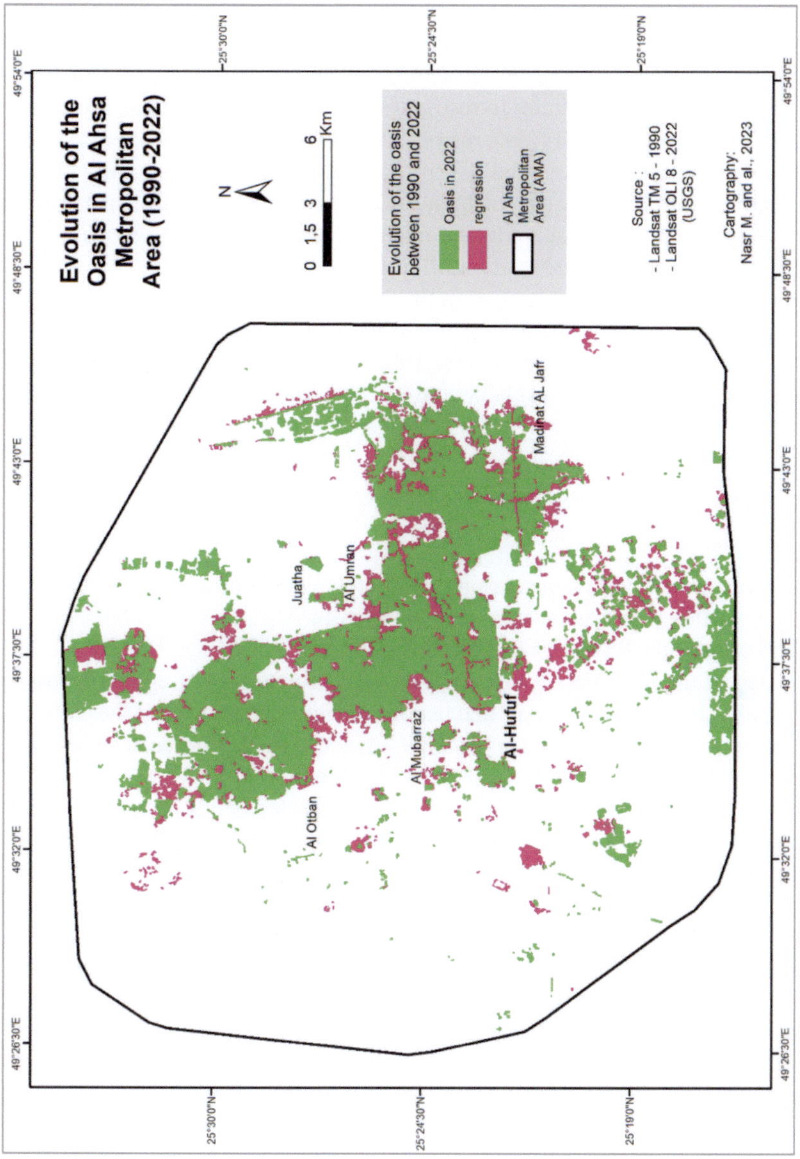

Fig. 9 Evolution of Al-Ahsa Oasis between 1990 and 2022. Cartography: Nasr et al. [26]

On the contrary of the ancient oasis, the analysis of the oasis' vegetation cover showed that a significant increase in vegetation cover occurred at the expense of bare soil between 1990 and 2022, estimated at 184 km^2 versus 211 km^2, which could be due to the inclusion of cultivated projects located in the northern and southern parts of the former oasis (plus 27 km^2). The vegetation cover class lost an area of about 3 km^2 to the bare soil class, in addition to the 4 km^2 areas occupied by the urban class.

In addition, as it is seen that the growth of urban areas is at the cost of Bare soil land and ancient oasis policies to restrict of this agricultural heritage should be prioritized. The results quantify patterns of land cover change in the metropolitan area. They demonstrate the potential of multi-temporal Landsat data to provide an accurate and cost-effective means of mapping and analyzing land cover changes over time, which can be used as inputs for land management and policy decisions.

Despite the adoption and integration of an integrative and global methodology, this study has some inherent limitations of the data. First, the 30 m spatial resolution of Landsat imagery hinders the accuracy and range of classified maps. Secondly, the lack of high-resolution imagery for the 1990s has limited accuracy, whereas for 2022, images like Sentinel are available to map large-scale changes.

5 Conclusion

"Land use and land cover change is one of the top agenda of the current world". In the context, this paper assesses LULC changes in Al-Ahsa Metropolitan Area in Saudi Arabia using RS data and GIS tools, and the LULC layers represent the digital image of the Al-Ahsa Metropolitan Area classified as built-up, agriculture, bare soil, and hills. The study reveals that major changes were detected in the built-up and agricultural land use classes. Our findings show that there have been significant changes in LULC between 1990 and 2022 when the built-up areas experienced a considerable increase. On the other hand, bare soil decreased, while agricultural land (oasis) simultaneously increased. The study showed that the increase in vegetation cover occurred in the northern part of the study area, while it was mainly concentrated towards the southern and northern areas. The irrigation and drainage project has contributed to improving the quality of the ecological environment of the oasis.

It would be advisable to coordinate the relationship between development and protection, harness the city's internal potential, achieve connotative growth, and avoid oil-stain expansion. In other words, the future urban areas of Saudi Arabia cities are already experiencing unprecedented spatial growth, which must be addressed adequately. With the continuous development of industrialization, urbanization, and salinization of agricultural land in Saudi Arabia, the degradation of the ecological environment is one of the most important environmental problems. Quantitative assessment and long-term analysis of the change in Saudi Arabia's environment can shed light on the driving factors of such change.

6 Recommendations

Several recommendations are given below regarding the design of the LULC for future use in the Al-Ahsa Metropolitan Area. This study has shown that urban areas will increase in the coming years due to overpopulation and the rapid increase in urban roads.

This situation will have different but dangerous effects on the territory of the Al-Ahsa Metropolitan Area due to various difficulties, like overcrowding, traffic, congestion, and a lack of facilities. It is recommended that people build outside the urban area, for example, in the south of the Al-Ahsa Metropolitan Area, by distributing the incentives available in the main city.

The current map should be updated using RS techniques and data. This will provide developers with a consistent, highly accurate LULC map and improve employment policy. The government is encouraged to use the capabilities of RS and GIS technology for mapping to provide adequate and reliable spatial information and valuable data to develop effective management and monitoring of LULC changes in the Kingdom of Saudi Arabia.

Acknowledgements This work was supported by the Deanship of Scientific Research, Vice Presidency for Graduate Studies and Scientific Research, King Faisal University, Saudi Arabia [Grant No. KFU 251350].

References

1. Abdelatti H, Elhadary Y, Babiker AA (2017) Nature and trend of urban growth in Saudi Arabia: the case of Al-Ahsa Province–Eastern Region. Resour Environ 7(3):69–80. https://www.researchgate.net/publication/316923401_Nature_and_Trend_of_Urban_Growth_in_Saudi_Arabia_The_Case_of_Al-Ahsa_Province_-_Eastern_Region
2. Al-Ahsa Municipality (2016) Al-Ahsa structural plan (Hofuf & Mubarraz). Al-Ahsa, Saudi Arabia
3. Al-Arfaj AM (2012) The recreational utilities and its impact on agricultural sustainability in Al-Ahsa Oasis. A paper presented on the sustainable agriculture forum in Al-Ahsa Oasis, December 2012, Al-Ahsa Chamber, Al-Ahsa Province, Saudi Arabia
4. Al-Shayaa MS, Baig MB, Straquadine GS (2012) Agricultural extension in the Kingdom of Saudi Arabia: difficult present and demanding future. J Anim Plant Sci 22:239–246
5. Al-Tahir AAS (1999) Al-Hassa: geographic study, 1st edn. A-Hussainy Modern Press, Al-Hassa, Eastern Province, KSA, p 1999
6. Almadini AM, Hassaballa AA (2019) Depicting changes in land surface cover at Al-Hassa oasis of Saudi Arabia using remote sensing and GIS techniques. PLoS ONE 14(11):e0221115. https://doi.org/10.1371/journal.pone.0221115
7. Almutairi A, Warner TA (2010) Change detection accuracy and image properties: a study using simulated data. Remote Sens 2(6):1508–1529
8. Alqarni S, Babiker A, Salih A (2018) Detection, mapping and assessment change in urban and croplands area in Al-Hassa Oasis, Eastern Region in Saudi Arabia using remote sensing and geographic information system. J Geogr Inform Syst 10:659–685

9. Auch RF, Wellington DF, Taylor JL, Stehman SV, Tollerud HJ, Brown JF, Loveland TR, Pengra BW, Horton JA, Zhu Z et al (2022) Conterminous United States land-cover change (1985–2016): new insights from annual time series. Land 11:298. https://doi.org/10.3390/land11020298
10. Chouari W (2021) Wetland cover change detection using multitemporal Landsat data: a case study of the Al-Asfar wetland, Royaume d'Arabie saoudite. Arab J Geosci 14:523. https://doi.org/10.1007/s12517-021-06815-y
11. Chouari W (2023) Spatiotemporal analysis of land cover changes in Al-Hubail Wetland (Kingdom of Saudi Arabia). J Indian Soc Remote Sens 51:585–599. https://doi.org/10.1007/s12524-022-01653-1
12. Coppin P, Jonckheere I, Nackaerts K, Muys B, Lambin E (2004) Review article digital change detection methods in ecosystem monitoring: a review. Int J Remote Sens 25:1565–1596
13. Deng JS, Wang K, Deng YH, Qi GJ (2008) PCA-based land-use change detection and analysis using multitemporal and multisensor satellite data. Int J Remote Sens 29:4823–4838
14. Dhieb M (2022) A bivariate dasymetric population map of Saudi Arabia. Curr Urban Stud 10:673–696. https://doi.org/10.4236/cus.2022.104040
15. Dhieb M, Yengui T, Nasr M (2022) Approaching the Tunisian human environment by using RS and the dasymetric method. In: Khebour Allouche F, Negm AM (eds) Environmental remote sensing and GIS in Tunisia. Springer Water, Springer, Cham. https://doi.org/10.1007/978-3-030-63668-5_2
16. Dubertret F, Le Tourneau FM, Villarreal ML, Norman LM (2022) Monitoring annual land use/land cover change in the Tucson metropolitan area with Google Earth Engine (1986–2020). Remote Sens 14:2127. https://doi.org/10.3390/rs14092127
17. Hansen MC, Egorov A, Potapov PV, Stehman SV, Tyukavina A, Turubanova SA, Roy DP, Goetz SJ, Loveland TR, Ju J et al (2014) Monitoring Conterminous United States (CONUS) land cover change with web-enabled Landsat data (WELD). Remote Sens Environ 140:466–484. https://doi.org/10.1016/j.rse.2013.08.014
18. Homer C, Dewitz J, Jin, SXian G, Costello C, Danielson P, Gass L, Funk M, Wickham J, Stehman S et al. (2020) Conterminous United States land cover change patterns 2001–2016 from the 2016 national land cover database. ISPRS J Photogramm Remote Sens 162:184–199. https://doi.org/10.1016/j.isprsjprs.2020.02.019
19. Lambin EF, Geist HJ, Lepers E (2003) Dynamics of land use and cover change in tropical and subtropical regions. Annu Rev Environ Resour 28:205–241. https://doi.org/10.1146/annurev.energy.28.050302.105459
20. Lark TJ, Mueller RM, Johnson DM, Gibbs HK (2017) Measuring land-use and land-cover change using the U.S, Department of Agriculture's Cropland data layer: cautions and recommendations. Int J Appl Earth Obs Geoinf 62:224–235. https://doi.org/10.1016/j.jag.2017.06.007
21. Latifovic R, Homer C, Ressl R, Pouliot DA, Hossian S, Colditz R, Olthof I, Chandra G, Victoria A (2012) North American land change monitoring system. In: Giri CP (ed) Remote sensing of land use and land cover: principles and applications. CRC Press, Boca Raton, FL, USA, pp 303–324
22. Lepers E, Lambin EF, Janetos AC, DeFries R, Achard F, Ramankutty N, Scholes RJ (2005) A synthesis of information on rapid land-cover change for the period 1981–2000. Bioscience 55(2):115–124
23. Li X, Zhou Y, Zhu Z, Cao W (2020) A national dataset of 30 m annual urban extent dynamics (1985–2015) in the conterminous United States. Earth Syst Sci Data 12:357–371. https://doi.org/10.5194/essd-12-357-2020
24. Lu D, Mausel P, Brondízio E, Moran E (2004) Change detection techniques. Int J Remote Sens 25:2365–2401
25. Municipality of Al-Ahsa (2018) Al-Ahsa creative city of crafts and folk art. A member of united nations educational, scientific and cultural organization (UNICECO) creative cities network since 2015. https://unesco.alhasa.gov.sa/En/Details.aspx?ArticleID=3138

26. Nasr M, Zenati H, Dhieb M (2022) Using RS and GIS to mapping land cover of the cap bon (Tunisia). In: Khebour Allouche F, Negm AM (eds) Environmental remote sensing and GIS in Tunisia. Springer Water, Springer, Cham. https://doi.org/10.1007/978-3-030-63668-5_7
27. Ojima D, Moran E, McConnell W, Stafford Smith M, Laumann G, Morais J, Young B (2005) Global land project, science plan and implementation strategy. IGBP report n°53 - IHDP report 19, 65 p
28. Olorunfemi JF (1983) Monitoring urban land use in developing countries—an aerial photographic approach. Environ Int 9(1):27–32
29. Parveen S, Basheer J, Praveen B (2018) A literature review on land use and land cover changes. Available from: https://www.researchgate.net/publication/327008642_A_LITERATURE_REVIEW_ON_LAND_USE_LAND_COVER_CHANGES
30. Sparlef L, Le Berre I, Gourmelon F (2011) Assessment of land use change in coastal areas using heterogeneous data—(Évaluation des changements d'occupation des sols en zone côtière à partir de données hétérogènes), Article 21(3):381–403
31. Stott A, Haines-Young R (1998) Linking land cover, intensity of use and botanical diversity in an accounting framework in the UK. In: Uno K, Bartelmus P (eds) Environmental accounting in theory and practice. Economy and environment, vol 11. Springer, Dordrecht. https://doi.org/10.1007/978-94-017-1433-4_15
32. Turner IIBL, Lambin EF, Reenberg A (2007) The emergence of land change science for global environmental change and sustainability. PNAS 104(52):20666–20671
33. United Nations Development Program (2003) Human development report: millennium development goals: a compact among nations to end human poverty, New York, USA
34. Villarreal M, Norman L, Wallace C, van Riper C, (2011) A multitemporal (1979–2009) land-use/land-cover dataset of the binational Santa Cruz Watershed. In: Open-file report 2011–1131; U.S. Geological Survey: Reston, VA, USA, p 26
35. Wickham J, Homer C, Vogelmann J, McKerrow A, Mueller R, Herold N, Coulston J (2014) The multi-resolution land characteristics (MRLC) consortium—20 years of development and integration of USA national land cover data. Remote Sens 6:7424–7441. https://doi.org/10.3390/rs6087424
36. Wu C, Du B, Cui X, Zhang L (2017) A post-classification change detection method based on iterative slow feature analysis and Bayesian soft fusion. Remote Sens Environ 199:241–255

GIS Contribution for Mapping of Flood-Hazard Areas in Arid Tunisia: Application to the Maknassy Basin (Lban-Ouadran Wadi Watershed, Tunisia)

Olfa Riahi

Abstract The floods in Tunisia pose a major risk to the environment, particularly soil and water resources. The country experienced several catastrophic events and that affected several regions, including the Maknassy Basin. Located in the communes that belong to Maknassy and Mazzouna, the basin of Maknassy (more than 1020 km^2) is drained from east to west by the Wadi Lban that represents itself a part of the watershed of the Lban- Ouadran wadi. The Maknassy Basin is a compact catchment area characterized by low slopes in the center and fairly steep slopes (always more than 15%) on the slopes that delimit it with a belt, often impermeable rock outcrops and relatively high drainage density. Thus, the physical characteristics of the Maknassy Basin make it an environment vulnerable to flood risk. This has led to a profound change in geomorphology of the interfluves and a change in the hydrological behaviour of wadis. These only increases the risk of flooding which can reach the disaster phase just like the situation in 1969 and 1989.In addition to material and infrastructure damage, hundreds of victims were also caused. This increases the risk (The danger or possibility of the occurrence of a hazard, affecting the human population, and the damage to the material and the environment) of flooding that can reach the disaster (It is the realization of increasing the risk of man-made losses on the basis of material and environmental losses) phase as was the case in 1969 and in1989 which have made a few hundred victims in addition to the material damage and in the infrastructure. This chapter focuses on the use and contribution of the GIS to the study and management of flood hazards in the Maknassy basin (the Lban-Ouadran wadi watershed, Tunisia) as part of a hydro-geomorphological study and with the aim of mapping to flooding hazard. The research methodology is based, in particular, on the analysis of the morphometric characteristics of the alluvial deposits, those of the river system and the terrain (slope and permeability) to identify the flood-hazard. The research methodology is based in particular on the analysis of the morphometric characteristics of alluvial deposits, those of the hydrographic network and the terrain

O. Riahi (✉)
Preparatory Institute for Literary Studies and Humanities of Tunis (IPELSHT), Tunis, Tunisia
e-mail: olfa.riahi@ipelsht.rnu.tn; olfariahi@gmail.com

F. Khebour Allouche (ed.), *Environmental Remote Sensing and GIS in MENA Region*, Earth and Environmental Sciences Library,
https://doi.org/10.1007/978-3-031-86574-9_14

(slope and permeability) to identify flood risks. Therefore, this analyse on the one hand shows the natural tendency of the Maknassy Basin to flood risk. On the other hand, the solution to this hazard is to protect the bottom of the cultivated basin (30% of the basin surface area), which is the most important agglomerations (such is the case of the Maknassy city, home to 4,215 people according to INS (National Institute of Statistics, Tunisia) estimates for 2024).

Keywords Flood · Hazard · Risk · Disaster · Lban-Ouadran Wadi · Maknassy · GIS · Tunisia

1 Introduction

Although they are generally located in arid climates, the countries of the MENA region can experience torrential rainfall. This results in exceptional flooding, which can be disastrous, causing damage to infrastructure (roads, housing, etc.), loss of water resources and run-off management systems, loss of soil through erosion of agricultural land, and loss of human life [3]. The fight against flooding in the MENA region requires scientific research into rural and/or urbanised catchment areas. This will make it possible to spatialise the flood hazard and provide decision-support tools for planners and decision-makers. Our study of the Maknassy basin (Oued Lban catchment) is part of this approach.. In Tunisia, flooding is a hazard that threatens water and soil resources [14]. The country has experienced several catastrophic events (1902, 1931–32, 1958–59, 1962, 1969, 1990, 2003, 2007, 2011, etc.) [18, 23] that have devastated several regions. Three of these events were devastating for central and southern Tunisia: those of 1932, 1969 and 1990 [2, 9]. The floods of 1969 were tragic, causing several hundred deaths and leaving several thousand people in misery, as well as material damage estimated at more than 30 million dinars [17, 25] and the bursting of the "stillborn" dam of the oued al Boua (Mazzouna region, central western Tunisia),) whose construction dates from the same year [26]. The 1990 floods caused more material damage, costing more than 90 million dinars, in addition to the hundreds of human lives lost [4, 7, 15]. This was despite the aridity that characterises the region. What's more, in arid climates, it is difficult to predict in advance the extent of climate-related hazards and the resulting consequences for lives and development.

Furthermore, the study of the morphometric characteristics of river impluviums draining floodplains [4, 7] is often combined with the analysis of human occupation features for the study and identification of flood-prone areas and the assessment of the degree of flood risk in wetland flood-prone areas. However, few studies have focused on the analysis of this environmental risk in arid areas where rainfall is scarce.

Furthermore, to deal with the risk of flooding, it is essential to implement sustainable water and soil management policies, to install hydro-agricultural schemes to mobilise soil resources and create a groundwater reserve, and to encourage environmentally-friendly farming practices. The fight against flooding in the MENA region requires scientific research into rural and/or urbanised catchment areas. This will make it possible to spatialise the flood hazard and provide decision-support tools for planners and decision-makers. Our study of the Maknassy basin (Oued Lban catchment area) is part of this approach. In this context, this chapter aims to study and map flood-related disasters (by analysing hydro-morphometric models, lithology and soil data), and to provide decision-support tools that can be used as a reference for risk management plan proposals.

The plan aims to protect and sustainably manage natural resources such as water and soil in the Maknassy basin (the catchment area of the Oued Lban-Ouadran). This is due to the use of GIS for the hydrological morphological diagnosis of its characteristics and its impact on the risk of flooding has adopted a method based on the identification and specialisation of alluvial deposits and arteries likely to generate floods according to the time of concentration of the flood, the peak flow of the flood, the runoff capacity and the jurisdiction.

2 Conceptual and Methodological Framework

2.1 Conceptual Framework

2.1.1 Hazards

Hazard is a natural phenomenon with uncontrollable occurrence and intensity [22], which can be characterized by three parameters: spatial frequency and temporal frequency. The result of a factor or process is at least partly beyond the control of human beings, and can only be transformed into a risk when there are human, economic, and environmental problems. In this context, the concept of hazard is of great significance in the study of major risks and is necessary in the field of land use planning. However, the assessment of the hazards (intensity, temporal proximity, frequency) of a given location does not predict possible damage or economic consequences in advance [22].

2.1.2 The Hydro-Morphometry of Impluviums and Its Impact on the Response and the Hydrographic Behavior of Wadis

Hydro-morphometric analysis is a tool of great interest for the prevention and the cartography of flood hazards (mapping, localization and specialisations the most vulnerable to the hazards of the floods is, on the one hand, a decision-making tool and, on the other, a tool for popularizing information to the local population [6, 8].

We used hydro-morphometric analysis to map the risk of flooding in the arid domain. It is based on two axes: the first is hydrological analysis. The second is the geomorphological study of the impluviums of the arteries 1. The latter allows the study of the form of alluvial deposits in the basin as a rainwater collection area.

Hydrology, on the other hand, allows the diagnosis of the architecture of the river system and its impact on the rate of concentration of runoff and, therefore, on the floods and consequently on the floods and their extent [10]. The combination of the results of analysis makes it possible to associate the impact of the form of the impluvia with the impact of the characteristics of the river system to the risk of flooding [11, 12].

Therefore, morphometric analysis is essential to assess the impact of the first-level watershed characteristics on runoff and the impact on erosion risk. In addition, the study of watershed characteristics focuses on the quantitative analysis of river systems [13, 16] to determine their hydrological behavior during floods. The influence of morphometric characteristics is explained in the Table 1.

In other words, the research of hydro-morphometry (see appendix n°1) of the Meknassy basin (Lban-Ouadran Wadi Watershed) involves analyzing the interaction between the morphometric parameters (see appendix n°1) of a catchment area and the quantitative characteristics of the river system (see appendix n°1)). The flood and its direct impact on flow characteristics (flow rate, concentration rate), especially when floods occur, and, therefore, the generation and severity of flooding regardless of the intensity and volume of precipitation.

2.1.3 Geomorphological Characteristics of Reliefs

The analysis of the relief structure includes two parameters: the slope value and the lithological properties of the geological structure (permeability/impermeability and resistance of rocks). The slope of the ground has a direct impact on the time of concentration of water flow and the power of floods. Therefore, the steeper the slope value, the more concentration and course time, and vice versa. In addition, the permeability, on the one hand, and the resistance of lithological outcrops, on the other hand, affect the wetting time. In fact, the stronger the permeability of lithological outcrops, the less precipitation will remain on the surface. In other words, the land is no longer impervious, and the river system is dense (Table 1).

In addition, on soft and soft substrates, the power of runoff is reduced. Because they are easy to transport, they form an alluvial charge, which reduces the capacity and competence of the flow, thereby reducing the rate of runoff and thus the rate of flood rise.

2.2 The Methodological Framework

2.2.1 The Relevance of GIS to the Flood Hazard Study

The goal of GIS is to realize the storage, management, update and development of geographic positioning, multi-scale, multi-theme and multi-date information [27]. In other words, the interpretation of the comparison and the temporal and spatial inquiry

Table 1 The impact of hydro-morphological characteristics on the behaviour of river basins (Riahi 2024)

			RVR	Hydrological behaviour					
				TVR	Concentration time	Peak flood flow	Speed of concentration	Capacity	Competence
Morphometric characters	Surface area	↑	↑	↑	↑	↓			
		↓	↓	↓	↓	↑			
	Compactness	↑			↓	↑	↑	↑	↑
		↓			↑	↓	↓	↓	↓
Characteristics of the river system	Drainage density	↑		↑	↓	↑	↑	↓	↑
		↓		↓	↑	↓	↓	↑	↓
	Hydrographic density	↑		↑	↓	↑	↑	↓	↑
		↓		↓	↑	↓	↓	↑	↓

RVR: Rainfall volume received
TVR: Total volume runoff
↑ Increase
↓ Decrease

of the information layer [28]. This makes it an important tool for multiple application areas, including environmental hazard research, risk assessment, and risk management. The products produced by the development of geographic information systems are indispensable tools for sustainable development decision-making support.

Therefore, using GIS to study flood risk is essential to better understand, prevent, model, and manage this risk. Therefore, the contribution of GIS is essential for flood risk mapping and flood risk modeling.

2.2.2 Study Stage

In order to study and map the hazards of flooding in the Maknassy basin (the Lban-Ouadran wadi watershed), we have taken several steps based on our work progress. We started by creating a GIS database. We integrated multi-date and multi- source data into each of them corresponding to a data layer. This information comes from the analysis of archival data (topography, geology, soil science, agriculture, etc.), Bibliographic data and data from the photo-interpretation of aerial photographs and Google Earth images (Landsat december 2020).

Subsequently, we managed and operated the GIS database by cascading, combining, calculating, and comparing the data layers of different hydro-morphometric forms, topography, lithology and soil characteristics in the basin of Maknassy. This allows us to map flood disasters. This can be used as a reference for the proposal of a risk management plan for the protection and sustainable management of natural resources, including water and soil, in the study area.

3 Study Area

The Maknassy basin (Fig. 1) has an area more than 1020 km^2 and is located between 34°47′23.56′ N 9°43′18.81′ E and 34°29′24.11′ N 9°11′12.47′ E. It represents a part of the upstream of the Lban-Ouadranewadi watershed. The latter, draining an area of 3610 km^2, is one of the rare wadis with exoreic flow of central-southern Tunisia. In the upper part, the watershed of the WadiLban-Ouadran takes the form of acatchment, with Hadej-BouHedma- BouDwaw in the south, jbelKbar, jbelTwema, and jbelGsayra in the northeast, and the Majoura and Meloussidjebels chain in the northwest [7]. The basin leads to another larger basin bordering Gouleb and Goudrarjebels in the north and opens to the east. It is located on two alluvial plains between the Bir Ali Ben Khlifa-Graïra plateau in the north and the Skhira plateau in the south [24, 29], which leads to the sea in Hchichina.

The catchment area of the upstream of the Lban-Ouadrane has a configuration of two sub-closed compartments or basin with flat bottoms to sub-flat surrounded by the north, east and south sides by dominant mountain reliefs except for a few narrow passages that allow the passage. As for its western limit, it is materialized by low-elevation hills [5].

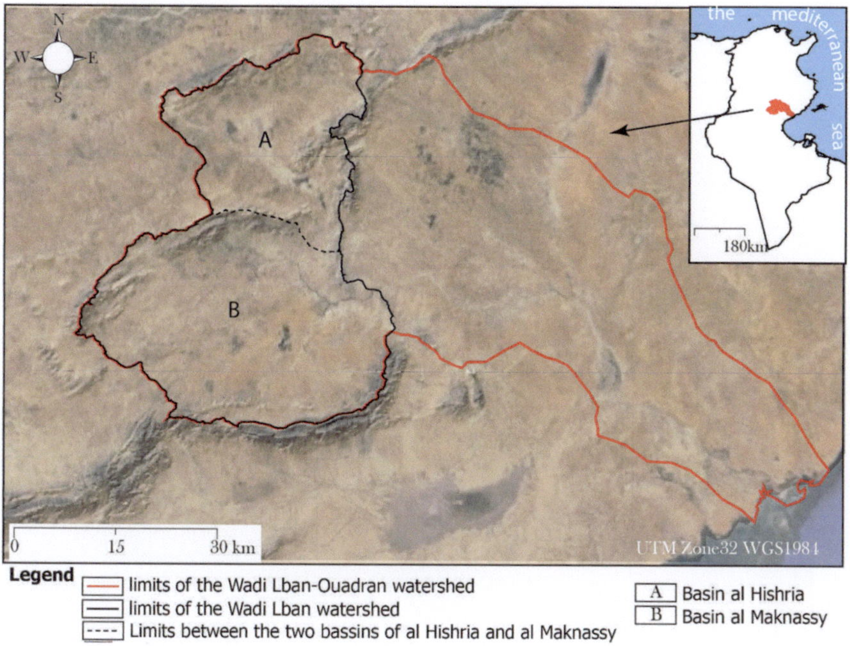

Fig. 1 The location of the study area (Riahi 2024@Google earth)

The first compartment is located in the northern part of the research site. It corresponds to the Hichriya Basin basin. This latter is bounded by jbel Kbar (569 m) on the north side, jbels Twema and Gsayra (510 m) on the northeast side, and jbel Boudinar (696 m) on the east side. In the south, the chain of al Mallousi separates the al Hichriya bowl from the Maknassy bowl. It has the particularity of spreading in a large area, namely sebkha an Njila. For this reason, we chose not to include the Al Hichriya Basin in our flood disaster research (Fig. 2).

In addition, the Maknassy Basin (the second compartment or basin of the upper partof the Lban-Ouadranwadi watershed) is located between the peak lines of jebel Méhiri, Zebbeus and Dribika el Hamra. The el Goussa chain, Bou Hedma-Bou Douaou borders the south and southeast. The ridges of the Majoura and Méloussi mountains form its western and northern borders. At the same time, the hydrological network of the Makenasy Basin is more organized and maintained at the level of the plain and surrounded by layers. Indeed, the Wadi Leben became important, especially after it merged with the Wadi Selam in Maknessy. It maintained a west-east orientation, and after receiving the largest tributary of the north western Jack Bells, its course bent to the southeast. Therefore, the wadi left its bed and lay on the plain [7].

Additionally, the topographic features of the Maknassy basin are diverse. Although its border is occupied by an almost continuous mountain zone, along the direction of the Atlas Mountains (north and south) and the submeridian (east).

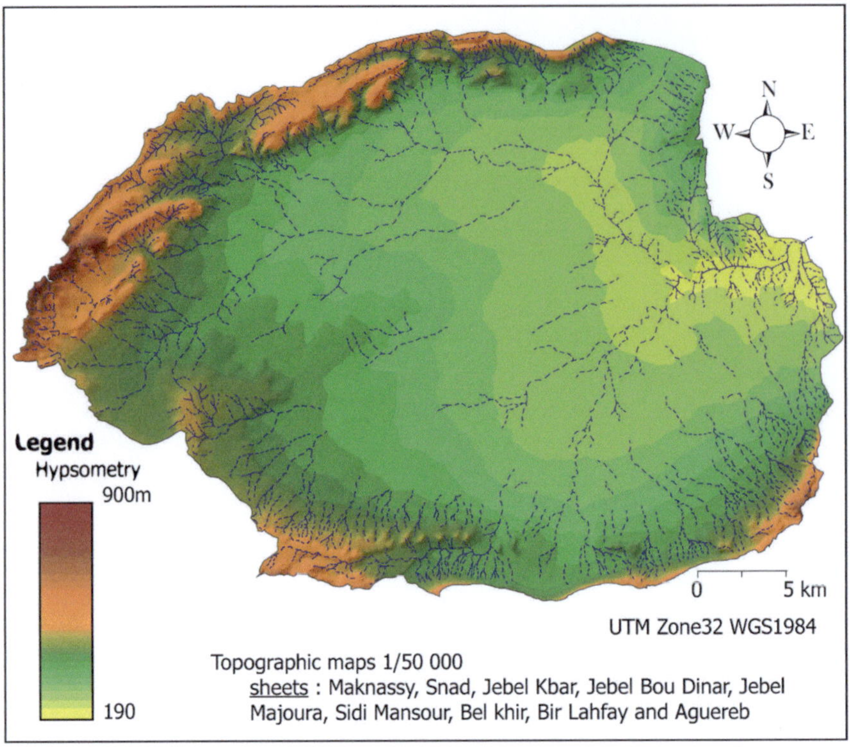

Fig. 2 Hypsometry of the Maknassy Basin (Riahi 2024)

The peak line of jbels always exceeds 600 m in height, separating the often elongated depressions, and the very wide plains largely mark the main part of the basin. However, the plateau is characteristic of the eastern piedmonts and the Ouled Amara region [5]. The two basin centers are characterized by low or even zero slopes over nearly 50% of the total area of the basins. The slope is generally greater than 25%, and it can reach 100% at the eaves (Figs. 2 and 3).

In addition, the topographical features of the Maknassy basin are diverse. Although its edge is occupied by an almost continuous mountainous zone, along the direction of the Atlas Mountains (to the north and south) and the Submeridian (to the east), the crest line of the jbels always exceeds 600 m in altitude, separating often elongated depressions. The crest line of the jbels always exceeds 600 m in altitude, separating often elongated depressions, and the extensive plains largely mark the bulk of the basin. However, the plateau is characteristic of the eastern foothills and the Ouled Amara region [5]. The two basin centers are characterized by low to zero slopes over almost 50% of the total basin area. The slope is generally greater than 25% and can reach 100% at the eaves (Figs. 2 and 3).

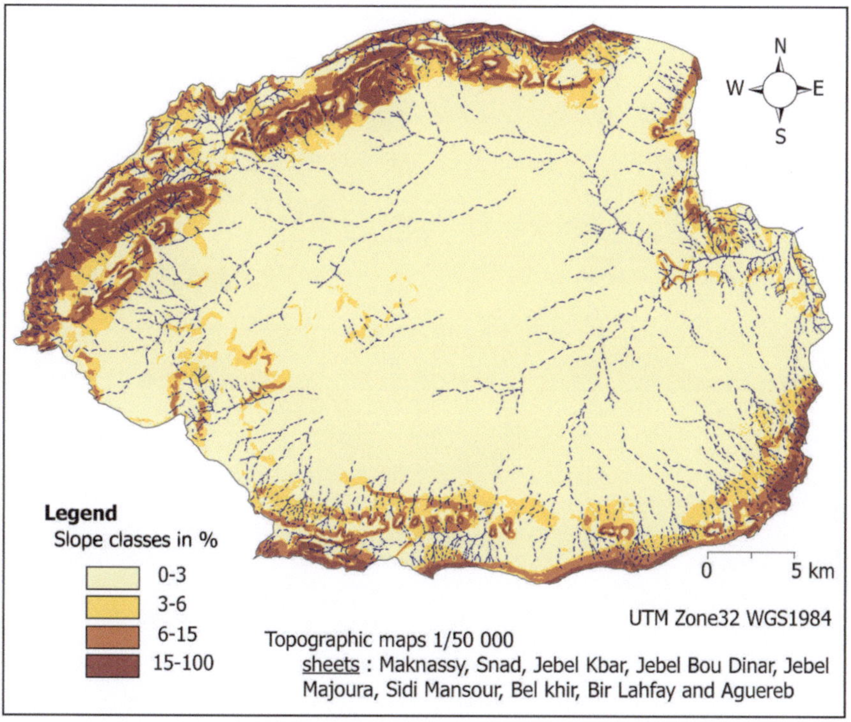

Fig. 3 Slope classes of the Maknassy Basin (Riahi 2024)

The basin system of the Maknassy watershed consists of a total of 42 second-level wadi watershed (according to the Hake flow classification (1957)) and 84 third-level sub-basins. The latter represents the impluviums within which precipitation water is concentrated.

Furthermore, impluviums represent the areas of the environment in which runoff is concentrated, which directly influence the behaviour of the hydrological network draining them (flow rate, concentration rate,...) and, therefore, the risk of flooding that they represent. The analysis of the morphometric characteristics of this part of the environment is of paramount importance for the study of the hazard of flooding. This analysis is based on the thresholds classification of the Gravelius (Kc) Compactness Index for Wadi watersheds. We have grouped the impluviums into three groups according to their compactness indices.

In addition, in the Maknassy Basin, the lithology of the study site is dominated by substrates with low water retention capacity, which is conducive to flooding during heavy rains. Therefore, the combination of these two features will only increase the risk of flooding by increasing the risk of severe flooding. Therefore, the physical characteristics of the Maknassy basin, especially the zonal and compact topography, are favorable factors for high flow, and its influence cannot be ignored [7].

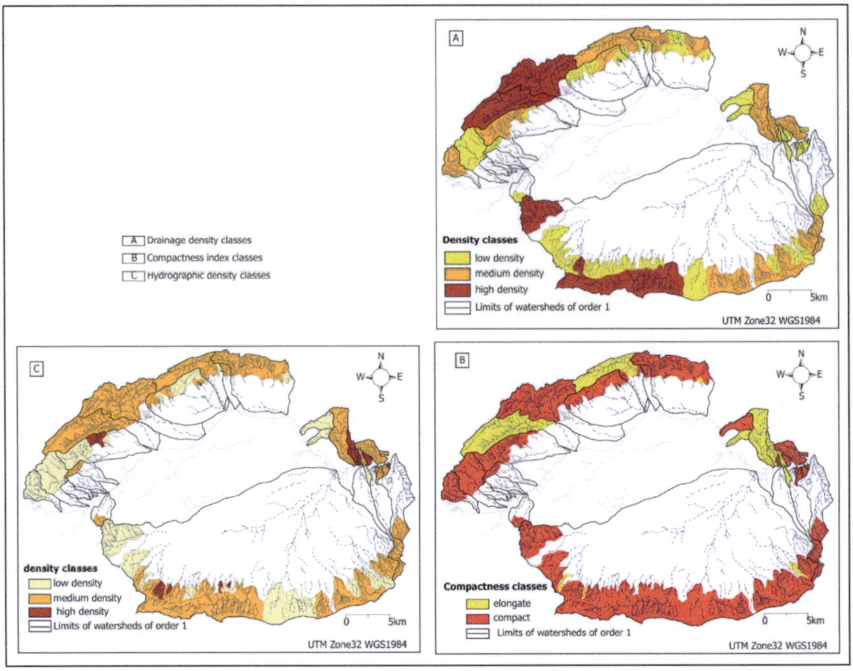

Fig. 4 The hydro-morphological characteristics of the watersheds (order 1) of the Maknassy basin (Riahi 2024)

4 Flood-Hazard Areas in the Maknassy Basin

The study of the flood hazard shows that the Maknassy basin is dominated by high-risk flood impluvium, which accounts for about 75% of the impluvium studied. Three classes of hazard are identified (Fig. 5).

4.1 *Low Hazard Impluviums*

Impluviums belonging to this hazard class account for only 15% of the impluviums studied and are located mainly on the west side of the basin. They are characterized by average inclinations of less than 3%.From a hydro-morphometrical point of view, this hazard class groups together three subcategories of impluviums (1st class in Fig. 5). The first and second groups are slender alluvial layers with medium to dense drainage density and low river system density. The third is compact, low-drainage and hydrographic impluvia. Lithology is dominated by permeable rock types. A slope value between 0 and 5% has been added to these characteristics. The combination of these characteristics produces a kind of hydrological behavior, which is characterized

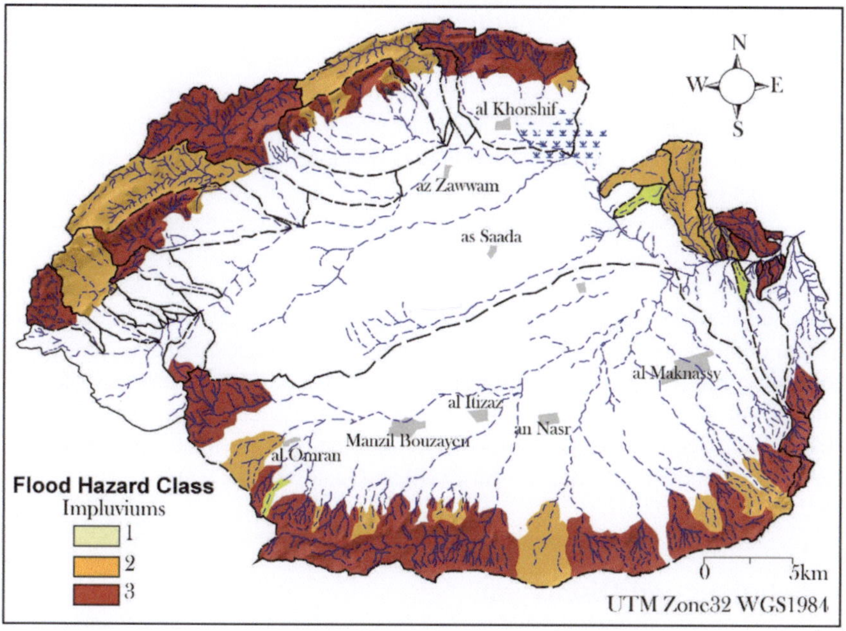

Fig. 5 Hazard in the maknassy basin Low chance Impluviums (Riahi 2024)

by a slower rate of runoff concentration and flow value, so the flood damage is the lowest in the alluvial under study.

4.2 Medium Hazard Impluviums

Two types of shapes characterize Impluviums in this category of hazards, which represent 20% of the impluviums studied (2nd class in Fig. 5). The first one is elongated with high drainage densities and medium hydrographic densities. The second is compacted with low drainage and hydrographic densities. These impluviums have, therefore, medium long concentration times and, thus a medium peak flood flow rate. On the other hand, runoff volume is important.

Moreover, they are characterized by the dominance of impervious outcrops. These characteristics make them the class representing an average degree of flood hazard and the source of the second flood wave during floods with the largest volume of water delivered.

4.3 *Very High Hazard Impluviums*

This category represents 65% of the impluviums studied. The hydrographic network drains through loose, impermeable substrates, which reduces the stability of the wadis [5]. and consequently floodwater overflow is frequent. In addition, this hazard class groups together compact impluviums with three types of density classes (3rd class in Fig. 5). The first is compact impluviums with low drainage densities and medium hydrographic densities. The second has medium drainage densities associated with high, medium and low hydrographic densities. Finally, the last group includes impluviums with high drainage densities and high, medium and low hydrographic densities. The combination of these characteristics leads to hydrological behavior marked by the shortest runoff concentration time and the highest peak flood flow. On the other hand, runoff is less important, which makes them the source of the first flood that transports the least amount of water. Therefore, they are the most severely flooded alluvial deposits in the Maknassy Basin (Lban-Ouadran wadi watershed).

5 Solutions and Discussion

Flooding has a direct impact on natural water and soil resources and their sustainability. In particular, they result in a loss of soil quantity and quality. Soil loss mainly affects areas where sediment is removed and transported, i.e. alluvial foothills and mountain foothills. In the foothills, soil loss in these parts of the catchment area is essentially quantitative. The various processes involved in uprooting and transporting materials counteract this. In the alluvial plain, on the other hand, soil loss is mainly due to the deposition of sterile particles from neighbouring mountainous terrain, as well as an increase in soil salinity.

In addition, flooding is a natural hazard that cannot be controlled in the Maknassy basin (Lban-Ouadran wadi catchment area). However, the vulnerability of the natural environment can be reduced by taking a number of measures to limit the damage caused by flooding in the Maknassy basin (the oued Lban-Ouadran catchment area). These measures must be taken by both the government and the local population. The government is responsible for building floodwater diversion and retention projects.

At the top of the piedmont, for example, flood diversion structures are urged to be built. Downstream, it is advisable to build floodwater collection and storage structures by building underground water supply structures to increase run-off, minimise the risk of salinisation of the soil downstream and create water reserves. Added to this is the need to disseminate information to the local population of the Maknassy basin in order to make them more aware of environmental problems, particularly the risk of flooding. In addition, the local population still has a responsibility to protect themselves and reduce the risk of flooding by avoiding building houses on small and medium-sized watercourses, particularly in the minor bed.

All in all, flooding is a scourge that threatens the economy in the MENA region, especially as the number of flash floods and people affected or killed has doubled over the last ten years [3]. This situation is due to a number of factors, including the fragility of the natural environment following the increasing trend towards aridification of the climate, and concrete construction (housing, infrastructure, etc.) in the middle beds of wadis, which do not allow water to infiltrate into the ground. In order to better understand and manage this risk, we have proposed in this chapter the study of the flood hazard as being a natural situation whose occurrence and intensity are uncontrollable [22].

The analysis and mapping of the flood hazard in the maknassy basin and in Tunisia in general, as well as in the MENA region, based on slopes, hydro-morphometric and lithological characteristics, makes it possible to preserve water resources, avoid water shortages, protect soil resources, avoid food insecurity and save human lives. The widespread use of this method of analysis will help prevent hazards from becoming risks or natural disasters, and thus avoid dramatic events such as those that occurred in Libya in 2023 (Dernah city), in Algeria in 2001 (Algiers city) and in Tunisia in 2018 (Nabeul city, Korbous city and Korba city).

However, it is also advisable to couple the results of our research method with the results of other approaches such as the geomatic method applied by Lahmer [19] in the study of the flood hazard in the Jendouba basin in the Middle Medjerdah watershed (north of Tunisia), or the method of field exploration and data coupling with geomorphological data applied [20] in the Bou Salem region (north of Tunisia), or the method of cross-referencing metrological data, geomorphological data, biogeographical data and anthropogenic data applied in the Tangier-Tetouan region in Morocco [21] and in cities in Lebanon [1].

6 Conclusion

In conclusion, our research has shown that the Maknassy basin (Lban-Ouadran Wadi watershed) is a natural environment at high risk of flooding. In fact, the hydromorphological characteristics of the impluviums discharged by the Maknassy basin (oued Lban-Ouadran catchment) are mainly determined by factors that make it a very high flood risk environment. Indeed, the morphology of the impluviums of the wadis draining the Maknassy basin (Lban-Ouadran wadi catchment) is dominated by high slope values. In addition, these conditions add a bedrock with low permeability and low resistance. Hydrological characteristics dominated by high compactness indices, drainage densities and very high hydrographic densities exacerbate this situation.

The combination of these factors has revealed the existence of three hazard classes, the two most sensitive of which (moderately and very sensitive) cover more than 65% of the impluviums of the order 1 wadis (on both the northern and southern slopes) of the Maknassy basin, which are at very high risk of flooding. This highlights the environmental and economic impact of the flood hazard and the real risk to natural resources, particularly water and soil, thus compromising the sustainability of development and the environment in the Maknassy basin.

To deal with this situation, a series of works and certain measures need to be taken as part of a participative approach. These measures must be taken by both the local population and the State. It is up to the State to build flood control infrastructure, such as soil protection works and water diversion and storage facilities. At the same time, it is highly advisable to invest in raising the local population's awareness of the environmental and economic problems associated with flooding. In addition, the residents of the Maknassy basin must take steps to reduce their susceptibility to flooding by adopting farming practices suited to their environment, and in particular by avoiding building in areas at risk.

7 Recommendations

A number of recommendations can be made both for future scientific research and for policy planners, decision-makers and stakeholders. With regard to future scientific studies, it is strongly recommended to:

- developing long-term, multi-source data collection systems to monitor flood hazards and their impact on water and soil resources and the rural economy;
- encourage multidisciplinary studies of flood hazards in rural catchment areas;
- involve the local population, as part of a participatory approach, in decision-making and the maintenance of flood protection facilities
- to disseminate information to the local population in catchment areas at high risk of flooding in order to convince them to rationalise the use of water and soil resources, and to educate them in the field of flood risk control

References

1. Abdallah C (2018) Risques (II): floods and human activities, Atlas of Lebanon: the new challenges, Presses de l'Ifpo, Beirut 2016. http://books.openedition.org/ifpo/10863. Accessed Aug 2023
2. Abderrahmane A (2013) The very strong risks rains in Central Tunisia: event of the event of 22 and 23 September 2009, alleas and climate risks. Coordination Henia L. and Helaoui, University of Tunis, pp 61–66
3. Banerjee A, Bhavnani R, Burtonboy-Catherine H, Hamad O, Linares-Rivas Barandiaran A, Safaïe S, Tewari D, Zanon A (2014) Natural disasters in the Middle East and North Africa : a regional overview. Global facility for disaster reduction and recovery (GFDRR). World Bank Group, Washington, D.C. http://documents.worldbank.org/curated/en/211811468106752534/Natural-disasters-in-the-Middle-East-and-North-Africa-a-regional-overview
4. Benaïchouche A (2013) Reorganization of hydrographic networks in the models of landscape evolution, at large spatial and temporal scales: study of the potential catches of the Meuse Paris, p 216

5. Bedoui C (2002) The maknassy basin: search for quaternary forms and deposits. Doctoral thesis of the FSHST, p 230
6. Ben Djoudi H, Hubert P (2002) The coefficient of compactness of Gravelius: critical analysis of a watershed index. Hydrol Sci-Journal-des Sciences Hydrologie 47(6): 21–930
7. Bourges J (1974) Overview of Tunisian south hydrology, exceptional observation network and raws. Dir.Des Water and Soil Resources Tunisia, p 164
8. Castany G (1982) Hydrology principles and methods. Bordas, p 236
9. Claude J, Pontanier R (1971) Some data on floods, morphological modifications and evolutions of the water and alluvials of the Leben-Ouadrane basin after the fall of the fall 1669, Dir. of water and soil resources Tunisia
10. Demeres S, Olsen T, Buffin-Bélanger T, Marchand J-P, Biron P-M, Morneau F (2014) The hydrogeomorphology applied to the Aléa of cold climate flood: the example of the Matane River (Quebec). Phys Geogr Environ V(8):67–88
11. Dubreuil P (1966) The physical and morphological characters of watersheds, their determination with acceptable precision. Office for scientific and technical research-overseas, hydrological service, p 31
12. Dubreuil P (1969) Practical complements to the rules of measurements of the morphological characteristics of the watershed. Tech Note Hydrol Serv ORSTOM 19:20
13. Eichert B-S, Kindler J, Schultz G-A, Sokolov A-A (1985) Calculation methods for hydrological studies regarding water development. Collection: studies and hydrology reports, Unesco, p 134
14. Fehri N (2011) Aggravation of flood risk in Tunisia, reflection elements. Rev Physio-Géo V(8):149–175
15. Hamza A (1990) Morphological dynamics and behaviour of land and water conservation infrastructures during the January 1990 rainfall in central and southern Tunisia, cahier of CRES. Géography 7:255–281
16. Humbert H (1990) Interest of drainage density for regionalizing hydrological data in the mountainous area. Hydrology in mountainous regions. I—Hydmlogical Measurements. Water Cycle 8(193):373–380
17. Kassab F (1979) The very strong rains in Tunisia. Publications of the faculty of arts and humanities of Tunis. Geogr Ser 5(11):234
18. Kallel R, Benzarti Z (1990) Directory of the main floods and floods observed in Tunisia over the last two decades, D.G.R.E., 16p
19. Lahmar L (2021) Multi-temporal and systemic approach to flood risk in the middle Medjerda river basin (northern Tunisia). Univ. Paris City, p 294
20. Lahmer (2005) Land use and flood risk in the Jendouba-Bou Salem plain. Univ. Tunis, p 150
21. Pateau M (2014) From hazard to natural risk: Tangier-Tetouan region example (Rif, Morocco). Geo-Eco-Trop 38(1):23–32
22. Observatoire territorial des risques d'inondation OTRI (2021) https://www.otri.fr/le-risque-inondation/alea-et-risques/. Accessed Aug 2023
23. Oueslati A (1999) Floods in Tunisia. Tunis, p 206
24. Pias J (1970) The floods of September and October 1969 in Tunisia, Part I: Pedological study. UNESCO technical report, UNESCO, Paris, p 36
25. Poncet J (1970) The climate catastrophe of the fall of 1969 in Tunisia. Ann Geogr 435:581–595
26. Riahi O (2017) The dynamics of the geomorphological landscapes and recent evolution of an arrived natural environment: the Hadej-Bou Hedma region (southern Tunisia) the doctoral thesis of the UNIV of Tunis-FSHST, 261p
27. Riahi O (2021) Contribution of GIS in the environmental monitoring of a tunisian biosphere reserve (Bou Hedma National Park). In: Khebour Allouche F, Negm A-M (eds) Environmental remote sensing and GIS in Tunisia. Springer Water, Springer, Cham, pp 303–327. https://doi.org/10.1007/978-3-030-63668-5_15

28. Riahi O (2021) Mapping environmental risk degradation under climate stress and anthropogenic pressure: case study of Abdeladim Watershed, Tunisia. In: Khebour Allouche F, Negm A-M (eds) Environmental remote sensing and GIS in Tunisia. Springer Water, Springer, Cham, pp 167–189. https://doi.org/10.1007/978-3-030-63668-5_9
29. Stuckmann G (1970) The floods of September and October 1969 in Tunisia, Part II: the morphological effects of the floods of September–October 1969 in central and southern Tunisia. UNESCO, Paris, p 35

Assessment of Underground Water Quality for Irrigation Proposes Using GIS and Remote Sensing Techniques

A. Gad and Rania M. Mohamed

Abstract Because of increased population growth in industrial areas, agricultural development, and current climate change, that causes changes in precipitation patterns, water quality and supply, have become major global issues. Water plays an important role in agricultural planning and reclamation. The main source of water in the study area is ground water (wells). Assessment of water quality, in the study area, is done by using **piper** diagram, while Evaluation of the sodicity and salinity hazards in accordance with USDA (1954). Geostatistical analysis resulted in the overlaying of the theme maps (EC, Na^+, Cl^-, HCO_3^-, and SAR) to create the IWQI map. The piper diagram demonstrated that the sources of such types of water had similar origins; the dominant cations are represented by Na^+ ions, while the dominating anions are represented by Cl^-. According to the USDA's categorization system, 75% of the wells' water is classified as salinity class C3, and about 17% of the wells are in the C4 class while only 8% of wells are in the C2. The outcome demonstrated groundwater samples are categorized as; severe restriction (SR), high restriction (HR), moderate restriction (MR), low restriction (LR) irrigation water (0.2%, 3.1%, 8.9% and 82.4%) respectively. Research results indicate that 5.4% of the samples collected from groundwater in the research region are categorized as irrigation water with no restrictions (NR). Thus (75%) of water in the study area is suitable for irrigating most crops except the sensitive ones. It is possible to control the low to medium SAR and moderate to high EC_w without applying any water or soil treatments. The issue can be resolved by leaching the extra salts from the root zone.

Keywords Under ground water · Water quality · Remote sensing · GIS · El-Minya · Egypt

A. Gad · R. M. Mohamed (✉)
Land Use Department, National Authority for Remote Sensing and Space Sciences (NARSS), Cairo, Egypt
e-mail: Rania.mansour327@gmail.com

F. Khebour Allouche (ed.), *Environmental Remote Sensing and GIS in MENA Region*, Earth and Environmental Sciences Library,
https://doi.org/10.1007/978-3-031-86574-9_15

1 Introduction

All living things, including humans, derive their sustenance from water. There are roughly 37 Mkm^3 of fresh water on Earth [30]; however, There is a significant need for new water resources, particularly in emerging nations, to meet the expanding needs for water [2]. Agriculture has made a significant contribution to meeting the growing population's food needs. Furthermore, it has promoted economic development in irrigated districts [16]. Many parts of the world rely heavily on groundwater as a source of fresh water. However, population increase and various economic activities have had a deleterious impact on groundwater quality [20]. Groundwater is a valuable source of freshwater. At the same time, anthropogenic activities, particularly industry, urbanization, population development, and excessive fertilizer application, are key causes of groundwater quality decline [19].

Groundwater in arid/semiarid locations is critical for water supply, irrigation, and ecosystem sustainability [12]. Groundwater is one of the most important natural and abundant resources, as well as the primary source of water supply around the planet. Water quality is an important issue for humans because it is intimately linked to human well-being [5]. Groundwater is significant not just because it is sustainable, but also because it is continuously efficient (Abou Zakhem and Hafez 2015). Water quality and supply have become major global issues as a result of rising population growth and industrial and agricultural development, and current climate change is contributing to changes in precipitation patterns [4]. Nowadays Egypt is focusing on fresh initiatives for land reclamation in order to establish new communities and produce enough important crops for self-sufficiency. West El-Minya is regarded as one of Egypt's most hopeful lands for upcoming changes. Mismanagement, exploitation, and poor water policy implementation will undoubtedly have an impact on groundwater quality and quantity, either directly or indirectly. As a result, ongoing investigations and the development of maps of potential groundwater in newly reclaimed lands are critical for long-term Managing groundwater. These investigations are expensive and necessitate an enormous amount of data [16]. Water production and management rely heavily on understanding the physicochemical distribution and depth of groundwater (Gupta and Sharma 2016). Many factors influence groundwater potential, Tolche [32] include depth of water, topography, lithology, geological formations, hydraulic parameters, lineaments and quality of the water. That is why the conclusions of groundwater potential investigations vary depending on the factors considered [18].

Water quality is often defined as the biological, chemical, and physical characteristics of a body of water until it is needed for a number of purposes, including drinking, industrial, recreational, and agricultural [17]. Improvements in data producing and access technologies, such as the internet, GPS, and remote sensing, are necessary for the continuous evolution of GIS ("Geographic Information System"), which calls for ever-more-efficient data processing and specialised solutions for unified models of environmental simulation [31]. Developing solutions to problems with water resources at the regional level, understanding the environment, reducing food contamination, figuring out water convenience, and assessing water quality may all be accomplished with the use of GIS [9]. Abdel-Fattah et al. [1] illustrated that, the water quality of the Bahr Mouise canal was classified into two classes based on the application of the irrigation water quality index (IWQI), while the water quality was classified into three classes using the PCA and AHC methods. Temporal variations in water quality were investigated, where the water qualities in winter, autumn, and spring (January, February, March, April, November, and December) were classified as class I (no restrictions) based on IWQI application, and the water salinity, sodicity, and/or alkalinity did not represent limiting factors for irrigation water quality. On the other hand, in the summer season (May, June, July, August, and October), the irrigation water was classified as class II (low restrictions); therefore, irrigation processes during summer may lead to an increase in the alkalinity hazard. Batarseh et al. [6] stated that The assessment Irrigation Water Quality Index (IWQI) and GIS-Zoning maps to evaluate groundwater resources for irrigation in Abu Dhabi Emirate, UAE. IWQI readings ranged from 8.78 to 77.44, with 52% of samples falling under severe irrigation limits. Groundwater can only be utilised for irrigation of salt-tolerant plants. The analysed wells were divided into four categories: heavy restriction (44%), moderate restriction (3%), low irrigation restriction (1%), and uncontrolled (0%). Wilcox and USSL diagrams verified IWQI results, indicating that untreated groundwater is usually unsuitable for irrigation. Waters classed as Permissible to Doubtful or Good to Permissible account for only 3%. Gaagai et al. [14] mentioned that, the IWQI values revealed that 33% of samples were severely restricted for irrigation, while 67% of samples varied from moderate to high restriction for irrigation, indicating that crops that are moderately to highly hypersensitive to salt should be watered in soft soils without any compressed layers. According to RIGW [25], groundwater in the El-Minya region ought to be regarded as a component of the regional aquifer systems of the Nile valley. There are two recognized aquifers: the Quaternary and the Eocene. The Quaternary aquifer covers a large area, particularly in the Nile Valley's western region. The aquifer is covered in Holocene Nile sandy clay and silt, which expand into a semi-permeable to impermeable layer. The Holocene Nile silt and sandy clay layer ranges in thickness from zero to sixteen meters, with an average thickness of nine to ten meters. Water is collected by the upper layer from subsurface seepage through irrigation canal beds and sides as well as return flows following agricultural land irrigation.

The Quaternary aquifer stretches far from north to south. It lies unevenly above the Middle Eocene Limestone and is confined to the east and west. According to RIGW [26], the aquifer's thickness decreases from 200 m below the Nile to a few meters west and east. According to Said [28], the Quaternary aquifer is underlain by the Middle Eocene limestone water bearing formation, it then sits on top of the Nubian limestone deposit.

The Samalut formation, which represents the Eocene limestone aquifer unit, is composed of firm, white, extremely fossiliferous limestone featuring marl and shale intercalations. The Eocene aquifer spans the extreme east and west margins of the research area. Because of its fractures, Eocene limestone is probably affected by a displacement network. This study aims to (i) identify the Piper's trilinear diagrams depict the classification of groundwater samples in the west El-Minya area. (ii) Evaluation USDA's classification of the salinity (C) and sodicity (S) and (iii) assess and classify the groundwater quality for irrigation using IWQI in the investigated area. The next section explains the methodology used to achieve these objectives.

The innovative aspect can be highlighted in the current chapter through the observed relation between increased population growth and geographic practices. Such relation could be confirmed by focusing on resulted impact of climate change, in precipitation patterns, water quality and supply, to be considered major global issues playing the important role in agricultural planning and reclamation. Innovation extends to employing the Piper diagrams in assessing water quality, where evaluation in terms of the risk associated with salinity and sodicity were considered. This article ended with an innovative conclusion confirming that leaching the excess salts from the root zone can solve the soil salinity problem.

2 Methodology

2.1 Study Area

The study area was chosen to represent part of promising sustainable soil and water resources. According to [33] the western El-Minya area Fig. 1 was chosen as a case study fitting the goal of the current investigation. In terms of geography, the study area is situated within latitudes 27°55′ and 28°34′ N and longitudes 29°44′ and 30°39′ E, west west of El-Minya Governorate Egypt, covering a total area of 572,900 ha, as shown in Fig. 1. The studied region is rounded beside the Nile River in the east, by a limestone plateau to the west, on the north by BeniSuef Governorate, and at the south by Assiut Governorate. El-Minya Governorate inclouds 9 districts, 54 primary villages, and 340 subordinate villages. According to Rania [20], El-Minya Governorate covers a total area of 3,227,900 ha, of which 3,043,900 ha are desert lands and the remainder is agricultural fields, built-up regions, and water bodies.

2.2 Climatic Conditions

El-Minya is mostly known for its chilly, rainless winters and scorching, dry summers.

From the meteorological station of El-Minya (https://www.ventusky.com/minya), the climate data were obtained within the period of 2002–2021 that is presented in Fig. 2.

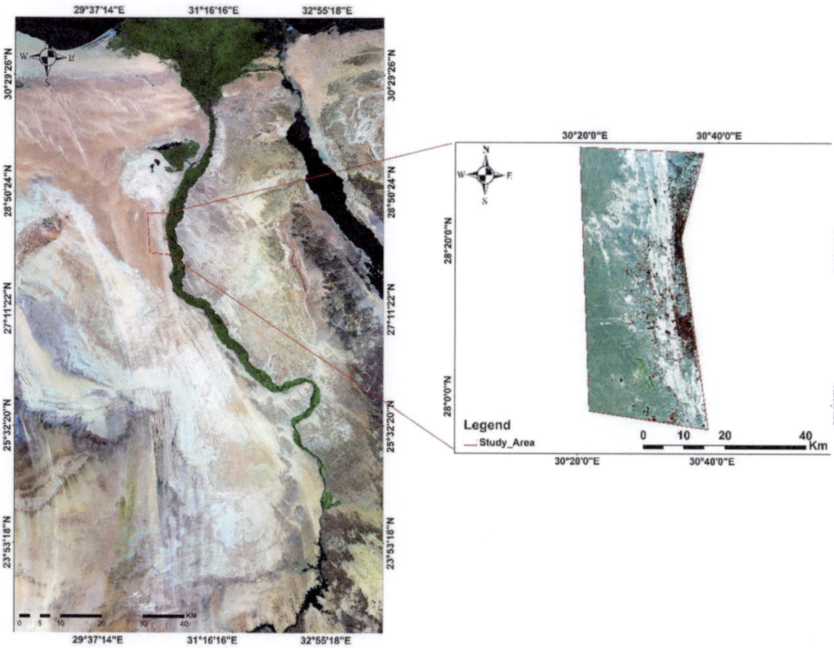

Fig. 1 Location of the case study area (Gad and Mohamed 2024)

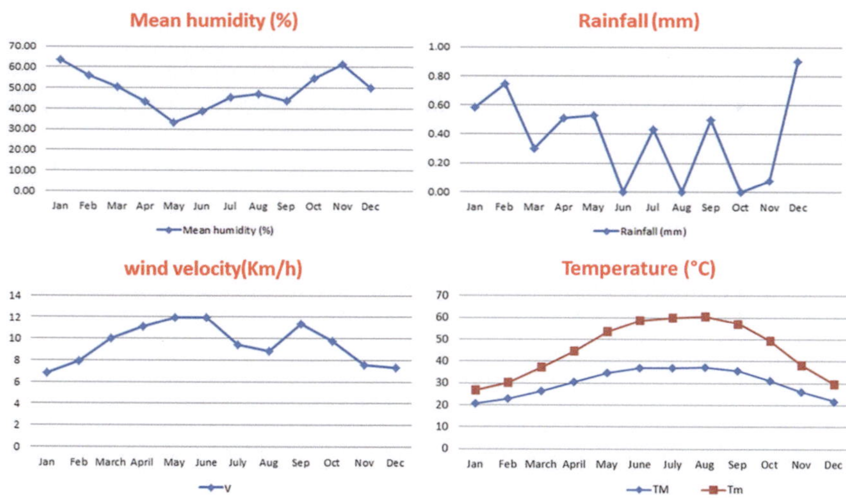

Fig. 2 Climatologically diagram of El-Minya Governorate from 2002 to 2021

2.2.1 Temperature

Figure 2 illustrates that the maximum and minimum temperatures have similar distribution curves. The lower temperature values are recorded from December to March period representing the rainy season, while the highest temperature values are recorded from May to October period, corresponding the dry season. In January, the lowest recorded temperature was 6.04 °C, while in August, the highest recorded temperature was 37.30 °C. Based on the meteorological data gathered, the study region's average maximum temperature was determined to be 30.36 °C, while its average lowest temperature was discovered to be 15.72 °C. The difference between mean summer and mean winter temperatures was more than 6 degrees Celsius. According to FAO 2006, the soil temperature regime in the study area could be described as thermic while the soil moisture regime as torric.

2.2.2 Rainfall

With the highest monthly rainfall of 3.08 mm recorded in November, the average monthly rainfall was quite low. June through October is considered the dry season.

2.2.3 Relative Humidity

Water vapour in the atmosphere was frequently corresponded to air temperature and is known as "humidity." The relative humidity is calculated using the amount of moisture required to saturate the air. Relative humidity ranged from 33.06% in June to 63.51% in October.

2.2.4 Wind Speed

As Fig. 2 illustrates, the maximum wind speed recorded was 20.66 km/h in May, while the lowest recorded was 6.78 km/h in January.

2.3 Geology of the Study Area

The West El-Minya region is mainly composed of Quaternary sediments (Wadi, Protonile, Prenile, Playa, Nile Silt, Sanddunes, Gravel, and Fanglomerate), Miocene (Dolomitic limestone and moghra Fm), Oligocene (tertiary Alkali Olivine Basalt), and Eocene limestone (Samalut, Minya, Hamra, and Rayan Fm.). The main lithostratigraphic units observed in the research region were described by Conco (1987). as depicted in Fig. 3. These units are Minia formation, it covers the middle latitudes of the wide plateau surrounding the oases of the western desert (Said 1960).

Samalut formation, it exhibits the first appearance of nummulites gizehensis banks in the stratigraphical sequence of Egypt. The formation display constant lithologic characters everywhere [8]. Rayan formation, it exhibits lithologic characteristics that enables its subdivision in to two subdivision. (a) An upper sequence (100 m thick). (b) Thick basal nummulitic limestone bed, corresponding to the muweilih [7] (Said 1960). Variegated sandstone containing fossil wood and vertebrate remains in the lower part of the formation, marly sandstone, yellow marl with gypsum veins, and conglomeratic nodular limestone with regular shale intercalation represent the majority of the Moghra formation. In the west, the structure gets thinner and shadier, reaching a thickness of only 57 m [28]. Fanglomerates formation, these deposits represent alluvial fans developed along the slopes and scarps west to the river Nile in upper Egypt. They are made up of cobbles and gravel derived from the sediments provided by the surrounded scarps. Playa deposits, these deposits consist of fine-grained clastic sediments foun in topography low areas. They are best developed at the foot of scarps and in depressions. Wadi sediments, wadi (dry valleys) of the mapped sheets are covered by alluvial gravel sand and silt, sand dunes, these dunes cover large areas of the western desert. According to Said (1960), these Eolian formations are partially remoulded sediments from braided streams. Much of the El-Minya is covered in dunes, primarily linear dunes. The limestone that makes up the Hamra formation is yellowish-white in colour, with some grit and sandy stone intercalations towards the top. The same locations record the infrastructure conglomerate. The whole succession is highly fossiliferous. The limestone beds are partly crystalline and display irregular bedding.

3 Methodology

3.1 Remote Sensing Imagery Data

3.1.1 Satellite Image

Landsat 8 OLI and Thermal Infrared Sensor (TIRS) images contain nine spectral bands, each with a spatial resolution of 30 m. Band 1 in the ultra-blue spectrum is important for aerosol and coastal researches. Band 9 is useful for identifying cirrus clouds. The resolution of Band 8 (panchromatic) is 15 m. There are nine spectral bands in Landsat 8 OLI and Thermal Infrared Sensor (TIRS) images, with Bands 1–7 and 9 having a spatial resolution of 30 m. Aerosol and coastal research benefits from the usage of band 1 (ultra-blue). Band 9 may be useful in identifying cirrus clouds. Band 8 (panchromatic) has a resolution of 15 m. At 100 m, thermal bands 10 and 11 can gather data and offer more accurate surface temperatures as displayed in Table 1.

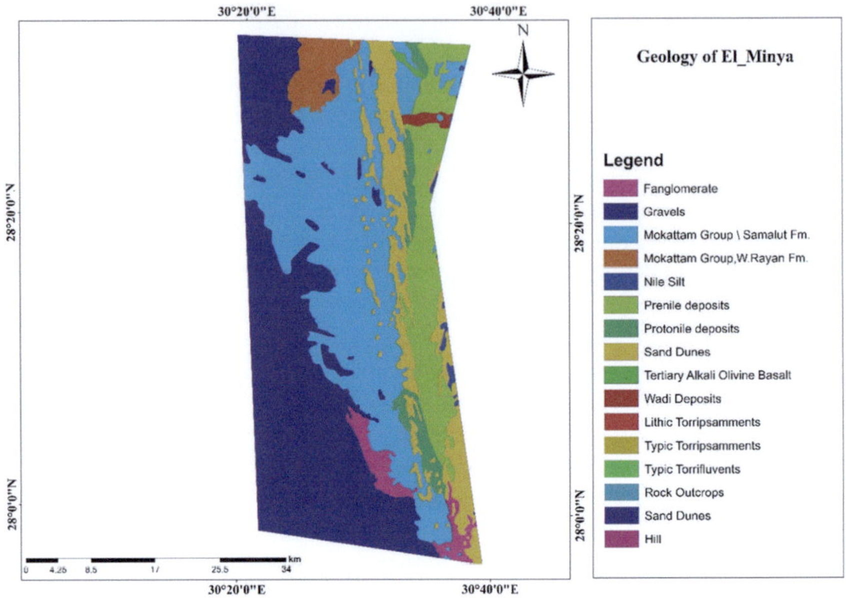

Fig. 3 Geology map of west El-Minya area [10]

Table 1 The band designations and their wavelength for the land sat satellites

	Bands	Wavelength (μm)	Resolution (m)
Landsat 8 OLI and TIRS	Band 1 is Ultra Blue (coastal/aerosol)	0.435 to 0.451	30
	Band 2 is Blue	0.452 to 0.512	30
	Band 3 is Green	0.533 to 0.590	30
	Band 4 is Red	0.636 to 0.673	30
	Band 5 is Near Infrared (NIR)	0.851 to 0.879	30
	Band 6 is Shortwave Infrared (SWIR) 1	1.566 to 1.651	30
	Band 7 is Shortwave Infrared (SWIR) 2	2.107 to 2.294	30
	Band 8 is Panchromatic	0.503 to 0.676	15
	Band 9 is Cirrus	1.363 to 1.384	30
	Band 10 is Thermal Infrared (TIRS) 1	10.60 to 11.19	100 * (30)
	Band 11 is Thermal Infrared (TIRS) 2	11.50 to 12.51	100 * (30)

The raw images from satellites may have minor mistakes in geometry. Sources of these errors may be scan error, mirror velocity and earth rotation. Consequently, it is crucial to correct these images and to rectify them to a definite map projection. This process is known as image geometric correction. The photos are normalised for any radiometric differences caused by the shift in the solar illumination angle resulting in a variation in pixel values on the basses of the geometric correction. The process of eliminating the influence of smoke, haze, dust, and greenhouse gas emissions is known as atmospheric correction [3]. Landsat Data Continuity Mission (LDCM) sensor (Landsat-8) data were used for the case study area remote sensing analyses in 2018. All additional digital image processing and analysis was carried out with the program Arc-GIS 10.3 and ENVI 5.2.

3.1.2 Radiometric and Atmospheric Correction

Radiometric correction is a precondition for generating high-quality scientific data, making it possible to discern between product artefacts and actual changes in Earth processes, as well as reliably construct land cover maps and detect changes [22].

The process of converting digital values obtained by a sensor into physically significant quantities, including surface temperature, brightness, or reflectance, is known as radiometric correction [21].

Atmospheric correction methods can be employed to reduce the effect of solar or viewing conditions, and sensor spectrum absorption induced by aerosols continues to provide corrective issues [26], as shown in Fig. 4.

3.1.3 Image Stacking

This study employed data from the blue to short-wave infrared portion of the spectrum due to the nature of the investigation. Spectrally omitted were the bands of the Thermal Infrared Sensor (TIRS), along with the bands of the cirrus and coastal regions. The panchromatic band was used to increase the reflectance bands up to a spatial resolution of 15 m. The bands were layered and spatially adjusted to fit the extracted west EL-Minya area.

Image subset. Because of the scene covers a much larger area than required, subset image file was selected to cover only the study area. Sub-setting refers to breaking out operation of large scene into one or more smaller areas. In this case it is helpful to expedite the processing process by shrinking the visual scene to just include the relevant area [13].

3.1.4 Image Post Processing

Purpose of the digital image processing is to obtain information, which is difficult to see from the raw data. Image processing can be divided into three categories: image

Fig. 4 Satellite data pre-processing steps

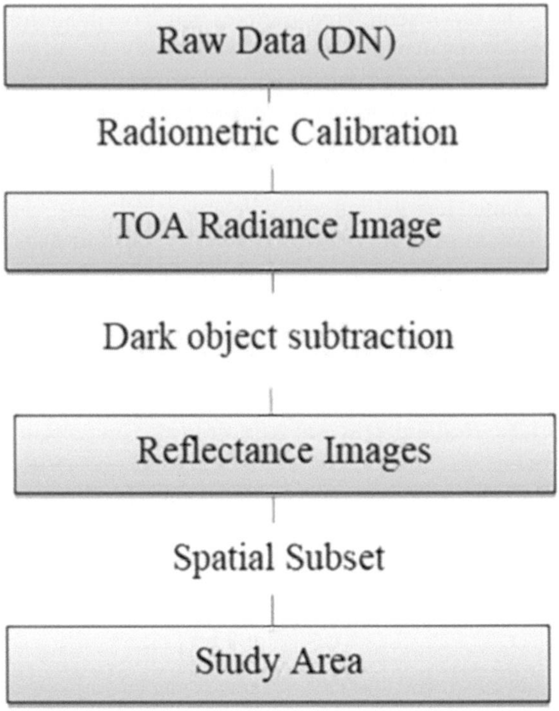

pre-processing, which includes methods for geometric corrections, data error and noise removal, and image enhancement, which improves pixel readings and data; and image classification, which groups the image's spectral content into distinct classes [27].

3.2 *Software*

This chapter is strongly methodology-oriented based on RS and GIS. Hence, the software utilized is a fundamental part of the whole process of this research, especially the methods and the analyses implemented.

The following software packages have been implemented during this research work:

ArcGIS map (ver. 10.8):

ArcGIS desktop is the type of GIS programs that is most commonly utilized, which is well-known worldwide. The Environmental System Research Institute Inc. produced it.

In this study, the elements of ArcGIS desktop-like ArcMap, Arc Catalogue, and Arc Toolbox were used to creating the geodatabase; handling, modifying, and storing data; georeferencing data from several sources; executing spatial multicriteria analysis and result data visualization; using geoprocessing tools for varied purposes; creating criterion maps and combining them; establishing a least-cost and cost-distance (ESRI Arc Map, version 10.8, User Manual, ESRI, Redlands, California, 2006). ENVI (ver.5.3): for digital processing of remotely sensed data (ENVI, Eastman, J.R., ENVI version 5.1 Guide to GIS and Image Processing Clark University Manual, 2009).

Microsoft excels (ver. 2016): to product tables, conduct calculation, and statistical analysis.

3.3 Geomorphologic Mapping

Wulder et al. [34] said that geomorphological features could have been detected by evaluating satellite images. Using a variety of topographic maps and a field survey, the landscape recovered from the visual interpretation of satellite photos over the digital elevation model (DEM) was analysed to determine the geomorphological components. This allowed for the acquisition of the natural 3D terrain.

3.4 The Field Investigation

In the current study, a number of twelve representative water samples were collected for laboratory analysis. Samples were chosen, on basis of dominant geomorphologic feature shown in (Fig. 5).

3.5 Analysis of Water Resource

The following determinations were carried out according to the methods described by Rowell (1995).

- Electrical conductivity (EC) using electric conductivity meter.
- Soluble Ca^{2+} and Mg^{2+} were determined by versinate method
- Soluble Na^{+} and K^{+} were determined by the Flame Photometer.
- Soluble $CO_2{}^{3-}$ and $HCO_3{}^{-}$ were determined by titration against HCl.
- Soluble Cl^{-} was determined by the Mohr's method.
- Soluble $SO_4{}^{2-}$ was calculated by the difference between total anions and cations.

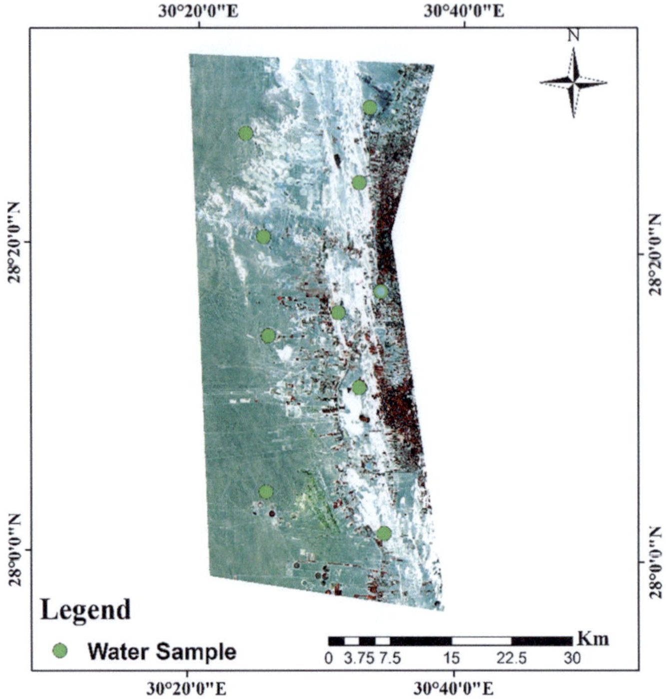

Fig. 5 Location map of water sample of the study area

3.6 *Assessment of Water Resources (IWQI)*

Groundwater samples were taken, moved to the lab, and kept at 4 °C in 1-L polythene bottles. The Irrigation Water Quality Index (IWQI) and the qi and wi of individual parameters were reported by Meireles et al. (2010).

Finally, geographic distribution maps of IWQIs and water quality metrics were created using the IDW (Inverse Distance Weight) technique. This was accomplished by utilising analytical results to obtain the statistical spatial distribution of the variables within a GIS framework. For all of the research, Excel 2010 and Arc GIS 10.8 were utilized.

Meireles et al. (2010) created the IWQI technique, which was initially used for agricultural water quality monitoring. To calculate each parameter's proportionate weight in this technique, the Ayers and Westcot Criteria (1985) and irrigation water quality data from the University of the California Committee of Consultants (UCCC) should be used.

The IWQI identifies key parameters that impact water quality for agriculture, such as EC, Na^+, Cl^-, HCO_3^-, and SAR. In the second step, the weight of water quality parameters, such as qi and wi, should be determined based on individual parameter values and the criteria proposed by Ayers and Westcot (1985).

It should be noted that in this framework, a lower number indicates bad water quality, and vice versa. The following equation (1) was used to calculate the value of qi.

$$qi = qimax - [(xij - xinf) * qimap)/xamp] \quad (1)$$

where xij is the observed value for the parameter; xinf is the corresponding value to the lower limit of the class to which the parameter belongs; qiamp is the class amplitude; and xamp corresponds to the class amplitude to which the parameter belongs. In this example, qimax represents the maximum value of qi for each class.

In this instance, the maximum value found when the water samples were examined was the upper limit, This was required to evaluate the final class's x amp. The values of (qi) were determined by the University of the California Committee of Consultants (UCCC) using factor amount, tolerance limit, and irrigation water quality criteria, as shown in Table 2.

The relative weights of each parameter are shown in Table 3. IWQI attributes for each class were displayed in Table 4 and the IWQI value derived from Eq. (2) using the previously established methodology.

$$IWQI = \sum_{i=1}^{n} (qi * wi) \quad (2)$$

In order to explain global variability in water quality, the formula uses the following notation: wi is the parameter's normalised weight; qi is the parameter's

Table 2 Parameter limiting values for quality measurement (qi) calculation

Qi	EC ds/m	Na^+ mq/l	Cl^- mq/l	HCO_3^- mq/l	SAR
85–100	0.2–0.75	2–3	1–4	1–1.5	2–3
60–85	0.75–1.5	3–6	4–7	1.5–4.5	3–6
35–60	1.5–3	6–9	7–10	4.5–8.5	6–12
0–35	<0.2	<2	<1	<1	<2
	>3.0	>9	>10	>8.5	>12

Table 3 Weights for the IWQI parameters

Parameters	Wi
EC ds/m	0.211
Na^+ mq/l	0.204
Cl^- mq/l	0.202
HCO_3^- mq/l	0.194
SAR	0.189
Total	1.00

Table 4 Restriction and characteristics of general IWQI

IWQI	Exploitation restriction	Recommendation for soil	Recommendation for plant
85 to 100	No restriction (NR)	Water has been used on practically every type of soil. Soil is at decreased danger of salinity/sodicity concerns	No toxicity risk for mast plants
70 to 85	Low restriction (LR)	Irrigation soil with a slight texture or moderate porosity can be used in this range. Soil leaching is indicated to prevent soil sodicity in heavy-textured soils	Evaluated risks for salt sensitive plants
55 to 70	Moderate restriction (MR)	Water in this range is best suited for soils with medium to high porosity properties. Moderate salt leaching is highly recommended to prevent soil damage	Plant with modest tolerance to salt can be grown
40 to 55	High restriction (HR)	This water level is suitable for usage in high-permeability soils with no impact layers. Irrigation is scheduled at a high frequency	Water with low Na^+, Cl^-, and HCO_3^- levels is suitable for irrigation of plants with moderate to high salt tolerance when combined with appropriate salinity control methods
0 to 40	Severe restriction (SR)	Under typical conditions, it is not recommended to use this range of water for irrigation	Only plants with great salt tolerance can expect waters with exceptionally low levels of Na^+, Cl^-, and HCO_3^-

quality, which ranges from 0 to 100 and represents a function of its measurements or concentration; and IWQI is a one-dimensional irrigation water quality index, which also ranges from 0 to 100. The restrictions on water consumption classes were outlined by Meireles et al. (2010) (Table 4).

4 Result and Discussion

4.1 The Main Geomorphologic Features of the Studied Area

Geomorphologic components could have been defined by describing satellite images. The geomorphic units were established by examining the topography with the aid of multiple topographic maps, a field survey, and visible interpretation of satellite pictures overlaid on the digital elevation model (DEM) to obtain the natural 3D terrain.

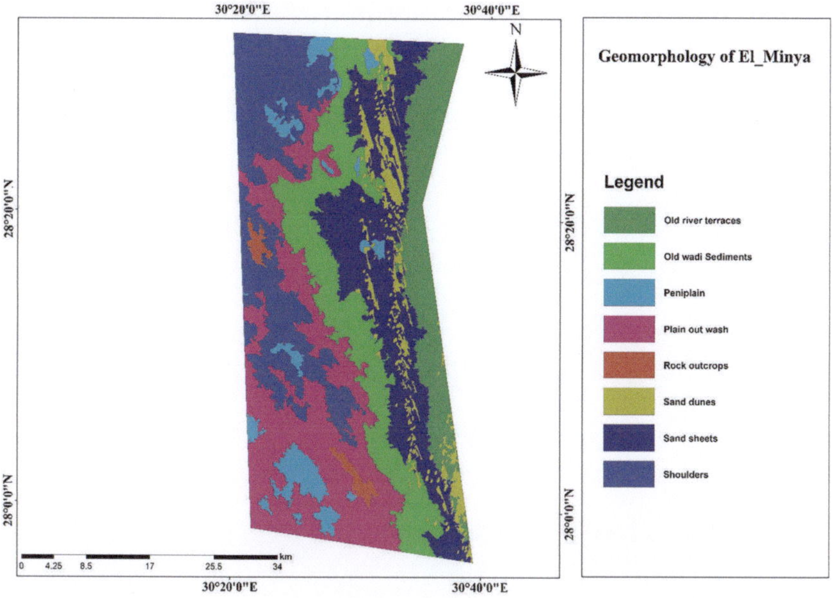

Fig. 6 Geomorphology of west El-Minya Governorate (Gad and Mohamed 2024)

The areas of geomorphic unity are revealed by data as follows: (1) Old wadi Sediments forming 5.4% of the total area, (2) Peniplain (forming 7.4% of the total area), (3) old river terraces (forming 7% of the total area), (4) Shoulders (forming 15.4% of the total area), (5) Sand sheets (forming 6.23% of the total area), (6) Rock outcrops (forming 0.84% of the total area), (7) Sand dunes (forming 2.29% of the total area), (8) Erosional Plains (forming 15.58% of the total area), (9) Hill (forming 1.45% of the total area), and (10) Plain out wash (forming 38.51% of the total area).The results are illustrated in Fig. 6.

4.2 Application of Piper Trilinear Diagram for Water Quality

The chemical characteristics of the research region and water were determined by plotting analytical results on a Piper diagram (Piper 1944). The two triangles on the Piper trilinear diagram represent cations and anions, respectively. The area in the shape of a diamond represents the combined positions of cations and anion. Based on the hydro-geochemical features of the water, we may identify it and draw conclusions from the combined single position of the diamond-shaped area.

The diamond-shaped region of the Piper diagram is divided into four main sections, each of which illustrates and clarifies a certain type of variation or domination of cations or anion. Figure 7 illustrates that the predominant cations in both waters are Na^+ ions, whereas the prominent anions are Cl^-. The origins of both waters are comparable.

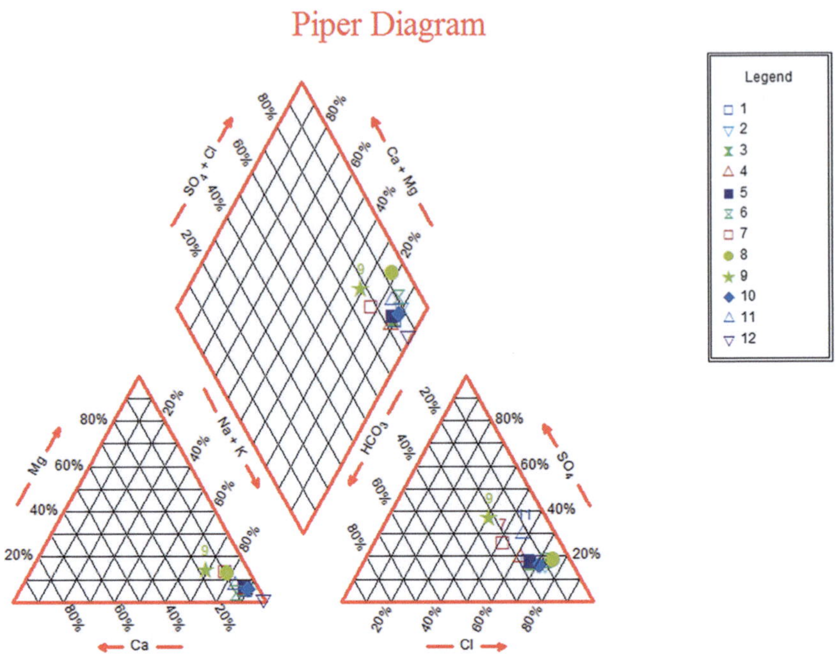

Fig. 7 Piper's trilinear diagrams depict the classification of groundwater samples in the research area

4.3 Classifications of the Ground Water for Irrigation According to USDA Method

The USDA's classification of the salinity (C) and sodicity (S) threats in the research region is shown in Table 5. The zone of moderate salinity hazard, where the total dissolved salt content in the water is moderate, is where most of the wells are located.

According to the USDA categorization system, 75% of the wells' water falls in the salinity class C3, 17% in the C4 class, and just 8% in the C2. On the other hand, nearly 80% of the wells have SAR values ranging between 2.5 and 18, which it is of medium quality for irrigation water. The C3-S1 class (medium salinity—low sodicity), C4-S2 class (high salinity—medium sodicity), and C2-S1 class (moderate salinity—low sodicity) are the salinity and sodic dangers and C4-S1 class (high salinity—low sodicity) 0.75% of Such water is rated as medium Salinity hazards and medium Sodicity hazards for irrigation, i.e. suitable for irrigating most crops except the sensitive ones. Low to medium SAR and moderate to high EC_w can be handled without using water or soil treatments. Excess salts can be leached from the root zone, which will solve the problem.

Table 5 USDA Salinity, sodicity classification of El-Minya area

Water classification	Sodicity hazard** (SAR)				Salinity hazard* EC (dSm^{-1})						Sample no
	Class S4	Class S3	Class S2	Class S1	Class C6	Class C5	Class C4	Class C3	Class C2	Class C1	
C3-S2			*					*			1
C3-S2			*					*			2
C3-S1				*				*			3
C3-S2			*					*			4
C3-S2			*					*			5
C4-S2			*				*				6
C2-S1				*					*		7
C4-S1			*				*				8
C3-S1				*				*			9
C3-S2			*					*			10
C3-S2			*					*			11
C3-S2			*					*			12

Note Low, moderate, medium, high, very high, and excessively high salinity hazards are distinguished (EC values <0.25, 0.25–0.75, 0.75–2.25, 2.25–4.00, 4.00–6.00, and >6.00 dSm^{-1}). Percentages of EC values <0.25, 0.25–0.75, 0.75–2.25, 2.25–4.00, 4.00–6.00, and >6.00 dSm^{-1} correspond to low, moderate, medium, high, very high, and extremely high salinity hazards

4.4 Irrigation Water Quality Index (IWQI)

The analytical data of groundwater samples are given in Table 6. The data revealed that, the studied water reaction (pH) ranged between 8.46 and 7.51 as shown in Figure 48. The water salinity values (EC_w) values are ranged between 0.67 to 3.87 dS/m. The soluble cations could be arranged in the descending order of $Na^+ > Ca^{2+} > Mg^{2+} > K^+$. The soluble anions could be arranged in the following descending order of $Cl^- > SO_4{}^{2-} > HCO_3{}^{2-}$. The dominant soluble salt in this water is sodium chloride. Also, data in Table 6 showed that the sodium adsorption ratio (SAR) values of the groundwater samples ranged between 2.06 and 15.08.

4.4.1 Spatial Distribution of EC and SAR in West West El-Minya Area

Figure 8a illustrated that, the salinity of the west west El-Minya ranged between 0.67 to 3.87 ds/m. The part most affected by salinity was located in the middle on the eastern side. On the other hand, salinity dropped both up and down from the salinity centre point. According to Piper's trilinear diagrams, Na^+ ions are the dominating cations, whereas Cl^- is the dominant anions. Therefore, the source of salinity may be due to the dominance of sodium chloride salt.

Table 6 Chemical analyses of the groundwater samples in the study area

Water sample	pH	EC dS/m	Cations (mmolc L^{-1})				Anions (mmolc L^{-1})			SAR
			Na^+	K^+	Ca^{2+}	Mg^{2+}	HCO_3^-	SO_4^{2-}	Cl^-	
1	8.32	1.16	9.22	0.16	0.62	0.48	1.80	1.96	7.30	12.43
2	7.98	1.51	13.10	0.15	0.73	0.82	1.64	2.30	11.20	14.88
3	7.98	1.62	13.60	0.18	086	0.90	2.62	2.64	10.90	2.06
4	7.83	1.30	10.80	0.22	0.71	0.63	2.40	2.50	8.10	13.19
5	8.29	1.72	13.67	0.19	0.90	0.96	2.63	2.80	11.10	14.18
6	8.31	2.72	20.18	2.40	2.60	0.98	2.44	4.60	19.30	15.08
7	8.46	0.67	4.70	0.11	0.60	0.80	1.62	1.82	3.80	5.62
8	7.51	3.87	26.48	3.16	3.48	4.76	2.40	6.80	28.50	13.05
9	7.68	1.07	5.66	1.18	1.68	1.36	2.30	3.70	4.20	4.59
10	7.62	1.55	12.90	1.15	0.73	0.82	1.960	2.36	10.80	14.65
11	8.08	1.33	14.50	1.05	1.63	1.32	1.62	3.80	7.56	11.94
12	7.78	1.57	10.85	2.41	0.87	0.62	1.95	2.44	9.63	2.57

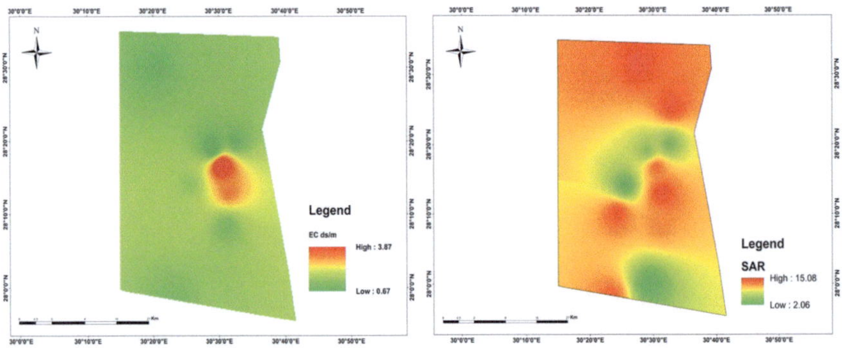

(a) (b)

Fig. 8 Spatial distributions of EC (A) and SAR (B) in west west El-Minya area

The distribution of SAR values in west west El_Minya is shown in Fig. 8B. SAR shows that all samples values range between 2.06 to 15.08. It is seen that samples of low SAR are mainly located in the center of the study area and the eastern lower part. While high SAR dominated the majority of the remaining region. This might be because of a larger rate of evaporation and the amount of rainfall received in a certain area which both had a significant impact on the quantity of minerals in the borehole. Heavy rainfall depletes mineral concentrations due to high penetration rates, but high evaporation rates enhance mineral concentrations, resulting in salt deposition in groundwater.

While Table 7 shows the computed characteristic of the irrigation water quality index, Table 6 shows the findings of the chemical analysis for each water sample.

Table 7 The water quality index values of groundwater samples in the study area

Sample no	EC ds/m		Na mq/l		HCO_3^- mq/l		Cl m q/l		SAR		IWQI	
	qi	qi*wi	qi	qi*wi	qi	qi*wi	qi	qi*wi	qi	qi*wi	Index	Class
1	81.83	17.27	29.03	5.92	83.29	16.82	56.93	11.04	10.93	2.07	53.12	HR
2	54.57	11.52	23.90	4.88	84.81	17.13	28.51	5.53	5.25	0.99	40.04	HR
3	53.86	11.37	23.24	4.74	75.49	15.25	26.28	5.10	100.00	18.90	55.35	MR
4	80.93	17.08	26.94	5.50	77.59	15.67	56.23	10.91	9.17	1.73	50.88	HR
5	53.22	11.23	23.14	4.72	75.40	15.23	26.04	5.05	6.87	1.30	37.53	SR
6	46.76	9.87	14.54	2.97	77.21	15.60	15.96	3.10	4.78	0.90	32.43	SR
7	100.00	21.10	85.00	17.34	85.00	17.17	100.00	19.40	76.74	14.50	89.51	NR
8	6.06	1.28	6.21	1.27	77.59	15.67	4.67	0.91	9.49	1.79	20.92	SR
9	82.42	17.39	84.09	17.16	78.54	15.86	84.65	16.42	80.81	15.27	82.10	LR
10	54.32	11.46	24.16	4.93	81.77	16.52	26.40	5.12	5.78	1.09	39.12	SR
11	80.74	17.04	22.05	4.50	85.00	17.17	56.70	11.00	12.07	2.28	51.98	HR
12	54.19	11.43	26.87	5.48	81.86	16.54	54.89	10.65	10.61	2.00	46.10	HR

According to the findings, groundwater in this area is most suited for plants that can withstand high salinity, such sugar beetroot. Waters with low concentrations of Na^+, Cl^-, and HCO_3^-, as well as soils with high permeability and an effective agricultural drainage system to prevent salt buildup, are also suitable. Severe restriction (SR) irrigation water was applied to the remaining groundwater samples (0.2%) in the research study area, indicating extremely low groundwater quality for irrigation. In summary, under normal circumstances, this groundwater should not be used for irrigation. In unusual instances, it may be used on occasion. It was discovered that 3.1% of the research area's groundwater samples were classified as high restriction (HR) irrigation water. If water with low Na^+, Cl^-, and HCO_3^- values is available, it should be used to irrigate plants with a medium to high salt tolerance (like tomatoes), using specific salinity control techniques. To get the best use in this situation, soils should be reached with high permeability and free of compacted layers. Additionally, the results show that 8.9% of the groundwater samples in the study area were categorized as moderate restriction (MR) irrigation water. This type of water is recommended for irrigating plants that can withstand moderate amounts of salt and are grown in soils that range from moderate to high permeable, with a suggestion of moderate salt leaching. Low restriction (LR) irrigation water, which may be used to irrigate any plant (such as bean) while avoiding salt-sensitive plants, makes up the bulk of groundwater in the research region (82.4%). This water is recommended for irrigation on light-textured or moderately permeable soils that leach salt.

It is advised against using it in soils that are high in clay because soil sodicity can occur in heavy texture soils. 5.4% of the groundwater samples in the study area are classified as No restriction (N.R.) irrigation water, meaning that they can be used to irrigate any type of plant without putting most plants at danger of toxicity. After conducting geostatistical analysis, the theme maps (EC, Na^+, Cl^-, HCO_3^-, and SAR) were superimposed to form the IWQI map that is depicted in Fig. 9. This map aids decision makers in managing sustainable agricultural growth and choosing suitable crops and irrigation systems by showing the distribution of groundwater quality for irrigation purposes in the research region based on IWQI values.

5 Conclusion

Groundwater in arid/semiarid environments is critical for water supply, irrigated agriculture, and ecosystem sustainability. Natural resources on irrigated land, such as in west El-Minya, Egypt, have a potential future for agricultural expansion projects if appropriate reclamation technologies and management practices are used. The findings and methods utilized in this study can be useful in analyzing subsurface water quality for irrigation applications. With the exception of sensitive crops, 75% of the water in the study area is suitable for irrigating most crops. It is possible to maintain low-to-medium SAR and moderate-to-high ECw without using any water or soil treatments. Leaching can be used to extract excess salts from the root zone, which will resolve the problem.

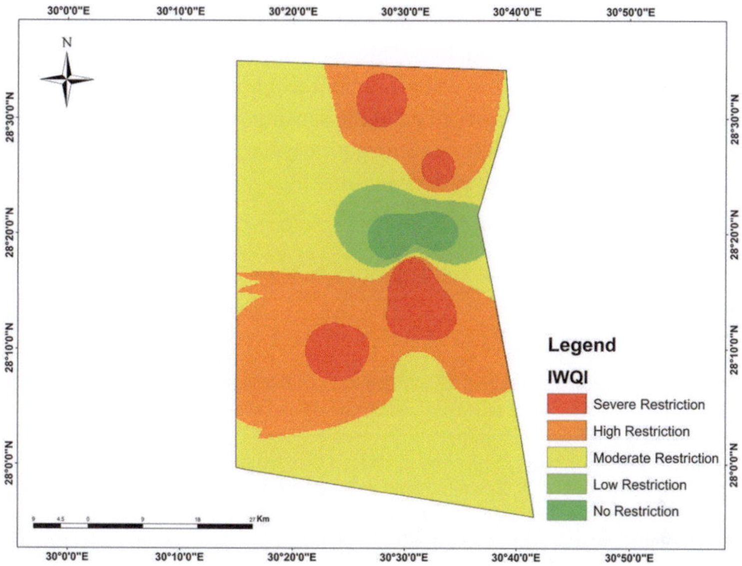

Fig. 9 IWQI map of west El-Minya area

6 Recommendations

Assessing groundwater quality for irrigation is crucial to ensure sustainable agricultural practices and protect crop health. Here are some recommendations for future studies:

1. Conduct regular and comprehensive testing of groundwater for a wide range of parameters, including pH, electrical conductivity (EC), total dissolved solids (TDS), major cations and anions (e.g., sodium, potassium, calcium, magnesium, chloride, sulfate, bicarbonate), heavy metals, and contaminants (e.g., nitrates, pesticides).
2. Implement long-term monitoring programs to observe trends and changes in groundwater quality over time. This can help in understanding the impact of agricultural practices, climate change, and other factors on groundwater quality.
3. Study the interaction between soil, water, and crops to understand how groundwater quality affects soil health and plant growth. This includes investigating soil salinity, sodicity, and the uptake of nutrients and contaminants by plants.
4. Promote integrated water management practices that combine the use of groundwater with surface water and other water sources. This can help in maintaining the balance and sustainability of water resources.
5. Educate farmers and stakeholders about the importance of groundwater quality and best practices for its management. Training programs can help in adopting sustainable irrigation and agricultural practices.

6. Advocate for the development and enforcement of policies and regulations that protect groundwater quality. This includes setting permissible limits for contaminants and promoting water-efficient technologies
7. Utilize advanced technologies like remote sensing AND GIS for real-time monitoring of groundwater quality and irrigation practices. These technologies can provide timely data and aid in decision-making.

References

1. Abdel-Fattah MK, Abd-Elmabod SK, Aldosari AA, Elrys AS, Mohamed ES (2020) Multivariate analysis for assessing irrigation water quality: a case study of the Bahr Mouise Canal, Eastern Nile Delta. Water 12(9):2537
2. Abdelhafez AA, Metwalley SM, Abbas HH (2020) Irrigation: water resources, types and common problems in Egypt. Technological and modern irrigation environment in Egypt: best management practices & evaluation, pp 15–34
3. Abdel-Hamid H. Inventory of soil salinity, alkalinity and water logging problems in the Nile Delta, Egypt. Doctoral dissertation, M.Sc. Thesis
4. Awodumi OE, Akeasa OS (2017) Gis applications for assessing spatial distribution of boreholes and hand dug wells in Boroboro community, Atiba Local Government, Oyo state. J Remote Sens GIS 6(208):2
5. Balakrishnan P, Saleem A, Mallikarjun ND (2011) Groundwater quality mapping using geographic information system (GIS): a case study of Gulbarga City, Karnataka, India. Afr J Environ Sci Technol 5(12):1069–1084
6. Batarseh M, Imreizeeq E, Tilev S, Al Alaween M, Suleiman W, Al Remeithi AM, Al Tamimi MK, Al AM (2021) Assessment of groundwater quality for irrigation in the arid regions using irrigation water quality index (IWQI) and GIS-Zoning maps: case study from Abu Dhabi Emirate, UAE. Groundw Sustain Dev 1(14):100611
7. Beadnell HJ (1905) The topography and geology of the Fayum Province of Egypt. National Print. Department
8. Bishay Y (1966) Studies on the larger Foraminifera of the Eocene of the Nile Valley between Assiut, Cairo and SW Sinai. Ph.D. Thesis, Alexandria Univ
9. Collet C (1996) Geographic information system needs and software
10. CONOCO, Coral (1987) Geological map of Egypt. Scale, 1.500, 000
11. Elewa HH, Qaddah AA (2011) Groundwater potentiality mapping in the Sinai Peninsula, Egypt, using remote sensing and GIS-watershed-based modeling. Hydrogeol J 19(3):613
12. El-Rawy M, Makhloof AA, Hashem MD, Eltarabily MG (2021) Groundwater management of quaternary aquifer of the Nile Valley under different recharge and discharge scenarios: a case study Assiut governorate, Egypt. Ain Shams Eng J 12(3):2563–2574
13. ESRI (Environmental Systems Research Institute) (2014) ArcGIS 10.1 help manual. Cell statistics, inverse distance weight (Spatial Analyst)
14. Gaagai A, Aouissi HA, Bencedira S, Hinge G, Athamena A, Heddam S, Gad M, Elsherbiny O, Elsayed S, Eid MH, Ibrahim H (2023) Application of water quality indices, machine learning approaches, and GIS to identify groundwater quality for irrigation purposes: a case study of Sahara Aquifer, Doucen Plain, Algeria. Water 15(2):289
15. Ibrahem S, Elalfy M, Hagras M (2020) Groundwater potentials of Eocene limestone aquifer in west-west El-Minya area, Egypt. Egypt J Desert Res 70(1):59–82
16. Ibrahim H, Yaseen ZM, Scholz M, Ali M, Gad M, Elsayed S, Khadr M, Hussein H, Ibrahim HH, Eid MH, Kovács A (2023) Evaluation and prediction of groundwater quality for irrigation using an integrated water quality indices, machine learning models and GIS approaches: a representative case study. Water 15(4):694

17. Khan F, Husain T, Lumb A (2003) Water quality evaluation and trend analysis in selected watersheds of the Atlantic region of Canada. Environ Monit Assess 88:221–248
18. Magesh NS, Chandrasekar N, Soundranayagam JP (2012) Delineation of groundwater potential zones in Theni district, Tamil Nadu, using remote sensing, GIS and MIF techniques. Geosci Front 3(2):189–196
19. Masood MU, Rashid M, Haider S, Naz I, Pande CB, Heddam S, Alshehri F, Elkhrachy I, Ahsan A, Sammen SS (2024) Exploring groundwater quality assessment: a geostatistical and integrated water quality indices perspective. Water 16(1):138
20. Nguyen TG, Phan KA, Huynh TH (2022) Application of integrated-weight water quality index in groundwater quality evaluation. Civil Eng J 8(11):2661–2674
21. Pons X, Arcalís A (2012) Diccionari the terminology of remote sensing. Enciclopedia Catalana and Inst. Cartog. Catalunya, Barcelona. ISBN: 978
22. Pons X, Pesquer L, Cristóbal J, González-Guerrero O (2014) Automatic and improved radiometric correction of Landsat imagery using reference values from MODIS surface reflectance images. Int J Appl Earth Obs Geoinf 1(33):243–254
23. Rania M (2020) Evaluation of irrigation water and soil quality for optimum agricultural use planning using GIS and remote sensing techniques. M.Sc. Thesis. Fac. Sci., Moshtohor. Univ. Banha, Egypt
24. RIGW (1989) Development and management of groundwater resources in the Nile Valley and Delta
25. RIGW (1997) Research institute for groundwater. Hydrogeological map of Egypt, Scale 1:1000, 000. Map Sheet of El_Minia
26. Roy DP, Wulder MA, Loveland TR, Woodcock CE, Allen RG, Anderson MC, Helder D, Irons JR, Johnson DM, Kennedy R, Scambos TA. Landsat-8: science and product vision for terrestrial global change research. Remote Sens Environ
27. Sabins FF (1997) Remote sensing: principles and interpretation. WH Freeman and Co., New York, USA
28. Said R (1983) Remarks on the origin of the landscape of the eastern Sahara. J Afr Earth Sci 1(2):153–158
29. Said R (2012) The geological evolution of the River Nile. Springer Science & Business Media
30. Sarath Prasanth SV, Magesh NS, Jitheshlal KV, Chandrasekar N, Gangadhar KJ (2012) Evaluation of groundwater quality and its suitability for drinking and agricultural use in the coastal stretch of Alappuzha District, Kerala, India. Appl Water Sci 2:165–175
31. Sweeney MW (1997) Geographic information systems. Wiley, Hoboken, pp 419–422
32. Tolche AD (2021) Groundwater potential mapping using geospatial techniques: a case study of Dhungeta-Ramis sub-basin, Ethiopia. Geol Ecol Landsc 5(1):65–80
33. Wolff P (1993) Land reclamation in Egypt. Der Tropenlandwirt-J Agric Tropics Subtropics 94(2):103–117
34. Wulder MA, White JC, Goward SN, Masek JG, Irons JR, Herold M, Cohen WB, Loveland TR, Woodcock CE (2008) Landsat continuity: issues and opportunities for land cover monitoring. Remote Sens Environ 112(3):955–969

Conclusions and Recommendations for "Environmental Remote Sensing and GIS in MENA Region"

Faiza Khebour Allouche

Abstract The MENA region is facing many challenges related to climate change, environmental degradation, urban sprawl, water scarcity, and hazards. However, for several years, multidisciplinary research has focused on using remote sensing and GIS to resolve these dares. The authors present reviewed chapters on strategies for rainwater management and the state of the art of freshwater fishing followed by the contribution of new technologies to identify the possibilities of developing these sectors. The review of the state of the art on marine fauna wealth and then its mapping has enabled the development of the marine fishing sector. Furthermore, exploring the GEE platform with its free access and the richness of its database makes it possible to evaluate and monitor the physical characteristics of the regions. Three main risks in coastal MENA regions are raised: flooding, heat islands and eutrophication. Researchers propose GIS spatial modeling and analysis approaches based on analyzing time series of satellite images to reduce these risks. Remote sensing and GIS applications in urban areas are presented in regions of the Maghreb. These chapters indicate the contribution of these tools to characterize the spatial organization of oases, variation in the spatiotemporal dynamics of plants, and the impact of man on urban expansion. Meanwhile applications in rural areas have dealt with flooding, degradation and variation in land use. The editors bring together experience, methods, models and results that show the way forward and support decision-making on sustainable resource management, landscape planning, risk assessment and monitoring in MENA regions and others. All this is necessary to develop strategies for a strengthened green revolution and an accelerated blue revolution.

Keywords Geo-information · Remote sensing · GIS · Google Earth · Radiometric indexes · Strategies · Planning · Monitoring · Flood · Degradation · Management · Environmental · MENA

F. K. Allouche (✉)
High Institute of Agronomic Science of Chott Mariem, Department of Engineering and Landscape Planning, Sousse University, Chott Meriem Sousse, Tunisia
e-mail: allouchekhebour@yahoo.fr

Lr GREEN TEAM (LR17AGR01), Carthage University, National Agronomic Institute of Tunis, Tunis Mahrajène, Tunisia

F. Khebour Allouche (ed.), *Environmental Remote Sensing and GIS in MENA Region*, Earth and Environmental Sciences Library,
https://doi.org/10.1007/978-3-031-86574-9_16

1 Update

In the following paragraphs, different studies concerning the environmental applications of remote sensing and GIS in MENA regions are presented. The methodological approaches and the diversity of the approaches used demonstrate significant gains for different users. Furthermore, new studies presented in this book highlight the importance of using GIS and remote sensing for sustainable and applicable territorial planning in the MENA regions. The brief results of the studies are introduced in this chapter.

1.1 Overview of Remote Sensing and GIS Applications and Strategies

Four chapters are identified related to remote sensing and GIS applications and strategies. The first study presents a pioneering approach to assess and understand rainwater harvesting (RWH) strategies, drawing from a case study from central West Jordan (Karak Plateau). By blending traditional fieldwork methods and modern remote sensing technology, particularly the analysis of very high-resolution (VHR) satellite imagery like GeoEye-1 data, firstly, efficient data collection and the compilation of a comprehensive, area-wide, inventory of RWH structures becomes possible. Integrating the latter with collected on-the-ground data, particularly regarding the dimensions of underground cisterns, facilitates the estimation of the total water storage potential available with the existing cisterns in the area. Additionally, outcomes from satellite imagery interpretation, including rooftop area and land use / land cover (LU/LC) information, allow for estimating the overall RWH potential and the rooftop rainwater harvesting (RRWH) potential for each settlement. The data enable a spatially explicit comparison between storage and harvesting potential, thus making the identification of discrepancies, untapped opportunities, and suitable actions for development possible. Moreover, the chapter explores spatial patterns in RWH structure distribution concerning potential relations to environmental factors and the historical development of the area. Overall, these analyses offer valuable insights into the development of RWH strategies and steps for their adoption and improvement in modern contexts. Notably, the findings are applicable beyond the studied region, while the methodological approach holds promise for addressing research questions beyond RWH assessment.

A review of the small pelagic fisheries' status in the Northeast of Tunisia before and after implementing of the National Strategy for the Development of Small Pelagic Fisheries (NSDSPF) was presented. Then, the Geographic Information System and the Vessel Monitoring System were used to map the fishing areas' evolution between 2003, when the national strategy for the development of bluefish capture (NSDSPF) was established, and 2021, revealing a great expansion in small pelagic fish harvesting. Results showed that the production of small pelagic fish in Tunisia has experienced several fluctuations over the last 28 years (1991–2019). Maps representing the pluriannual variation of the small pelagic fishing areas show a large concentration of fishing units around Kelibia during the first years of the implementation of the NSDSPF. However, the period between 2010 and 2017 was marked by an increase in this activity, then stagnation from 2020 to 2021, with a focus on the fishing areas in the offshore areas. This is mainly due to the development of motorization and the modernization of fishing units targeting small pelagic fish in the eastern region of Tunisia. Nevertheless, the seasonal monitoring of small pelagic fish harvesting activities in the eastern region of Tunisia showed a clear variability between seasons in 2021. The greatest concentration was observed during the summer, with the remoteness of the new fishing areas and the sea state factor becomes primary for this fishing activity because the maneuvring of the seine requires good weather conditions. Despite the modernization of the equipment, the development of motorization, and the increase in the size of the vessels, the weather and sea state factors remain decisive for this fishing activity, especially in the offshore areas.

The evolutionary trends of the available freshwater fishery resources conducted in two tunisian reservoirs show that the yield of the study species is moderate in Siliana (23.72 ind/1000 m^2 and 4.7 kg/1000 m^2), while yields in Kasseb Reservoir are found to be quite low (123 ind/1000 m^2 and 18 kg kg/1000 m^2). The statistical analysis demonstrates that Siliana Reservoir possesses a high abundance of mullet population. Indeed, this reservoir is endowed with more effective habitat conditions than Kasseb, favouring the growth and survival of rudd and roach. In addition, the spatial distribution of different species and catch rates in these two reservoirs shows that the studied populations are diverse. However, some ecological niches in the dams are underexploited, which requires certain measures to be undertaken, such as boosting fishing operations, seeding Mugilidae fry and introducing new species.

The fourth chapter presents an application of a powerful cloud-based platform for managing and analyzing Earth observation data "Google Earth Engine" in Algeria. The findings from the case study reveal that variations in NDVI reflect spatial and temporal changes attributed to drought and forest wildfires in Algeria. The NDSI calculations has effectively identified saline depressions in the North of Algeria, serving as a valuable tool for assessing soil salinity.

1.2 Geospatial Technologies and Risks

Using the normalized difference water index, four main bodies of water are identified in Tunisia, especially, on the north side it is the lake, sebkha of Monastir, the Sebkha of Moknine on the south-east side and to the west it is the Sebkha of Sidi El Heni. The characterization of the eutrophication phenomena was detected by applaying the normalized difference chlorophyll index who four degrees of eutrophication are established (i) NDCI $\geq$ 0.45 presents the class with a very high degree of eutrophication; NDCI = [0.4–0.45] relates to the class with a high degree of eutrophication; NDCI = [0.35–0.4] corresponds to the average degree of eutrophication; NDCI =

[0.3–0.35] represents the class with a low degree of eutrophication and NDCI = [0.3–0.25] refers to the class with a very low degree of eutrophication. According to physical criteria, results showed that this phenomenon persists in hydromorphic soils and in the proximity of areas occupied by irrigated annual crops. In addition, active sludge, aerobic stabilization and prolonged aeration contribute to increasing the process. However, natural lagoons like Lake Moknine play an important role in de-eutrophication. Marine, physical and anthropogenic parameters contribute to accentuating or weakening the phenomenon.

Flood susceptibility mapping in Morocco is done using different physical parameters integrated on a GIS tool. Resultats show that the FR, WF, and WoE can be regarded as efficient and effective methods for flood susceptibility maps in a GIS environment. Another objective of the current study was to determine the most important causative factor in the study area. Based on the Wf, the order of factors was Lithology, SPI, TWI, stream density, land use, and slope have a high influence on flood occurrence, while the TPI and aspect, elevation, and NDVI are the medium important factors, in addition, the distance to a river, TRI, distance from road, rainfall, and curvature have a minimum influence in flood hazard.

The seventh chapter shows a predominance of heat islands that emphasizes the state of the city and confirms a dominance of the urban mineralized character, creating a large urban heat island area. However, cold islands are poorly represented, with just 21%. Vegetated surfaces outside the administrative boundaries of Belouizdad, play also an important role in heat mitigation and should be taken into account in the estimation of urban heat islands. These cool islands must be preserved and maintained, because in contrast with impermeable surface, vegetated area represent a cooler surface temperature and could thus contribute to mitigating the phenomenon of the urban heat island. However, the projected map shows an option for reducing UHIs, if heat island mitigation measures are implemented.

The stochastic modeling and prediction of four annual and monthly stream flow time series conducted in northern Algeria led to the building of some mathematical models for both of annual and monthly time lags and the extent of random variability was clearly highlighted. It presents a case of study of flood susceptibility mapping using statistical methods such as frequency ratio, weight of evidence, and weighting factor. The statistical methods used are based on the mathematical relationship between observed flood and controlling factors. However, the innovative aspect of this chapter is the use of the integration of remote sensing data and GIS to produce a detailed hazard maps, then the incorporation of citizens in data collection for providing the information about flood locations. For the annual scale, the average variability explained by the models is close to 30% against nearly 60% for the monthly scale. This means that the random variability within the data is about 70% for the annual scale, against only 40% for the monthly one. This result shows the magnitude of the risk incurred if the obtained ARIMA and SARIMA models are used for simulation or forecasting. Actually, the internal dependence structure of the data is reinforced by the introduction of the seasonal effect for the monthly data, which decreases the total random variability.

1.3 Urban Applications

In Maghreb, vegetation cover, a crucial factor for the sustainability of urban ecosystems, is undergoing significant change in cities. After the French colonial period giving key consideration to public vegetation in the new towns built such as in Sousse, Algiers or Marrakesh, the importance attached to greenery in cities is not clearly defined and today consideration of urban vegetation in Maghrebian spatial planning policy is still nascent.

By taking a spatialized approach involving satellite remote sensing results show an overall decline in vegetation cover in the cities of Algiers, Sousse and Marrakech between the 1970s and the 2020s with two distinct periods; a swift and fairly steady decline in greenery until 2000, slowing significantly and even reverses slightly after this date. The spatial distribution of vegetation across the cities has also changed, much more uniform in 1973 than during the later periods where greenery became increasingly concentrated within clearly defined areas. A neighborhood-based analysis of the vegetation cover makes it possible to connect the contrasting spatial distribution and variation observed with the public spatial and urban planning policies since the 1970s, as well as since the colonial period and their implications in terms of socio-environmental justice and socio-spatial segregation generated by the spatial organization of vegetation in Algiers, Sousse and Marrakesh. Today, the proportion of green space per capita shows that these cities have an overall average (of 3.1 m^2 per capita for Marrakesh), falling short of the minimum 10 m^2 per capita recommended by UN-Habitat. In the three cities, the parks and gardens inherited from the French colonial period were largely neglected after independences, and almost completely ignored by the urban planning and cities management departments, resulting in a lack of care for intra-urban green spaces. For a long time, urban planning departments considered the development of parks and gardens in urban areas as an element in the well-being and quality of the urban environment, to be secondary to the primary functions of traffic, housing and commerce. However, we have noticed the same renewed interest in urban greenery in these cities since the 2000s, in the form of new laws and regulations and rehabilitation projects of French colonial gardens and parks.

The spatial distribution of green spaces in Boufícha’s city shows that their amount and distribution are severely limited. Therefore, it is advised to establish additional green areas, such as via managing the urban park in Boufícha’s northern. GE images were helpful in detecting the LULC changes that have taken place from the year 2004 to 2019 in the study area. The city’s population was surveyed to learn about their objectives and suggestions for park design, as well as to better comprehend the relationship between the people and the nature. Thus, it enabled the implementation of environmental governance on a local level. Thus, the survey’s results served as the foundation for the urban park development concept, which is significant for the area’s growth and the enhancement of the local population’s quality of life. Consequently, it is a component of Boufícha’s sustainable development initiative.

A diachronic study is applied on the bay of Algiers using Geographic Information System and imagery to assess the expansion of the mineral element often to the detriment of the plant element, and the necessity opening the bay of Algiers and its infrastructures onto the sea. Three dates have been selected for this study (2001, 2011 and 2022). For the Bay of Algiers, in view of its potential, its population and its central place it occupied in the economy and business, the question is: is the city opened up to the sea? Does it have any prospects and ensure sustainability? To this end, we chose three key sites (*El Kettani*, the *Sablettes promenade* and the *Mohammadia business district*) that underwent dramatic change between 2001 and 2022, to take a critical, analytical look at coastal development. Results show a constant expansion of the mineral element to the detriment of the plant element between 2001 and 2022. In total, Algiers Bay has lost 2.55 km^2 of vegetation, representing a loss of more than 6.5% of the bay's total surface area by 2022. An analysis of a few examples of projects shows that they do not always meet the needs for a Mediterranean city, an open city and sustainability concepts. The Bay of Algiers is a highly sought-after area of great potential and attractiveness whose development needs to be reviewed. The bay's coastline needs real, well-thought-out planning, based on concepts of sustainability and openness to the sea, if it is to become a modern, attractive and green Mediterranean city.

1.4 Rural Applications

LULC change analysis of Al-Ahsa Metropolitan Area in KSA over the last 32 years shows a notable urban sprawl at the expense of the ancient oasis. Since 1970, the cities of Hofuf and Mubarraz have experienced rapid population growth due to socio-economic development and the significant expansion of the productive sectors. According to Al-Ahsa Municipality (2016), the population of the two main cities, covering an area of about 23,000 km^2, rose from 444,970 in 1992 to 768,000 in 2016, and led to increased built-up areas. The objective of this chapter is to map the recent evolution of land use and land cover (LULC) in Al-Ahsa Metropolitan Area (Saudi Arabia) between 1990 and 2022. To understand this complex theme, an integrative, global, and hierarchical methodological approach, combining multi-source and multi-date data was adopted. This approach involves the creation of a prototype land-use map based on the strict rules of graphic semiology. The main cartographic results present the various changes in land use and land cover.. The application aims to highlight the importance of change detection methods for understanding urbanization patterns and assessing their impacts on the oasis ecosystem. Post-classification comparison of land-use maps revealed a growth in built-up areas to the detriment of the old oasis center and bare soil. The Remote Sensing (RS) and Geographic Information System (GIS) approaches are highly effective and offer real added value in visualizing land occupation and land usage. All these documents are indispensable and highly important for rational planning.

The GIS contribution for mapping flood-hazard areas in arid Tunisia noted that the Maknassy basin (Lban-Ouadran Wadi watershed) is a natural environment at high risk of flooding. In fact, the hydro-morphological characteristics of the impluviums discharged by the Maknassy basin (oued Lban-Ouadran catchment) are mainly determined by factors that make it a very high floodrisk environment. Indeed, the morphology of the impluviums of the wadis draining the Maknassy basin is dominated by high slope values. In addition, these conditions add a bedrock with low permeability and low resistance. Hydrological characteristics dominated by high compactness indices, drainage densities and very high hydrographic densities exacerbate this situation. The combination of these factors has revealed the existence of three hazard classes, the two most sensitive of which (moderately and very sensitive) cover more than 65% of the impluviums of the order 1 wadis (on both the northern and southern slopes) of the Maknassy basin, which are at very high risk of flooding. This highlights the environmental and economic impact of the flood hazard and the real risk to natural resources, particularly water and soil, thus compromising the sustainability of development and the environment in the Maknassy basin.

The research conducted to the use of remote sensing in monitoring and evaluation of land degradation processing in west El-Minya, Egypt showed that the rest of the groundwater samples (0.2%) in the study area are classified as severe restriction (S.R.) irrigation water (i.e., very poor groundwater quality for irrigation purposes). Also (3.1%) of the groundwater samples, in the study area, are classified as high restriction (H.R.) irrigation water and (8.9%) of the groundwater samples are classified as moderate restriction (M.R.) irrigation water. Most of the groundwater samples of in the study area (82.4%) are classified as low restriction (L.R.) irrigation water. The results show that(5.4%) of the groundwater samples, in the study area, are classified as No restriction (N.R.) irrigation water. Thus (75%) of water in the study area is suitable for irrigating most crops except the sensitive ones. The low to medium SAR and moderate to high EC^{w} can be managed without any water and soil treatments. Leaching the excess salts from the root zone can solve the problem.

2 Conclusions

Emphasizing practical mapping technologies based on advanced remote sensing data and GIS tools, this book provides methods and approaches for assessing, modeling and monitoring risks, resources and improving strategies for suitable environments in MENA regions. Written by 50 + leading researchers working at the forefront of this critical area, it offers case studies from a variety of MENA regions, highlighting the subtle requirements of each. In a very practical way, it demonstrates the experience, utility, and models for developing strategies to reach sustainable development goals.

The multitude of capabilities and insights provided by GIS, has further elevated its importance in providing researchers, conservationists and policy makers to understand freshwater ecosystems and to improve the accuracy and efficiency of stock assessment for developping sustainable fisheries strategies. The application of radiometric indices as environmental indicators is essential in various studies, as discussed, to evaluate and monitor the coastal eutrophication that threatens the MENA regions. Furthermore, the combination of radiometric indices with other physical factors such as slope and lithology contributes to detect flood risk and monitor vegetation. In an urban context, the use of RS and GIS helps decision makers to understand the links between the contrasting spatial distribution of the urban vegetation cover and its variation with the public, spatial and urban planning policies and to assess LULC dynamics, mineral and planted surfaces and UHA too. In addition, these approaches are highly effective and offer real added value to manage natural ressources, supply chain management and increase food security in MENA region.

3 Recommandations

The materials presented in the chapters of "Environmental Remote Sensing and GIS in MENA region" emphasises the recommendations:

- The adoption of rainwater harvesting (RWH) practices is recommended since it aligns with sustainable local water resource management. However, it is also essential to recognize that it alone cannot serve as a panacea. Integrating time-tested RWH methodologies with innovative approaches is crucial and it should be complemented by other strategies such as water reuse and efficiency enhancements to realize its full potential and efficacy.
- The replacement of the old units with new powerful ones allows fishermen to operate in under-exploited marine areas. Besides, the establishment of a biological rest period for the small pelagic fisheries and a revision of the regulatory framework are strongly recommended. In addition, it is crucial to continue exploring new fishing areas for small pelagic fish stocks to ensure the durability and sustainability of this activity.
- Launch regular monitoring and evaluation of GIS-based interventions in order to adjust strategies based on results and to improve Tunisian freshwater resource management practices continuously.
- Use GEE time series to detect and analyze the spatial and temporal evolution of different phenomenon in various scales.
- Setup a coordinated target-focused monitoring program or experiment that uses in situ measurements in combination with UAV remote sensing data to detect eutrophication. This may improve future monitoring and management of coastal ecosystem health as well as other ecosystems that are subject to regular monitoring.
- For flood susceptibility mapping there are recommendations to follow for good accuracy such as to explore the various factors contributing to flood susceptibility, including topography, land use/land cover, soil characteristics, hydrological conditions, climate patterns, and human activities; choosing the high performance

models and analyzing the most contributing factors in flood susceptibility and the redundancy factors.

- Implement heat island mitigation strategies based on planting vegetation; reorganize gradually the urban fabric by demolishing old buildings and replacing them with efficient and ecological solutions and use innovative, permeable materials for paving roads and parking for reducing UHI risk.
- The predictive flow performance of such prediction models depends on the internal dependence structure of the data. Hence, because of seasonality that reinforces this dependence structure, the monthly SARIMA prediction models produce better predictions than the annual ARIMA models. The accuracy of the obtained results need to be confirmed by the applicability extent to a larger number of locations in northern Algeria.
- Apply a very high-resolution images to refine the spatial and temporal variations of objects or phenomenon
- Giving plants an important place in landscaping projects and try to integrate ecological connectivity concepts into development projects
- Decicion makers are encouraged to use the capabilities of RS and GIS technology for mapping to provide adequate and reliable spatial information and valuable data to develop effective management and monitoring of LULC changes.
- Build floodwater diversion structures and storage facilities downstream of the piedmonts
- Geographic information systems and remote sensing provide the required information for planners and decision-makers to help in organizing land resources data and water quality maps.
- It is possible to control the low-to-medium SAR and moderate-to-high EC^{w} without applying any water or soil treatments. The issue can be resolved by extracting the extra salts from the root zone by leaching.

MIX
Papier aus verantwortungsvollen Quellen
Paper from responsible sources
FSC® C105338

If you have any concerns about our products,
you can contact us on
ProductSafety@springernature.com

In case Publisher is established outside the EU,
the EU authorized representative is:
Springer Nature Customer Service Center GmbH
Europaplatz 3, 69115 Heidelberg, Germany

Printed by Libri Plureos GmbH
in Hamburg, Germany